D1189728

Concrete Materials and Practice

Concrete Materials and Practice

L. J. Murdock
D.Sc. (ENG.), Ph.D., F.I.C.E.
K. M. Brook
B.Sc. F.I.C.E., F.I.H.E.
B.Sc., F.I.C.E., F.I.H.E.

A Halsted Press Book

John Wiley & Sons
New York

First published 1948
Fifth edition 1979 by
Edward Arnold (Publishers) Ltd

Published in the U.S.A.
by Halsted Press, a Division
of John Wiley & Sons, Inc.
New York

Library of Congress Cataloging in Publication Data

Murdock, Leonard John.
Concrete materials and practice.

"A Halsted Press book."
Bibliography: p.
1. Concrete. 2. Concrete construction.
I. Brook, Keith Malcolm, joint author. II. Title.
TA439.M8 1979 620.1'36 78-27476
ISBN 0-470-26639-2

Printed in Great Britain

Preface to First Edition

RESEARCH both in Great Britain and abroad has provided an extensive knowledge of the fundamental properties and characteristics of concrete and it is possible in the laboratory to design and prepare concrete of predetermined quality. In the field, however, varying conditions, lack of knowledge, relatively inefficient plant, errors, quantities of material involved, changes in the basic materials, and other difficulties, cause variations in quality of the concrete which can only be minimized by careful control, education, and experience.

The object of this book is to provide a broad outline of the science of concrete making, particular attention being given to the translation of the knowledge gained by testing and research into practice in the field.

The properties of concrete are described early in the book since it is important to recognize that changes in the constituent materials, and the manner in which they are combined and used, may improve some properties at the expense of others, and a compromise must always be made to obtain the special qualities required as economically as possible.

The basic materials and the technique of concrete making under various conditions and for many purposes are discussed in some detail. The method of testing, the design of mixes to meet various requirements, and the methods of control of quality on the site, which are included, have been found by experience to give satisfactory results.

Brief mention is made of the design of reinforced concrete mainly to enable the concrete engineer to recognize errors in bending and fixing the reinforcement. Similarly, examples of formwork are included on account of the importance of careful design and construction to the final result.

The author is indebted to Messrs. G. Wimpey and Co., Ltd., for permission to publish information based on experience on many sites large and small, as well as a number of photographs and charts. Considerable use has been made of the results of researches carried out at the Building Research Station and the Road Research Laboratory, and due acknowledgement is made to the Directors of these establishments for access to this work. Acknowledgment is also made to the Cement and Concrete Association for their generous help in providing photographs; to Mr. R. V. Chate of the Reinforced Concrete Association and Dr. N. Davey of the Building Research Station for their valuable comments and criticisms of the original manuscript; and to Dr. W. MacGregor for reading the proofs.

Preface to Fifth Edition

THE preface to the first edition has been reprinted as it still describes the main aims of this book. The principles of concrete production do not change, but there have been changes in technology, in plant and equipment and, of considerable importance, there is the increased dependence on ready-mixed concrete. The book has been revised to take account of these changes, including a new chapter on ready-mixed concrete. More attention has also been given to concrete construction overseas and a new chapter has been introduced on concreting in hot weather.

This edition is presented in metric terms, with Imperial equivalents shown only where it is felt that it will assist the reader in establishing 'dimensions' in terms to which he has been accustomed.

For the fifth edition, Mr. K. M. Brook has become joint author. He is responsible for concrete technology and its application in George Wimpey & Co. Limited. He has a wide experience of laboratory research and of concrete practice on site, covering structures in all their varied forms and in many parts of the world. His knowledge of site organization and practice makes an invaluable contribution.

The authors are indebted to Mr. G. F. Blackledge for much practical comment.

The authors are conscious of the value of close cooperation between those responsible for quality control on site, the management, the design and planning engineers, the research engineers and the scientists. They have been fortunate in having such a broad wealth of knowledge so readily available in George Wimpey & Co. Limited. In particular they are appreciative of the considerable assistance from Mr. R. C. Baker in revising the chapter on reinforcement, from Mr. J. R. Illingworth on the chapter on formwork, and from Mr. H. Holland for his interest in the design of concrete mixes, including the preparation of Figs. 7.2 and 7.3.

L.J.M.
K.M.B.

1978

Acknowledgments

Grateful acknowledgment is made to the following who have given permission for the use of photographs and information:

Cement and Concrete Association. See Preface to First Edition and Figs. 4.1, 11.3, 11.7, 12.1, 13.6, 17.1, 17.2, 17.3, 17.4, 17.5, 17.7, 17.14, 17.15, 18.10, 18.16, 19.12, 19.13, 21.1, 21.3, 21.6, 21.7, 23.1.
Construction Industry Research and Information Association. Figs. 14.3, 18.1, 18.2, 18.3.
Institution of Civil Engineers. Figures and information reproduced from Papers to which reference is made in the Bibliography.
Concrete Society. Frontispiece.
The Controller, HM Stationery Office. Figs. 2.5, 2.9, 3.8, 4.8, 4.9, 5.1, 6.2, 11.2, 11.4, 11.5, 13.1, 24.1.

Acrow Ltd. Figs. 18.5, 18.6.
E. P. Allam Ltd. Figs. 13.5, 23.3.
American Concrete Institute. Fig. 3.7, also parts of Figs. 3.6, 3.14, 9.11, 12.3.
Associated Portland Cement Manufacturers Ltd. Fig. 4.2.
Benford Ltd. Figs. 8.7, 9.8, 9.10.
British Lift Slab Ltd. Fig. 18.18.
British Steel Piling Ltd. Fig. 22.3.
Cement and Steel Ltd. Fig. 19.11.
Columbus Dixon Ltd. Fig. 25.4.
Concrete Ltd. Figs. 23.4, 25.2.
Concrete Development Co. Ltd. Figs. 20.1, 23.4.
Concrete Pump Co. Ltd. Fig. 9.14.
Franki Piling Co. Ltd. Fig. 22.2.
S. V. Gardner. Fig. 19.15.
Ham River Grit Co. Ltd. Fig. 3.1.
Holland & Hannen and Cubitts Ltd. Fig. 20.2.
Mac Mix Ltd. Fig. 10.1.
Nordberg Manufacturing Ltd. Fig. 3.3.
Frederick Parker Ltd. Fig. 3.2.
Pegson Ltd. Fig. 23.6.
PSC Equipment Ltd. Fig. 20.3.
Ready Mixed Concrete Ltd. Fig. 4.10.
Road Machines (Drayton) Ltd. Fig. 9.13.
Rocoman Plant Ltd. Fig. 9.17.

Welding Industries Ltd. Fig. 23.7.
Winget Ltd. Fig. 9.7.
George Wimpey & Co. Ltd. Numerous illustrations.

Extracts from British Standards and from British Standard Codes of Practice are given by permission of the British Standards Institution, 2 Park Street, London W1, from whom official copies of the Standards and the Codes can be obtained.

Contents

Twickenham Bridge: an example of the pleasing appearance obtainable by bush-hammering gravel concrete, and by the use of inset lines of tiles to mask the lift planes. Concrete which has stood the test of the passing years.

1

Introduction

THE USE of some form of cement to bind together stones, gravel and other material has been practised from very early times. It is probable that fire was found to change limestone to a form of quicklime which became very hot on slaking with water and then slowly set, like the lime mortar commonly used in brickwork a few years ago. Some of the limestone would have had clay intrusions producing lime with hydraulic properties, which was found to be more durable. The Egyptians certainly used lime mortar in the construction of their pyramids, and analyses have shown that they possessed much practical knowledge of the subject.

The Greeks and Etruscans also used cementitious mortars, but a Roman, Vitruvius, is the first author on the subject whose writings have descended to us. His works show that, although the Romans adopted a different scientific phraseology, they knew as much about the chemistry of cementitious limes as the scientists of the eighteenth century. Vitruvius's work was not, in fact, superseded until M. Vicat made his researches in France in the early part of the nineteenth century.

One of the most notable examples of Roman work which still remains is the Pantheon, the dome of which is 43·5m (142 ft 6 in.) in span and cast in solid concrete. Wooden boards were used as shuttering, and the marks of these can still be clearly seen on the face of the concrete. The calcareous cements used by the Romans were either composed of suitable limestones burnt in kilns, or were mixtures of lime and pozzolanic material which combine to form a fairly hard weather-resistant concrete. Pozzolana was found naturally as volcanic ash or tuff in the district around Pozzuoli in Italy, or was produced artificially by grinding clay tiles and potsherds. Progress until the latter part of the eighteenth century appears to have been small except in the spread of knowledge of methods of manufacturing hydraulic limes. The choice of materials became less a matter of chance and in 1796 the term Roman cement came into use.

Smeaton experimented to ascertain the cause for some limes having the power to act under water, for the foundations of his structures, including the Eddystone lighthouse built in 1756. His researches led him to the use of a blend of Blue Lias lime and clay. What is perhaps remarkable is that he failed to discover that burning to a somewhat higher temperature, so that the materials vitrified, produced a material which could be ground to provide a much stronger and more weather-resistant cement.

Joseph Aspdin of Leeds first used the term Portland cement in 1824 to describe a patent cement formed by heating a mixture of finely divided clay and limestone or chalk in a furnace to a temperature sufficiently high to drive off all the carbon dioxide. It was called Portland cement because the appearance of the resulting concrete resembled Portland stone.

It was not until 1845 at Swanscombe that Isaac C. Johnson invented the cement which was the prototype of our modern Portland cement, by increasing the temperature at which the limestone and clay were burnt until they formed a clinker. From then onward many refinements and technical changes produced a gradual improvement in the quality of cement, accompanied by a steady expansion in the uses to which concrete was applied. At about the turn of the century the structural possibilities of embedding iron or steel in some form in this concrete were generally recognized. Since concrete itself is weak in tension, it was realized that steel could provide tensile resistance. Fortunately the coefficients of thermal expansion of the two materials are very similar and rusting of the steel is inhibited by the lime in the cement.

Improvements in Portland cement production and in the technique of concrete construction have greatly accelerated in recent years, and a great deal of research has been done throughout the world, notably in Britain, the USA, France and Germany. Concrete, as it is known today, is a building and structural material, the properties of which may be predetermined by careful design and control of the constituent materials. These constituent materials are, a hardened binding medium or matrix formed by a chemical reaction between cement and water, and the aggregate to which the hardened cement adheres to a greater or less degree. The aggregate may be gravel, crushed rock, slag, artificial lightweight aggregates, sand or other similar material.

The aggregate, the cement and the water are all mixed together, and in this state are plastic and workable, properties which permit moulding into any desired shape. Within a few hours of the preparation of the mixture, the cement and water undergo a chemical combination, generally referred to as hydration, which results in hardening and the development of strength. The strength development process may continue indefinitely under favourable moisture and temperature conditions, with a general improvement in the quality of the concrete.

Selection of constituent materials

For concrete to be suitable for a particular purpose, it is necessary to select the constituent materials and to combine and use them in such a manner as to develop the special properties required as economically as possible. For instance, the properties required in an engineering structure are, generally, weather resistance and the strength characteristics on which the designer has based his calculations; in roads, concrete must be strong and wear-resistant to sustain the pounding action of heavy and fast-moving traffic; in water towers and reservoirs it must be impermeable; in foundations it may be required to resist the disintegrating action of sulphates in the soil. Similarly, sea structures must be resistant to the chemical attack of sea salts, as well as to abrasive action. Concrete floors, in addition to abrasion by foot traffic and all types of vehicles, are often subjected to chemical attack by various agencies. Other properties which may be required in certain circumstances include lightweight, thermal insulation, special architectural finishes, and impedance to X-rays or the products of nuclear fission.

Fig. 1.1. A pleasing example of concrete construction for a modern office building. The structure is of in-situ reinforced concrete and is clad with smooth white precast panels

Some special types of construction such as piling, dock work, mass concrete dams, tunnel work, and so on, call for a special technique which will be described later. Similarly, the laying of concrete floor toppings such as granolithic, the moulding of cast stone, precast concrete staircase units, concrete railway sleepers, fence posts, paving flags, kerbs, garden ornaments, and other items require individual modifications in procedure. Lightweight concrete and open-textured concrete form another class of concrete product to be treated separately.

The selection of materials and the choice of method of construction is not easy, since many variables will affect the quality of the concrete produced, and both quality and economy must be considered. As an example of the necessity for compromise, concrete of the same strength may be produced by a lean mix which is difficult to place, or by a somewhat richer and more plastic mix which is easy to place. In such a case it is probable that the increased cost of cement in the richer mix will be offset by the saving in cost of placing the concrete. Additionally, the finish of the richer mix will in all probability be better. Other properties of the concrete are, of course, affected by an increased cement content, and must be considered according to their relative importance.

Limitations of concrete

Concrete has limitations and certain disadvantages which should be realized by both the designer and the builder, since recognition of these characteristics will go far towards the elimination of costly difficulties in construction, and of cracking and other structural weaknesses that detract from the appearance, serviceability, and life of the structure.

The principal limitations and disadvantages are:

(1) *Low tensile strength.* Members which are subjected to tensile stress must, therefore, be reinforced with steel bars or mesh.

(2) *Thermal movements.* During setting and hardening the temperature of the concrete is raised by the heat of hydration of the cement, and it then gradually cools. These temperature changes can cause severe thermal strains and early cracking. Hardened concrete expands and contracts with changes in temperature at roughly the same rate as steel. Expansion and contraction joints must be provided in many types of structure if failures are to be prevented.

(3) *Drying shrinkage and moisture movements.* Concrete shrinks as it dries out and even when hardened, expands and contracts with wetting and drying. These movements necessitate the provision of contraction joints at intervals if unsightly cracks are to be avoided.

(4) *Creep.* Concrete gradually deforms under load, the deformation due to creep not being wholly recoverable when the load is removed. Creep is particularly important in relation to prestressed concrete. Creep and shrinkage are difficult to separate in measurements of deformation during testing.

(5) *Permeability.* Even the best concrete is not entirely impervious to moisture, and contains soluble compounds which may be leached out to a varying extent by water. Joints, unless special attention is given to their construction, are apt to form channels for the ingress of water.

Impermeability is particularly important in reinforced concrete where reliance is placed on the concrete cover to prevent rusting of the steel.

Variability in quality of concrete

A list of variables likely to occur during the manufacture of concrete is given below, and while this list is not exhaustive it does give some indication of possible sources of variation in strength and other properties.

(A) *Materials*

- (1) Cement: Quality and rate of hardening.
- (2) Fine aggregate:
  - (a) Grading, affecting workability
  - (b) Moisture content, affecting water/cement ratio
  - (c) Silt, affecting strength
  - (d) Cleanliness, affecting strength and durability
- (3) Coarse aggregate:
  - (a) Grading, some effect on strength
  - (b) Moisture content, affecting water/cement ratio
  - (c) Cleanliness, affecting strength and durability
- (4) Water: Quantity affecting most properties, quality affecting setting, strength, durability, etc.
- (5) Admixture (if used): Modification of properties of the concrete will depend upon type and amount of admixture used

(B) *Methods of batching and mixing*

- (1) By volume:
  - (a) Bulking of sand affects proportions
  - (b) Accuracy of measurement
- (2) By weight:
  - (a) Moisture content of aggregate
  - (b) Accuracy of measurement

(3) Loss of materials entering mixer
(4) Efficiency of mixer

(C) *Methods of construction*
(1) Compaction: Air voids reduce strength
(2) Curing: Necessary to increase strength and improve other properties
(3) Weather conditions during placing and curing of concrete

Workmanship

In concrete construction it is not enough for the concrete engineer to supervise and criticize the work, without justifying his comments or attempting to educate the foremen, gangers, and workmen. Quality of workmanship is one of the most important factors in the production of good concrete structures, and the key to good workmanship is knowledge and an interest in the work being done. Most workmen are keen to do a job well, subject to the conflicting interest of speed since this affects their earnings, and if the reasons for doing a job in a certain way are explained to them, they will often make an effort to carry out the suggestions made.

The concrete engineer must, therefore, learn to deal tactfully and to encourage an interest in improvements of method in the men actually making and placing the concrete. He must also see that the quality of the work complies with the requirements of the particular purpose for which it was intended, while studying economy and the smooth progress of the work. The concrete engineer must bear in mind that he is responsible for maintaining a reasonable and uniform standard of concrete quality, while in the interests of economy it is obviously unnecessary to produce concrete having a strength far in excess of that required.

2

Properties of Concrete

THE characteristics of concrete should be considered in relation to the quality required for any given construction purpose. The closest practicable approach to perfection in every property of the concrete would result in poor economy under many conditions, and the most desirable structure is that in which the concrete has been designed with the correct emphasis on each of the various properties of the concerete, and not solely with a view to obtaining, say, the maximum possible strength.

Although the attainment of the maximum strength should not be the sole criterion in design, the measurement of the crushing strength of concrete cubes or cylinders provides a means of maintaining a uniform standard of quality, and, in fact, is the usual way of doing so. Since the other properties of any particular mix of concrete are related to the crushing strength in some manner, it is possible that as a single control test it is still the most convenient and informative.

The testing of the hardened concrete in prefabricated units presents no difficulty, since complete units can be selected and broken if necessary in the process of testing. Samples can be taken from some parts of a finished structure by cutting cores, but at considerable cost and with a possible weakening of the structure. It is customary, therefore, to estimate the properties of the concrete in the structure on the basis of the tests made on specimens moulded from the fresh concrete as it is placed. These specimens are compacted and cured in a standard manner given in BS 1881: 1970[1] as in these two respects it is impossible to simulate exactly the conditions in the structure. Since the crushing strength is also affected by the size and shape of a specimen or part of a structure, it follows that the crushing strength of a cube is not necessarily the same as that of the mass of exactly the same concrete.

Crushing strength

Concrete can be made having a strength in compression of up to about 80 N/mm^2 (12 000 lb/in^2), or even more depending mainly on the relative proportions of water and cement, that is, the water/cement ratio, and the degree of compaction. Crushing strengths of between 20 and 50 N/mm^2 at 28 days are normally obtained on the site with reasonably good supervision, for mixes roughly equivalent to 1:2:4 of cement:sand:coarse aggregate. In some types of precast concrete such as railway sleepers, strengths ranging from 40 to 65 N/mm^2 at 28 days are obtained with rich mixes having a low water/cement ratio.

The crushing strength of concrete is influenced by a numbers of factors in addition to the water/cement ratio and the degree of compaction. The more important factors are:

(1) *Type of cement and its quality.* Both the rate of strength gain and the ultimate strength may be affected.

(2) *Type and surface texture of aggregate.* There is considerable evidence to suggest that some aggregates produce concrete of greater compressive and tensile strengths than are obtained with smooth river gravels (see p. 55).

(3) *Efficiency of curing.* A loss in strength of up to about 40 per cent may result from premature drying out. Curing is therefore of considerable importance both in the field and in the making of tests. The method of curing concrete test cubes given in BS 1881 should, for this reason, be strictly adhered to.

(4) *Temperature.* In general, the rate of hardening of concrete is increased by an increase in temperature. At freezing temperatures the crushing strength may remain low for some time.

(5) *Age.* Under normal conditions concrete increases in strength with age, the rate of increase depending on the type of cement. For instance, high-alumina cement produces concrete with a crushing strength at 24 hours equal to that of normal Portland cement concrete at 28 days. Hardening continues but at a much slower rate for a number of years.

Variations in the crushing strength of site-made concrete are discussed in Chapter 27, p. 415).

The above refers to the static ultimate load. When subjected to repeated loads concrete fails at a load smaller than the ultimate static load, a fatigue effect. A number of investigators have established that after several million cycles of loading, the fatigue strength in compression is 50–60 per cent of the ultimate static strength.

Tensile and flexural strength

The tensile strength of concrete varies from one-eighth of the compressive strength at early ages to about one-twentieth later, and is not usually taken into account in the design of reinforced concrete structures. The tensile strength is, however, of considerable importance in resisting cracking due to changes in moisture content or temperature. Tensile strength tests are used for concrete roads and airfields.

The measurement of the strength of concrete in direct tension is difficult and is rarely attempted. Two more practical methods of assessing tensile strength are available. One gives a measure of the tensile strength in bending, usually termed the flexural strength. BS 1881 : 1970[1] gives details concerning the making and curing of flexure test specimens, and of the method of test. The standard size of specimen is 150 mm × 150 mm × 750 mm long for aggregate of maximum size 40 mm. If the largest nominal size of the aggregate is 20 mm, specimens 100 mm × 100 mm × 500 mm long may be used.

A load is applied through two rollers at the third points of the span until the specimen breaks. The extreme fibre stresses, that is, compressive at the top and tensile at the bottom, can then be computed by the usual beam formulae. The beam will obviously fail in tension since the tensile strength is much lower than the compressive strength. Formulae for the calculation of the modulus of rupture are given in BS 1881 : 1970.

Test specimens in the form of beams are sometimes used to measure the modulus of rupture or flexural strength quickly on the site. The two halves of the specimen may then be crushed so that besides the flexural strength the compressive strength can be approximately determined on the same sample. The test is described in BS 1881 : 1970.

Values of the modulus of rupture are utilized in some methods of design of unreinforced concrete roads and runways, in which reliance is placed on the flexural strength of the concrete to distribute concentrated loads over a wide area.

More recently introduced is a test made by splitting cylinders by compression across the diameter, to give what is termed the splitting tensile strength. Details of the method are given in BS 1881 : 1970. The testing machine is fitted with an extra bearing bar to distribute the load along the full length of the cylinder. Plywood strips, 12 mm wide and 3 mm thick are inserted between the cylinder and the testing machine bearing surfaces top and bottom. From the maximum applied load at failure the tensile splitting strength is calculated as follows:

$$f_t = \frac{2P}{\pi l d}$$

where f_t = splitting tensile strength, N/mm^2
P = maximum applied load in N
l = length of cylinder in mm
d = diameter in mm.

As in the case of the compressive strength, repeated loading reduces the ultimate strength so that the fatigue strength in flexure is 50–60 per cent of the static strength.

Shear strength

In practice, shearing of concrete is always accompanied by compression and tension caused by bending, and even in testing it is impossible to eliminate an element of bending.

Deformation under load

When concrete is loaded, deformation occurs which increases as the load increases, as with steel and other materials. However, whereas steel deforms elastically at loads below the elastic limit, so that the specimen returns to its original size when the load is removed, concrete deforms partly as a result of elastic strain, and partly as a result of plastic strain or creep. This is illustrated in Fig. 2.1 which shows the stress–strain curve for a continuously increasing load. The relationship for an elastic material, such as steel, is given by the straight line OA, while that for a partly plastic material, such as concrete, is given by the line OB. On unloading a concrete specimen stressed to the point B, the elastic strain disappears, but the plastic strain represented by OX remains. The effect of repeated loading is also shown in which the deformation

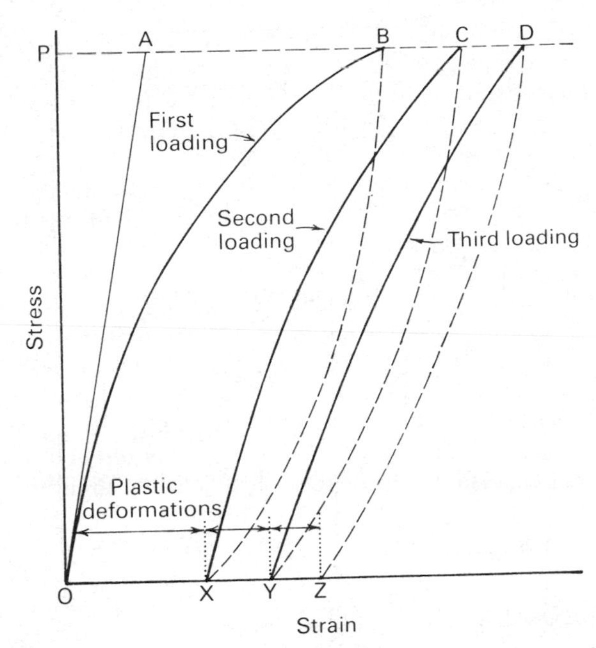

Fig. 2.1 Stress–strain curves for concrete subjected to repeated loading

due to plastic strain XY, YZ diminishes at each repeated loading, although the total plastic deformation or creep OX, OY, OZ is continually increasing. The strain has been exaggerated for clarity; within the normal working range the stress–strain curve for concrete is for practical purposes a straight line.

When the load is increased beyond the working range, the stress–strain curve deviates considerably from a straight line, indicating that stress and strain are no longer proportional for practical purposes. This limit of proportionality may be anything from 25 to 75 per cent of the ultimate strength, with 40 per cent as an average value.

When concrete is subjected to a constant sustained load, the deformation produced by the load may be divided into two parts; the elastic deformation which occurs immediately the load is applied, and the plastic flow or creep which begins when the load is applied but continues to increase without further increase in load for the whole time the specimen is loaded. This continuing deformation with time under constant load is illustrated in Fig. 2.2, being represented by line AB. If at time X the load is removed, there is an immediate elastic recovery BC equal to the elastic deformation OA. Plastic recovery, which is much smaller than the plastic flow or creep AY, then follows slowly.

Modulus of elasticity

The common measure of the elastic property of a material is the modulus of elasticity, which is the ratio of the applied stress to the deformation per unit length resulting from the application of that stress.

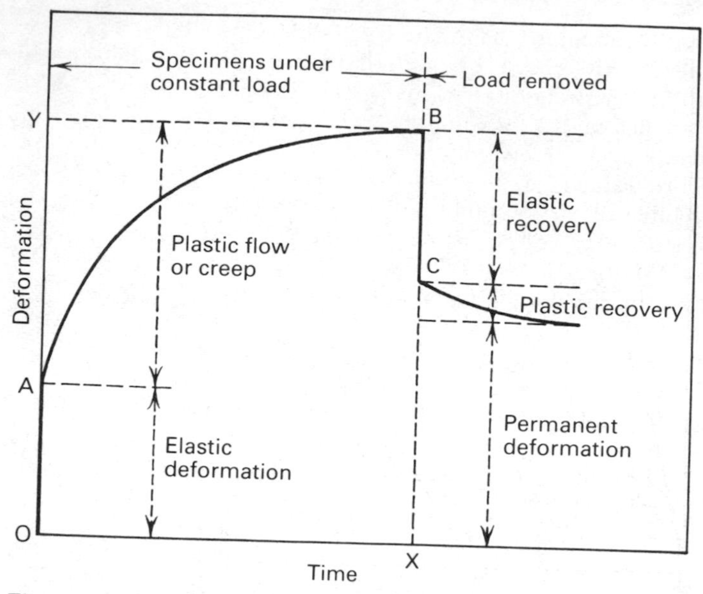

Fig. 2.2. Illustration of deformation of concrete under load

Then,
$$E = \frac{f}{e} = \frac{WL}{Ax}$$

where E is the modulus of elasticity

$f = \frac{W}{A}$ is the applied stress, W being the load and A the cross-sectional area

$e = \frac{x}{L}$ is the deformation per unit length or strain, x being the deformation under load W, and L the length of the member.

A, x and L are measured in the same units.

Concrete is not a truly elastic material, as already explained, and the stress–strain curve for continuously increasing loading is of the type shown at OB in Fig. 2.1, the curvature being exaggerated for clarity. If there was no plastic deformation, the curve would become a straight line OA, being the tangent to the curve OB at the origin. The true modulus of elasticity or instantaneous modulus could be calculated from the tangent OA, but for practical purposes it is usual to determine the modulus at an arbitrarily selected stress, applied at a defined rate, which takes into account movement due to creep. The modulus is then

$$\frac{\text{OP}}{\text{PB}} = \frac{f}{e}$$

It is necessary to load at a defined rate since concrete has more time to creep at a slower rate of loading. Since the creep of concrete is influenced by a number of factors, the modulus of elasticity is also affected.

The modulus of elasticity is not directly related to the other properties of concrete, although higher strengths are usually accompanied by higher values of E. For ordinary concretes the modulus of elasticity lies in the range 25 to 36 kN/mm^2. As concrete ages, its modulus of elasticity increases, a fact of some importance since a restrained concrete member already in a state of strain due, for instance, to cooling after the initial hardening, becomes subjected to an increasing tensile stress. However, this effect is offset to some extent by the effects of creep as described below. If completely restrained the stress at any time would be given by the usual formula, $f = Ee$.

Poisson's ratio

When subjected to compression, concrete contracts longitudinally and expands laterally. The ratio of the lateral strain to the longitudinal strain is known as Poisson's ratio, and within the normal working range of loading may be taken to have a value of 0·2.

Creep

Some mention of plastic flow or creep has already been made in connection with the modulus of elasticity. Creep is a non-elastic deformation under load, believed to be due to closure of internal voids, viscous flow of the cement–water paste, crystalline flow in aggregates, and the squeezing of water from the cement gel under load. In practice, creep and drying shrinkage usually occur simultaneously and are often confused; exceptions are dams, tunnels and other underground works, and massive structures where little or no drying out occurs. Creep is useful in that it enables a readjustment of stress to take place when local stresses of high intensity might otherwise result in failure of the structure. Whether creep is beneficial or not depends on circumstances. For instance, when the concrete is reinforced with steel, creep generally leads to a transfer of load from the concrete to the steel. This may relieve high stresses in the concrete but may also result in overstressing of the steel.

In prestressed concrete, creep has an adverse effect resulting in a reduction in the tension in the prestressing wires or cables; it is necessary to allow for this by an increase in the initial pretensioning (see also p. 302).

Creep reduces the tendency to crack in restrained concrete members by relieving stresses due to shrinkage. The higher the percentage of reinforcement the greater is the restraint and hence, the tendency to develop cracks, but the restraint has a beneficial effect in that, while cracks are induced at more frequent intervals, they are usually so fine that they pass unnoticed, and are of no importance (see also p. 284).

The creep of unreinforced concrete is approximately proportional to the ratio of stress to strength of concrete within the normal range used in design, but as failure is approached the rate of creep increases rapidly.

The rate of creep is dependent on the following factors apart from the stresses imposed:

(1) *Strength.* An increase in strength usually leads to a reduction in creep,

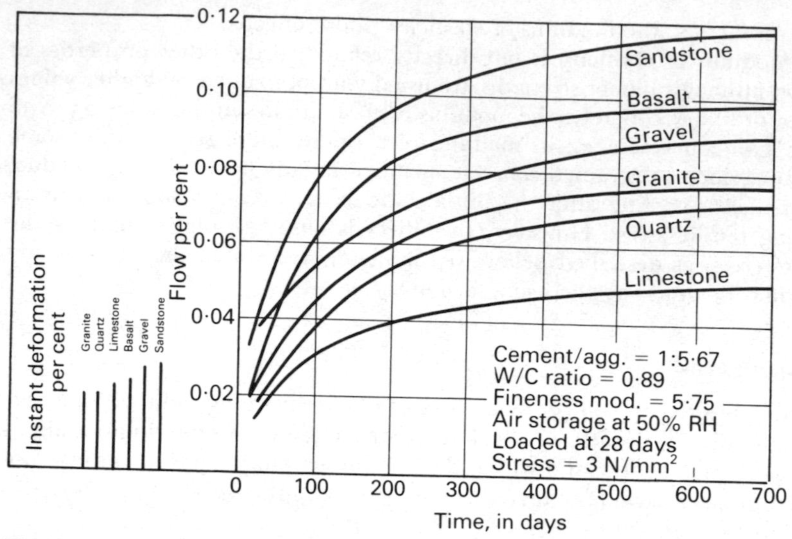

Fig. 2.3 Flow of concrete—effect of mineral character of the aggregate

since the stress/strength ratio is reduced. This assumes new conditions are not imposed which also increase the stress, for example shrinkage of a restrained member.

(2) *Cement.* There is some difference of opinion on the influence of cement type on creep. Generally it is thought that the type of cement is not of great importance, except that creep is related to the stress/strength ratio for the concrete, and strength development is, of course, related to cement type.

(3) *Proportions of mix.* Creep decreases as the water/cement ratio and the volume of cement paste decrease. The movement in a very wet mix may be as much as twice that in an otherwise similar dry mix.

Neville [2] has suggested that the creep of concrete depends mainly on the creep of the cement paste and is reduced by the coarse aggregate. Newman [3] suggests that the ultimate creep (C_c) of concrete can be predicted from the equation:

$$C_c = C_p(1 - V_a)^n$$

where C_p is the ultimate creep strain of cement paste, being of the order of $2500–3000 \times 10^{-6}$ mm/mm for a stress level of half the compressive strength

V_a is the solid volume fraction of total aggregate

n increases from 1·7 to 2·1 with the time under load.

The solid volume of the aggregate includes air voids within the aggregate particles but does not include voids between separate particles (see also p. 52).

(4) *Aggregate.* Creep increases as the aggregate becomes finer; and it is generally greater when porous aggregates are used.

The type of aggregate has a marked influence on creep. R. E. Davis and H. E. Davis [4] have given values for the creep of concrete made with different aggregates, as shown in Fig. 2.3.

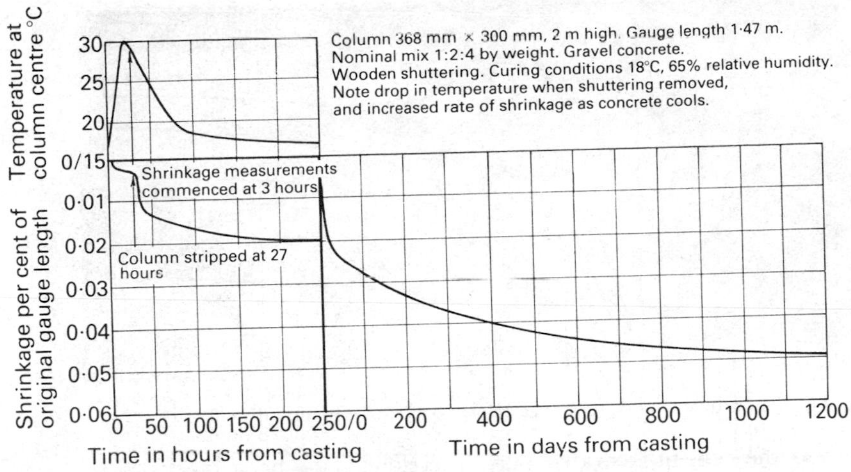

Fig. 2.4 Example of shrinkage of a concrete column from three hours after casting, with indication of temperature rise to be expected

(5) *Curing*. Creep decreases as cement hydration proceeds, so that concrete kept continuously wet creeps less than that cured in air. Alternate wetting and drying, however, leads to an increase in the magnitude of creep.[5]

(6) *Age*. The rate of creep decreases as the concrete ages. The creep at 1 year may be twice that at 28 days, but a further increase in creep of 20 per cent may take 5 years. The ultimate creep after 20 or 30 years may be taken as 30 to 40 per cent above that at 1 year.

Drying shrinkage and moisture movements

Two stages occur in the shrinkage of concrete as it is placed in practice:

(i) An initial shrinkage in the concrete while it is still in a fluid or plastic state, due to the reduction in the volume of the system cement plus water amounting to about one per cent of the absolute volume of the dry cement[6]; to the loss of water by seepage through the forms; to absorption by the forms, or in the case of roadwork by the formation; and to evaporation.

This shrinkage may be masked by the thermal expansion which accompanies the temperature rise resulting from the chemical reaction in the setting of the cement.

(ii) A further drying shrinkage as the concrete hardens and is allowed to dry.

This may be accompanied by a thermal shrinkage as the concrete cools.

The combined effects are shown in Fig. 2.4, which gives the results of tests made on a column. This diagram shows the effects of both temperature changes in the column and drying shrinkage. While the measurements were commenced at three hours after casting, the point at which the plastic concrete can be considered to have become rigid may not have occurred until some hours later. Expansion in the concrete mass due to the rise in temper-

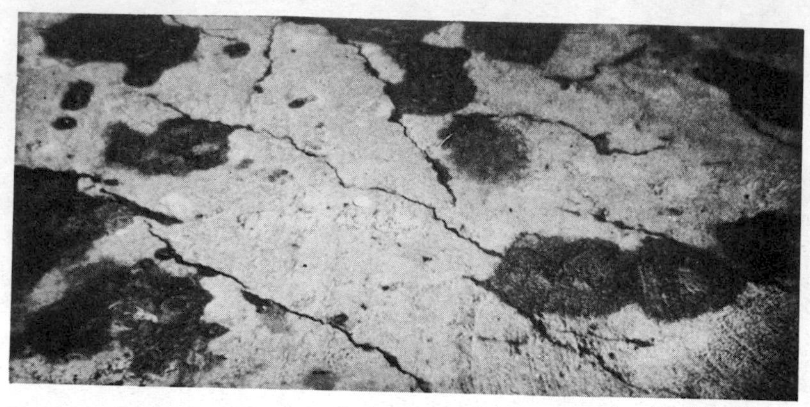

(*Crown copyright*)

Fig. 2.5 Cracks due to initial shrinkage before concrete has hardened

ature as the cement hydrated may, therefore, have been taken up to some extent by plastic movement. When the concrete became rigid, however, further expansion or contraction of the concrete would be measured as a movement of the rigid structure.

Much depends on the temperature of the concrete at the time it changes from a plastic to a rigid state. If, for instance, this occurs at the peak shown in Fig. 2.4, that is at about 30°C, then the shrinkage, which will be important from the practical point of view, will be the thermal shrinkage as the concrete cools from 30°C to the surrounding temperature of 18°C, in the test quoted, and, added to this, the drying shrinkage as the concrete dries out.

The shape of the curve on the left-hand side of Fig. 2.4 indicates the occurrence of thermal shrinkage after the temperature peak has been passed.

The extent of the shrinkage which occurs before the concrete sets is dependent on the efficiency of the curing arrangements and on the extent to which water can be absorbed by the shuttering or formation. It can, therefore, be minimized by taking proper precautions. An example of a concrete road cracked by this initial shrinkage shortly after placing is shown in Fig. 2.5 (see also plastic cracking, p. 19).

Dirty and seriously oversanded aggregates which require a high water/cement ratio to obtain satisfactory workability give concrete which is particularly prone to initial shrinkage cracking. Loss of water and consequent shrinkage of the concrete before or during setting is a fairly common cause of cracking in structural concrete, especially at changes of section. Sometimes, incipient cracking occurring at this early stage becomes more obvious later.

The shrinkage which occurs as the concrete hardens and dries out is due to shrinkage of the cement gel and is partly irreversible to the extent of 0·3 to 0·6 of the drying shrinkage[5] for ordinary concrete mixes, the lower value being more common. It is only partly irreversible because on subsequent wetting it expands, although not sufficiently to recover its original volume. Further drying and wetting produce a practically reversible shrinkage and expansion. These movements are illustrated in Fig. 2.6.

The shrinkage measurements in Fig. 2.4 were commenced at three hours so that virtually all the initial drying shrinkage illustrated diagrammatically in

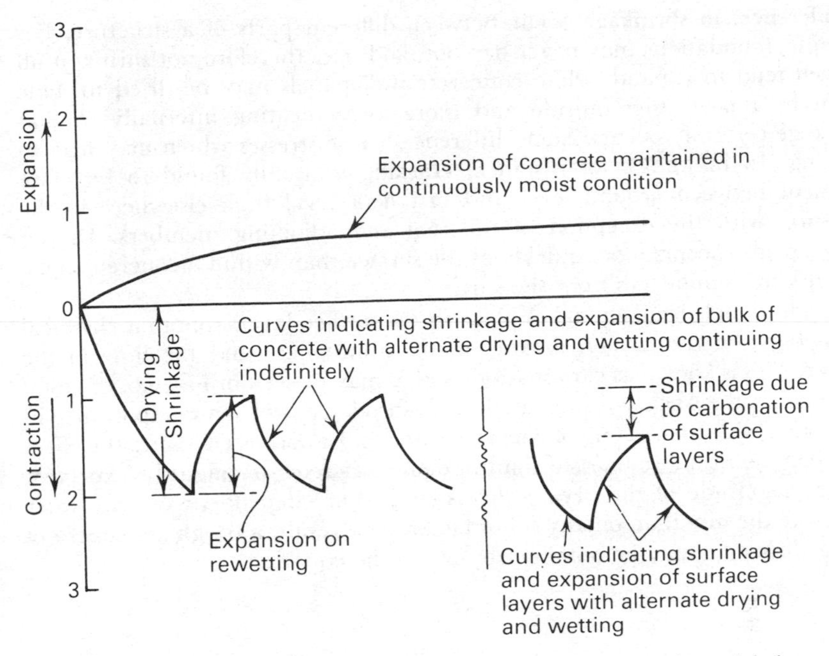

Fig. 2.6 Drying shrinkage and moisture movements with alternate wetting and drying

Fig. 2.6 was included. Shrinkage test procedures, however, vary considerably. For instance, the Building Research Station in their Digest No. 35[7], specify that after 2 days moist-curing the specimens should be dried in air for 26 days, then soaked in water for 4 days. The drying shrinkage is then measured after drying the specimens at 50°C for a further 3 weeks or so until they attain constant length. By this method the original drying shrinkage appears to be missed and what is measured is the smaller shrinkage on the second cycle, see Fig. 2.6.

The American method, ASTM C157-75[8] is quite different. The specimen is first measured wet about 24 hours after mixing. It is then cured for a further 27 days in lime-saturated water and its length again measured. After that the specimen may either be stored again in lime-saturated water or in air at 23 ± 1·1°C and a relative humidity of 50 ± 4 per cent. The expansion in water or the shrinkage in air is measured at specified intervals up to 64 weeks. It will be noted that this test measures the initial drying shrinkage and, since the temperature is much lower than in the BRS test, an entirely different result is obtained (see also thermal effects, p. 20).

The variation in methods of measuring shrinkage makes it very difficult to compare the results obtained from different laboratories. While a standard method of measurement would be useful for comparative purposes, the values obtained would not necessarily be a reliable guide to shrinkage in the structure, where conditions have such an important influence.

Continued immersion in water from the time of casting produces a slight expansion of the order of a quarter of the drying shrinkage, but on drying the concrete shrinks as before.

Differences in shrinkage occur between different parts of a structure. For example, foundations may never dry out, and may therefore not shrink at all or even tend to expand, while concerete wall panels may be dried to some extent by the weather outside and more so by heating internally so that shrinkage certainly occurs. Such differences cause stresses which may lead to cracking. In buildings, for instance, cracking is usually found to be more prevalent between ground level and first floor level than elsewhere in the structure, with the exception of the roof and adjoining members. Drying shrinkage also occurs more quickly at the surface than within a concrete mass, since drying commences from the surface.

An additional shrinkage takes place in the surface layers due to a chemical action between the carbon dioxide in the atmosphere and the lime in the cement. This is known as carbonation, and as may be seen in Fig. 2.6, the total shrinkage at the surface is increased. This shrinkage is of some importance in connection with the crazing of concrete surfaces. In porous concretes the effect of carbonation is more marked than in dense concretes, as might be expected.

The magnitude of the drying shrinkage is more dependent on the water content of the mix than on any other factor. As a result, a rough assessment of drying shrinkage may be obtained by use of the equation

$$S = \frac{1{\cdot}68W - 400}{10\,000}$$

Where S = drying shrinkage, per cent reduction in length
W = water content expressed in kg/m^3 of concrete.

This formula takes account of the cement content of the mix, because of its interdependence on the water/cement ratio. Other factors which influence drying shrinkage are the type of cement; the size and shape of specimen; the surrounding atmospheric conditions; age and the amount of reinforcement.

With so many factors, refinements in methods of assessing drying shrinkage are of little practical importance, and there is not much point in attempting other than a rough assessment by the simple equation above, or from Fig. 2.7 which for some purposes may be more convenient.

The mineral character of the aggregate affects the drying shrinkage of concrete in which it is used. This has been attributed in the past to greater compressibility or to pronounced moisture movements in the aggregate itself. Carlson[9] showed greater shrinkage with the more easily compressible aggregates, while Davis and Troxell[10] found that a concrete containing sandstone shrank twice as much as an otherwise similar concrete. More recently it has been suggested[7][11][12] that some dolerites of Scottish origin exhibit high volume changes on wetting and drying and that concrete made from these aggregates has a considerably higher drying shrinkage than that of concrete made from non-shrinking aggregates, e.g. quartzite gravels, limestones and granite. However, aggregates classified by this particular test as having high drying shrinkages have been used in Scotland for many years. There have been a few reports of undue movements and lack of durability, but the evidence is not conclusive that the aggregate was the major cause.

The presence of clay in aggregates or as a coating on aggregates leads to

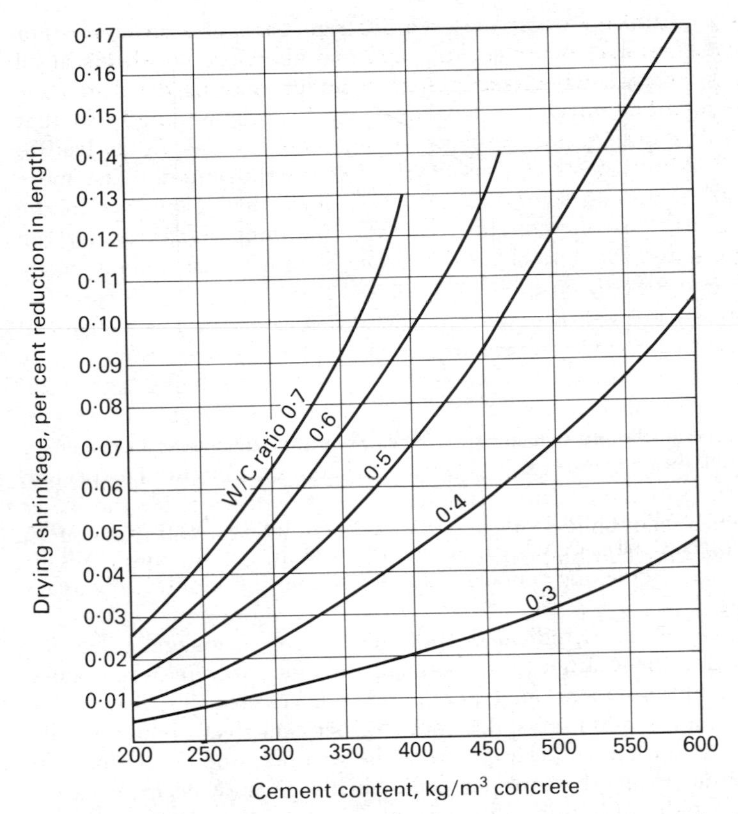

Fig. 2.7 Average values for drying shrinkage of concrete specimens stored in air

increased shrinkage. Clay coatings, in addition to being undesirable owing to poor bond, can increase shrinkage by up to 70 per cent[13].

The rate at which drying shrinkage occurs decreases as the size of specimen increases. For instance, with a small specimen 75 mm square the drying shrinkage under constant temperature and humidity conditions is virtually complete in less than a month, whereas a concrete section, say 1 metre square, continues to shrink for several years under the same conditions.

A further illustration of the influence of size of specimen and age on shrinkage, is given by Atherton[14] who measured the shrinkage in the members of a turbo-generator foundation, and compared this with the shrinkage of a 1 m × 100 mm × 100 mm laboratory specimen. The readings were taken in a 1·8 m (5 ft) thick wall at the steam end of the turbo-alternator block. The shrinkage in the 1·8 m thick wall had reached 0·04 per cent after about 18 months, by which time the smaller specimen had reached 0·06 per cent. The thick wall was still shrinking after 8 years, and had then reached over 0·06 per cent, whereas the smaller specimen was by then shrinking at a much slower rate, and the final shrinkages of the two were about equal. These differences arise because concrete normally only dries from exposed surfaces.

L'Hermite[5] has observed that desiccation may reach a depth of 75 mm in

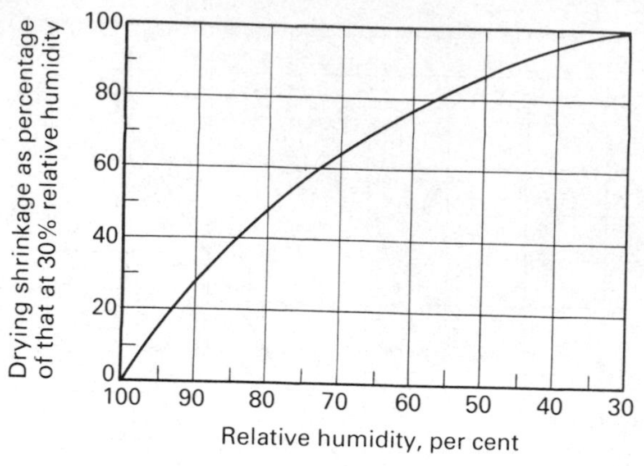

Fig. 2.8 The effect of the relative humidity of ventilated storage conditions on the drying shrinkage of concrete

one month, but only 700 mm after 10 years. This drying of the surface layers gives rise to internal stresses which, if not relieved by creep, may lead to cracking. In practice cracking is not usually noticed until 6 months to 2 years after concreting.

Lea and Davey [15] have indicated that the rate of shrinkage is less for concrete exposed to the weather (near London) than for concrete stored under the constant conditions mentioned above. They give shrinkage values for specimens 3000 mm × 500 mm × 800 mm (0·4 per cent steel), including the seasonal variation of the member exposed to the weather, due to wetting and drying, amounting to nearly 0·01 per cent. Shrinkage may be expected to occur far more rapidly in hot dry atmospheres. An indication of the influence of the relative humidity of the atmosphere is given in Fig. 2.8, based on recommendations by the Comité Européen du Béton [16]. The diagram is plotted to show the drying shrinkage at any relative humidity as related to that at 30 per cent humidity, since the actual coefficient of shrinkage depends on other factors, as already indicated.

Slow drying of concrete as compared with quick drying tends toward reduced cracking because of the compensating effects of extensibility and plastic flow which are thus given time to take effect, and because of increased strength derived from the longer presence of moisture in the concrete.

In normal reinforced concrete drying shrinkage results in a compressive stress in the bars, which themselves resist and reduce the overall drying shrinkage in the reinforced concrete member. It is generally assumed that the drying shrinkage of a reinforced concrete member is half that of an unreinforced member, although the actual value is obviously dependent on the amount of reinforcement used.

The Comité Européen du Béton suggests [16] that the influence of the reinforcement may be taken into account by multiplying the shrinkage coefficients by the term

$$1 - \frac{10 \times A_r}{A_c}$$

where A_r is the cross-sectional area of the longitudinal reinforcement and A_c is the cross-sectional area of the member.

In prestressed concrete, it is necessary to make an allowance for the relief of stress in the steel resulting from creep in the concrete and the steel, and drying shrinkage in the concrete. A good average value for the loss of prestress in the steel is 140 N/mm², although in poor concrete it may be as high as 360 N/mm²[(17)]. Other authorities assess the loss of prestress at between 12 and 18 per cent of the initial steel stress[(18)].

Shrinkage and moisture movements are of importance because they result in cracking of the concrete unless the stresses developed can be relieved by creep or, in the case of large areas, by suitably placed joints. Thus floors, roads, roofs and long walls are subdivided into panels or sections to control cracking due to drying shrinkage and moisture changes, as well as thermal movements.

Plastic cracking

Plastic cracks usually develop before the concrete hardens and normally form with 30 minutes to 2 hours of placing and finishing the concrete. They are due to either shrinkage or settlement or a combination of both.

Plastic shrinkage cracks. Plastic shrinkage cracks on horizontal surfaces may be identified and distinguished from other types of cracks by the following features:

(1) They are roughly straight and often more or less parallel.

(2) They are often, but not always, at an angle of about 45°.

(3) They may vary in length from a few inches to several feet, and seldom extend across the full width of the unit under construction.

(4) They appear within about two hours of placing and finishing the concrete and may be seen soon after, or sometimes before, the water sheen disappears from the surface of the concrete, but they are often not noticed until the day after construction.

(5) They can occur in both unreinforced and reinforced horizontal slabs, either on the ground or suspended.

The principal cause of their occurence is an excessively rapid evaporation of water from the concrete surface. When the rate of evaporation exceeds the rate at which bleeding water would naturally rise to the surface, plastic shrinkage cracks are likely to occur.

The ambient temperature, the relative humidity, the wind speed of the air, and the bleeding characteristics of the concrete are all factors influencing the formation of these cracks. The depth to which these cracks penetrate into the concrete can vary from about 25 mm to the full depth of the section.

Plastic shrinkage cracks are unlikely to affect significantly the structural performance or durability of the concrete section, particularly where it is reinforced. Furthermore, it has been found that plastic cracks tend to retain their original shape and do not increase in length or width with time. The principal concern with this type of cracking is the possible corrosion of embedded reinforcement.

Cracks finer than 0·15 mm on external paved areas are unlikely to permit ingress of moisture to the reinforcement and can usually be ignored. The risk of moisture penetration increases for crack widths between 0·15 mm and 0·5 mm and it is prudent for such cracks to be sealed using a low viscosity polymer or latex emulsion. Cracks wider than 0·5 mm present a risk of dirt and moisture ingress and possible spalling due to frost action, and in this case the line of the crack should be routed out to a width of no more than 20 mm and to a depth of 15 mm and then sealed. Where the slab is to receive a sand-cement screed the risk of corrosion is negligible and hardly warrants sealing of the cracks. Only in the rare and exceptional case of a crack extending across the full width of a ground floor or paving slab, such that it may act as a movement joint, should it be necessary to consider replacing the slab or bay.

Reference is made to measures which may be taken to reduce the risk of plastic shrinkage cracking in Chapters 12 and 16.

Plastic settlement cracks. This type of crack is normally, but not always, confined to the surface of relatively deep sections and they are almost always located over the line of top reinforcement. However, they have been seen in slabs only 150 mm thick.

After the concrete has been compacted and finished the solid particles will settle slightly until prevented by the hardening which results from the chemical reaction between the water and the cement. This settlement displaces water which rises to the surface where it is seen as a sheen. This is the bleeding referred to previously. The settlement of the particles is only very slight and if no reinforcement or change of section is present there is not likely to be any problem. If, however, reinforcement is present and this is fixed in position the settlement of the material directly over the reinforcement will be much less than the settlement at either side and this results in the surface of the concrete being stretched so that a crack, directly over the line of the reinforcement may occur. This type of crack is undoubtedly aggravated by drying conditions. Similarly a change of section may give rigid support preventing the slight settlement of the concrete compared with that adjoining.

Thermal effects

The thermal coefficient of expansion and contraction is the change in a unit length per degree of temperature. Its value varies a little with the richness of the mix, the water content and the type of aggregate, but for all practical purposes it may be taken as $1{\cdot}0 \times 10^{-6}$ per °C for normal aggregate concretes and $0{\cdot}8 \times 10^{-6}$ per °C for lightweight aggregate concretes, for unrestrained movement. In estimating the movement likely to occur in large masses of concrete it is usually safe to assume a value of one-half that given above, the other half being absorbed by creep without a serious danger of cracking.

Thermal expansion and contraction are not often uniform throughout a mass of concrete. The chemical combination of cement and water is accompanied by the generation of an appreciable quantity of heat, which can only escape by conduction to the surface. It follows that the larger the mass, the higher the internal temperature at an early age compared with that at the surface, with a consequent difference in thermal shrinkage as the mass cools.

This leads to the development of tensile stresses conducive to cracking.

When the temperature of concrete is raised by heating, or even by the sun's rays, moisture is lost from the cement gel and a further drying shrinkage results. After the initial thermal expansion, heated concrete therefore tends to shrink again to some extent as this further drying gradually occurs. As a result, the stress conditions in concrete that is subject to temperature changes are complex. Examples are tanks for hot liquids, roads, roofs, and heated floors, particularly when adjoining parts of the structure are not heated so that a thermal gradient is also present. Warping and cracking in these circumstances are not uncommon, and considerable ingenuity in designing expansion joints may be necessary if they are to be avoided (see page 180).

Both shrinkage and creep are affected by temperature. Ross[19, 20] has made extensive studies which have provided much useful data on these relationships.

Durability

The useful life of concrete is ordinarily limited by the disintegrating effects of:

(1) Weathering by the action of rain and frost, and by the expansion and contraction resulting from alternate wetting and drying.

(2) Chemical attack by such agencies as sea water; moorland, marsh and other waters; industrial chemicals and wastes; sewage; animal and vegetable oils, fats, grease; milk; and sugars.

(3) Wear by abrasion from foot and vehicular traffic, by wave action, and by water-borne and wind-borne particles.

Weathering by rain and frost action is chiefly a function of watertightness or impermeability, since leaching and attack by the carbonic and other acids present in rainwater, and disruption by frost action, depend on the penetration of water into the surface. The mechanism of rain and frost action and the effect of chemical attack by the various agencies listed are not in themselves properties of concrete, and will be described more fully in a later chapter.

The quantity of cement used in concrete does not affect the resistance to weathering to a very large extent unless the cement content is not sufficient to fill the interstices between the aggregate. Glanville[21] investigated the effect of cement content on permeability and found it was of no great advantage to use more than about 300 kg of cement per m^3 provided only sufficient water was used to make the mix workable. To obtain the same degree of impermeability, very wet mixes required a higher cement content, say 400 kg or more of cement per m^3.

Permeability

Concrete has a tendency to be porous due to the presence of voids formed during or after placing. First, it is usually necessary, in order to obtain workable mixes, to use far more water than is actually necessary for chemical combination with the cement. This water occupies space, and when later it dries out, leaves behind air voids. In addition to these initially water-occupied

Fig. 2.9 Efflorescence resulting from the percolation of water through a defective construction joint, bringing with it chemical salts

voids, there is always a small percentage of entrained air voids. Secondly, there is a decrease in absolute volume of the cement and water after chemical reaction and drying out, so that the set cement paste, when dry, occupies less volume than the fresh paste, whatever the water/cement ratio.

If proper care is taken, Portland cement can be made sufficiently impermeable for most purposes without the addition of any special materials. From the above analysis of the cause of voids, it is clear that for the most dense and least permeable concrete, the water/cement ratio should be reduced to a minimum consistent with adequate workability for thorough compaction without segregation. A compromise is, therefore, required between a low water/cement ratio and adequate workability, and the best conditions will depend on the type of structure and the method of compaction.

Other factors influencing permeability are:

(1) The soundness and porosity of the aggregate.

(2) Age. The permeability decreases with age, the decrease being greater for wet mixes than for drier ones. Glanville, in Building Research Technical Paper No. 3[21], has shown that after a year there is only a small difference between the permeability of mixes of normal consistency and very wet mixes.

(3) The grading of the aggregate should be chosen to give the most work-

able concrete with the least water. Harsh gradings, especially as regards the sand, should be avoided.

(4) Curing has an important influence, and it is necessary to keep the concrete moist particularly during the first few days.

Joints are the weakest point in a concrete structure, and an example of weeping through a construction joint is given in Fig. 2.9[22]. If impermeability is important, it is wise to provide some form of mechanical seal, as indicated on p. 182, since it is difficult to avoid local seepage due to segregation, poor compaction or some other occurrence. Whether such measures are considered necessary or not, it is essential that every effort should be made to obtain a good bond.

Permeability is very difficult to measure accurately, and tests are not often made outside the research laboratory.

Resistance to abrasion

Resistance to abrasion is directly related to crushing strength and in general it is safe to assume that concrete having the highest crushing strength has the greatest resistance to abrasion. Occasionally tests are made on the wear of paving flags, by exposing them to abrasion by steel balls for 48 hours, and finding the loss in weight. The method is described in British Standard 368 : 1971 for Precast Concrete Flags.

A tough igneous rock aggregate is often preferred to harder and more brittle flint aggregate in paving concrete.

Autogenous healing

Many tests have shown that the finer cracks in fractured concrete will completely heal under moist conditions. Apparently when the fracture occurs, unhydrated cement is exposed, and this in the presence of moisture hydrates and sets.

REFERENCES

1 BS 1881 : 1970, *Methods of Testing Concrete*, British Standards Institution.

2 NEVILLE, A. M., Creep of concrete as a function of its cement paste content, *Mag. Concr. Res.*, **16**, No. 46, March 1964. Cement and Concrete Association.

3 NEWMAN, K., Properties of concrete, *Structural Concrete*, **2**, No. 11, Sept./Oct. 1965. Reinforced Concrete Association.

4 DAVIS, R. E. and H. E., Flow of concrete under the action of sustained loads, *J. Amer. Concr. Inst.*, March 1931.

5 L'HERMITE, R., Volume changes of concrete, *4th International Symposium on the Chemistry of Cement*, Washington D.C., 1960.

6 SWAYZE, M. A., Early concrete volume changes and their control, *Proc. Amer. Concr. Inst.*, **38**, April 1942, pp. 425–440.

7 *Shrinkage of Natural Aggregates in Concrete*, Building Research Station Digest No. 35, June 1963. H.M.S.O.

8 ASTM Designation C157–75, *Standard Test Method for Length Change of Hardened Cement Mortar and Concrete*, ASTM Standards, Part 14; 1976, p. 111.

9 CARLSON, R. W., Drying shrinkage of concrete as affected by many factors, *Proc. Amer. Soc. Civ. Engrs*, **38**, Pt. II, 1938, pp. 419–437.

10 DAVIS, R. E., and TROXELL, G. E., Volumetric changes in Portland cement mortars and concretes, *J. Amer. Concr. Inst.*, **25**, 1929, pp. 210–260.

11 SNOWDEN, L. C. and EDWARDS, A. G., The moisture movement of natural aggregate and its effect on concrete, *Mag. Concr. Res.*, **14**, No. 41, July 1962. Cement and Concrete Association.

12 JENKINS, R. A. SEFTON, PLOWMAN, J. M. and HASELTINE, B. A., Investigation into the causes of the deflection of heated concrete floors, including shrinkage, *Structural Engineer*, **43**, No. 4, April 1965, pp. 105–117.

13 POWERS, T. C., Causes and control of volume change, *Portland Cement Assoc. J. Res. and Development Laboratories*, **1**, No. 1, Jan. 1959.

14 ATHERTON, A., in Discussion on RAE, F. A., Design and construction of a reinforced concrete foundation block for a 200-MW turbo-generator, *Proc. Inst. Civ. Engrs*, July 1963, pp. 379–380.

15 LEA, F. M. and DAVEY, N., The deterioration of concrete in structures, *Paper No. 5722*, Inst. Civ. Engrs, 1949.

16 COMITÉ EUROPÉEN DU BÉTON, *International Recommendations for the Design and Construction of Concrete Structures*, 1970. Published by the Cement and Concrete Association.

17 FLUCK, P. G. and WASHA, G. W., Creep of plain and reinforced concrete, *J. Amer. Concr. Inst.*, Title No. 54–59, **29**, No. 10, April 1958.

18 MAGNEL, GUSTAVE, Creep of steel and concrete in relation to prestressed concrete, *J. Amer. Concr. Inst.*, **19**, No. 6, Feb. 1948.

19 ROSS, A. D., The elasticity, creep and shrinkage of concrete, *Proc. Conference on the Mechanical Properties of Non-metallic Brittle Materials*, London, April 1958. Butterworths Scientific Publications.

20 ROSS, A. D., ILLSTON, J. M. and ENGLAND, G. L., Short and long-term deformations of concrete as influenced by its physical structure and state, *International Conference on the Structure of Concrete, 1956, Paper* No. H.1. Cement and Concrete Association.

21 GLANVILLE, W. H., *The Permeability of Portland Cement Concrete*, Building Research Technical Paper No. 3, revised edition. HMSO.

22 DAVEY, N., The surface finishing of concrete structures, *J. Inst. Civ. Engrs*, April 1942.

3
Aggregates

MOST factors pertaining to the suitability of aggregate deposits are related to the geological history of the surrounding region. The geological processes by which a deposit was formed or by which it was subsequently modified are responsible for its size, shape and location, the type and condition of the rock, the grading, rounding, and degree of uniformity, and a number of other factors which are pertinent to the question of utilization.

The most important properties of aggregates are the crushing strength and the resistance to impact: other important properties include the size and shape of the particles, which can affect the bond with cement paste; the porosity and water absorption characteristics affecting the resistance to frost attack and chemical attack; and the immunity from shrinkage.

Sand and gravel

The most widely used and economical aggregates in the UK are sand and gravel. Natural sand and gravel deposits occur as superficial (drift) deposits either laid down by rivers or as glacial and fluvioglacial spreads left behind when the ice sheets melted.

River deposits are still the most common and generally the most satisfactory because they have a more consistent grading as a result of a sorting action by the rivers, the individual pieces are usually rounded or irregular, and abrasion caused by river transportation and deposition leads to a partial elimination of the weaker particles. In general, river gravels are fairly uniform in thickness and deposits are worked varying between 1 and 6 m. The Thames and Trent valleys are major river sources of high quality gravels.

Glacial deposits, although of widespread occurrence, are not all economically workable and are more likely to be less uniform in quality, although thicknesses of 10 m are not uncommon. Major workings in these gravels are in the Vale of St. Albans, East Anglia, the Midlands and part of North England. The Lancashire and Cheshire deposits are mainly sand, while in Clwyd glacial deposits are a major source of gravel. In Scotland high quality gravels occur north of the Midland Valley, but some gravel deposits in the Southern Uplands have adverse effects on the shrinkage properties of concrete.

Gravel and sand are also dredged from river estuaries, mainly sand from the Bristol Channel and Liverpool Bay which is used with crushed rock in concrete. Production from the North Sea and English Channel has increased considerably over the last 10 years. Although marine-dredged aggregates have been used to a limited extent for many years, this increased production and use has given cause for some concern over acceptable limits for shell and salt contents.

Crushed rock

Various types of rock, when crushed, are suitable for use as aggregates in concrete.

(1) *Limestones* are sedimentary rocks composed chiefly of calcium carbonate. The harder and denser types of limestone, particularly the carboniferous types found in areas such as Derbyshire and the Mendips, are very suitable for concrete. Less hard and dense are the oolitic limestones represented by Portland stone, Bath stone, Clipsham stone, etc., which tend to be used more for architectural cast stone. They would be unsatisfactory as a road aggregate because, in addition to poor resistance to wear, they would be liable to become saturated and disintegrated by frost action. Chalk is normally unsuitable because it is too soft.

Dolomitic limestones contain a large proportion of magnesium carbonate in addition to calcium carbonate and, provided that they are hard and dense, are suitable as aggregate for concrete.

(2) *Igneous rocks*. These include granites, basalts, dolerites, gabbros and porphyries. Some confusion results from the commercial use of the term 'granite' to cover not only most of the igneous rocks, but also some types of limestone. Other local names include: whinstone, toadstone, and greenstone for dolerites and basalts; and elvan for porphyries.

Granite is hard, tough and dense and is an excellent aggregate for concrete. Geologically, granite is coarsely crystalline in structure, free from bedding planes, and is composed of quartz, felspar, mica, and hornblende. In Cornwall, particularly St. Austell, granite sands are found. These have been formed by disintegration of the felspar and hornblende into china clay, leaving the quartz crystals as a sand. In the selection of such material, care should be taken to ensure that the sand is washed free from china clay and that excessive amounts of mica are not present (see p. 46).

Basalts are igneous rocks similar to granite, but with a much finer grain structure due to more rapid cooling when they were formed. They are excellent aggregates.

Dolerites have a fine crystalline grain structure and contain more felspars and such materials as augite and olivine than the granites. Some dolerites, when used as aggregate for concrete, may cause cracking and disruption of the concrete. It has been found that the deleterious rocks expand and contract with changes in humidity. Mineralogical examination has shown that such aggregates contain decomposed ferro-magnesium silicates, such as olivine. The commonest mineral found in these circumstances is montmorillonite. In some parts of the world, the disintegration of some igneous rocks may be far more serious; in Australia for instance, rock which looks as hard as Cornish granite may become a heap of clay in a matter of months simply by exposure to the weather.

(3) *Sandstones*. When hard and dense, most sandstones are suitable as aggregates. The best are those which are composed of quartz grains cemented with hydrated iron oxide or amorphous silica, known geologically as ferruginous and siliceous sandstones. Calcareous sandstones cemented with calcium carbonate are more liable to attack by the carbonic and sulphuric acids present in the atmosphere. Imperfect cementation of the constituent grains

makes some sandstones friable and very porous and they are then unsuitable aggregates.

Sandstones range from the harder types composed of very close grains, through the softer and more loosely grained grits, to the sandy shales, in which the presence of clay causes softness, friability and high absorption.

(4) *Shales* are usually poor aggregates, being soft, weak, laminated and absorptive. In addition, the normal flat shape of the particles makes compaction of any concrete in which they are used, very difficult.

(5) *Metamorphic rocks* are variable in character. Marbles and quartzites are usually massive, dense, and adequately tough and strong, providing good aggregates. However, some schists and slates are often thinly laminated and, therefore, unsuitable.

Artificial aggregates

Blastfurnace slag is probably the most widely used of the artificial aggregates, more so on the Continent and in the USA than in Britain. It is produced in blast furnaces during the production of pig iron and consists of the silicates and alumino-silicates of lime. Slags vary in hardness, and any which 'dust' or disintegrate due to the presence of too much ferrous iron, should be avoided. Many unsound slags may be detected by immersing them in water for a period of two weeks, during which time they should show signs of disintegration. Laboratory methods of testing for soundness of blastfurnace slag are given in BS 1047: Part 2: 1974[1]. These are: (i) a chemical analysis for total sulphur, acid-soluble sulphate, silica, alumina, lime, magnesia, iron oxide, and titanium oxide; (ii) tests for stability including iron unsoundness; an analytical test for falling, dusting or lime unsoundness; and a microscopic test; (iii) a water absorption test. A minimum bulk density of 1250 kg/m^3 is also required.

Clean broken-brick of good quality provides a satisfactory aggregate, the strength and density of the concrete depending on the type of brick; engineering and allied bricks when crushed make quite good concrete of medium crushing strength. In using second-hand bricks it is essential that all plaster should be removed, otherwise the calcium sulphate present is liable to prevent or delay setting, and to cause disintegration in a short time. Bricks containing soluble sulphates in excess of 0·5 per cent should also be avoided. Brick aggregate should be saturated before use, because of its relatively high absorbency. Porous types of brick should not be used as aggregate in reinforced concrete work owing to the danger of penetration of moisture which may lead to corrosion of the steel reinforcement.

Glass is not a reliable aggregate, since most varieties possess strong alkali-aggregate reactivity.

Lightweight aggregates

Some aggregates, such as clinker, breeze, pulverized fuel ash (fly ash), pumice, foamed slag and wood waste, are used in the manufacture of lightweight non-structural concrete. Artificial and processed aggregates, notably expanded clays and shales, foamed slag, and sintered fly ash pellets, are being used on an increasing scale for lightweight structural concrete.

Fig. 3.1 Vibrating screens and overhead storage hoppers at a gravel pit in the London area

All these aggregates are discussed in Chapter 24 dealing with lightweight concrete.

Production of sand and gravel aggregates

Sand and gravel are obtained either from pits or by dredging from the bottom of rivers or from the sea bed. Pits are termed 'wet' or 'dry', according to the water level in the vicinity, although some pits which would otherwise be wet are dried out by pumping. The deposits are usually covered with overburden which has to be removed by scrapers, mechanical shovels, draglines or by sluicing. The overburden consists of the soil and stones mixed. Removal is not always straightforward owing to the presence of faults and dips in the gravel seams, and unless care is taken to remove these materials, occasional loads of dirty aggregate occur.

The sand and gravel are then excavated by mechanical shovels in dry pits; by draglines in both wet and dry pits; and by means of crane and grab, or suction pumps, in wet pits. Gravel seams often overlie beds of clay, and it is especially necessary that the inclusion of lumps of clay in the gravel should be avoided, since even with vigorous washing the clay lumps are extremely difficult to remove. The danger of the inclusion of clay lumps is greater in wet pits than in dry pits where draglines or grabs are used, because the driver is unable to see the bottom, and, therefore, to control the depth to which he digs.

The mixed gravel and sand is usually conveyed from the pit to the screening plant by conveyor belt or in light trucks, or by pipeline when the gravel is raised by a suction pump. Conveyor belts are commonly used to raise the materials from ground level to the level of the washing and screening plant.

Screening and washing of the gravel and sand is carried out either by means of cylindrical rotating screens or by the more modern vibrating screens. A recent improvement to vibrating screens has been made by carefully designing

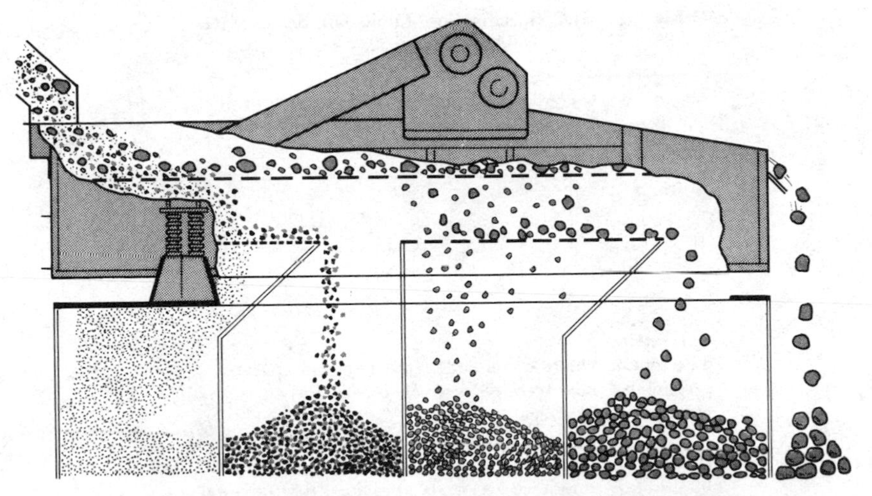

Fig. 3.2 Screening action of a Parker 'Powerflow' horizontal vibrating screen

Fig. 3.3 Overhead classifiers used for improvement of the grading of sands at a gravel pit in the London area

them to produce a resonant effect. The sand and dirty water are usually carried away together and are then dewatered by another unit. From the screening unit, the sand and gravel are fed by gravity into suitable hoppers. Oversize material goes to a crusher and is then returned for rescreening. A typical arrangement of screening plant is given in Fig. 3.1. The screening action of a horizontal vibrating screen is illustrated in Fig. 3.2. The sand at some pits is treated further to modify its grading, by means of classifiers, such as those seen in Fig. 3.3.

Table 3.1 GLC Specification Limits for Sea-dredged Aggregates

Nominal size of aggregate mm	Allowable percentage by weight of dry aggregate of shell as calcium carbonate
40	2
20	5
10	15
sand	30

Salt content
The sodium chloride content of the fine and coarse aggregate shall not exceed respectively 0·10% and 0·03% by weight of dry aggregate.

If either aggregate exceeds the above limits, the material shall still be considered acceptable provided that the total sodium chloride concentration derived from the aggregates is not greater than 0·32% by weight of the cement in the mix.

Marine-dredged aggregate, although contributing increasingly to the overall production figures, is only commercially workable in a few areas of limited extent. Marine-dredged material is, however, usually quite clean apart from the presence of sea shells and salt. Washing with frequently changed fresh water will reduce the salt content but the disposal of such wash water can present problems at some processing plants.

The increasing use of sea-dredged aggregates in concrete created some concern about the effects of shell and salt content and in 1968 the Greater London Council introduced specification requirements relating to shell content and salt content, as shown in Table 3.1.

Other authoritative sources, such as CP110 and the Property Services Agency (PSA) of the Department of the Environment have introduced limits which differ somewhat from those shown in Table 3.1.

One method used for obtaining aggregates from the bed of the North Sea, 8 km from the coast in depths up to 30 m, is by the use of a steel suction tube about 600 mm diameter. The steel tube is pivoted over the side of the ship until the lower end rests on the sea bed. The aggregate is then pumped up the tube through a large impeller and into the open hold of the ship. Using this method, about 1500 tonnes of aggregate can be loaded in about 1 hour.

On discharging the aggregate at the plant, the material is processed in the normal manner.

In hot desert countries, such as those in the Middle East, aggregate supplies are usually a serious problem. Apart from the shortage of good quality natural material in most areas, many of the available sources are contaminated by the presence of deleterious salts. This applies to both sand, which is usually abundant but with poor grading and particle shape, and to sources of limestone, which in addition may also be of a friable and porous nature.

Selective winning and processing of both sand and coarse aggregate can reduce the amount of sulphates, chlorides, silts, clays and dust to some extent, but more care than can usually be achieved is needed, even following intensive and detailed examination of potential sources.

Further information can be obtained from Fookes and Collis [2] and also in Chapter 26.

Although these desert countries consist very largely of sand, it is often quite difficult to find a satisfactory source near the site, since most of the sand is too fine, and frequently contains a considerable proportion of silt. Usual sources are old river valleys or wadis, in which the sand is dug from pits. Since the fine sand and dust on the desert surface is continually drifting in the wind, particularly during sand storms, sand pits and stockpiles soon become silted up, causing increased difficulty in obtaining a uniform and satisfactory grading, and cleanliness.

Any question of washing the aggregate may be quite impracticable due to the absence of water, except perhaps for meagre supplies of brackish water, which almost invariably contains appreciable quantities of sulphates and chlorides, and would therefore not be effective in producing a salt-free material.

Production of crushed stone aggregates

The preparation of crushed stone aggregates is not as easy as that of gravel. The overburden must first be removed, the method depending on the formation of the top surface of the rock. The rock is next blasted and conveyed in lumps in trucks, overhead skips, or by belt conveyor to the crushing plant. A primary crusher of the studded roll, gyratory, jaw, or cone type reduces most of the material to less than 75 mm diameter. The crushed rock is taken, usually by belt conveyor, to the primary screen which separates the material above and below the 75 mm size. That passing the primary screen is conveyed to the final screens, which sort it out into the various sizes required. Material retained on the 75 mm primary screen is passed through secondary crushers and again returned to the screens.

Six types of crushers are used for the production of crushed stone aggregate. These are:

(1) Jaw crushers consisting of one or more swinging jaws operating against a fixed jaw, the distance apart and the length of movement determining the ultimate size of the crushed stone. The single toggle type is used for the softer stones, while a double toggle arrangement gives a more powerful thrust to crush harder stones.

(2) Gyratory crushers in which a crushing head is rocked by an eccentric on the inclined revolving shaft which carries it.

(3) Disc crushers comprising one stationary and one revolving saucer-shaped disc which open and close as the material passes through.

(4) Hammer or impact crushers of various types. Hammer crushers are subject to considerable wear and maintenance costs are liable to be heavy.

(5) Roll crushers in which the crushing is effected by feeding the material between the toothed, serrated or corrugated surface of one roll and the similar

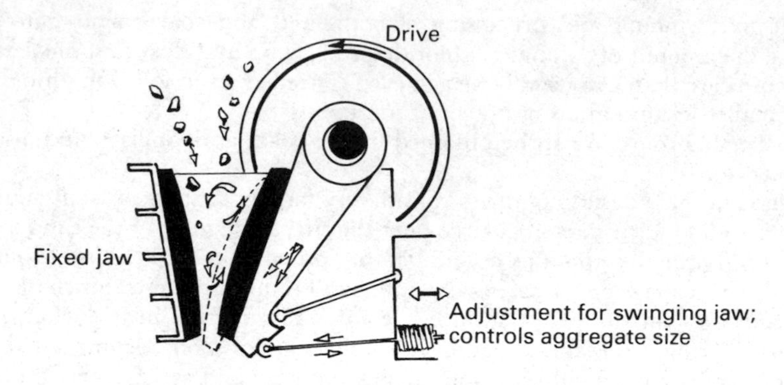

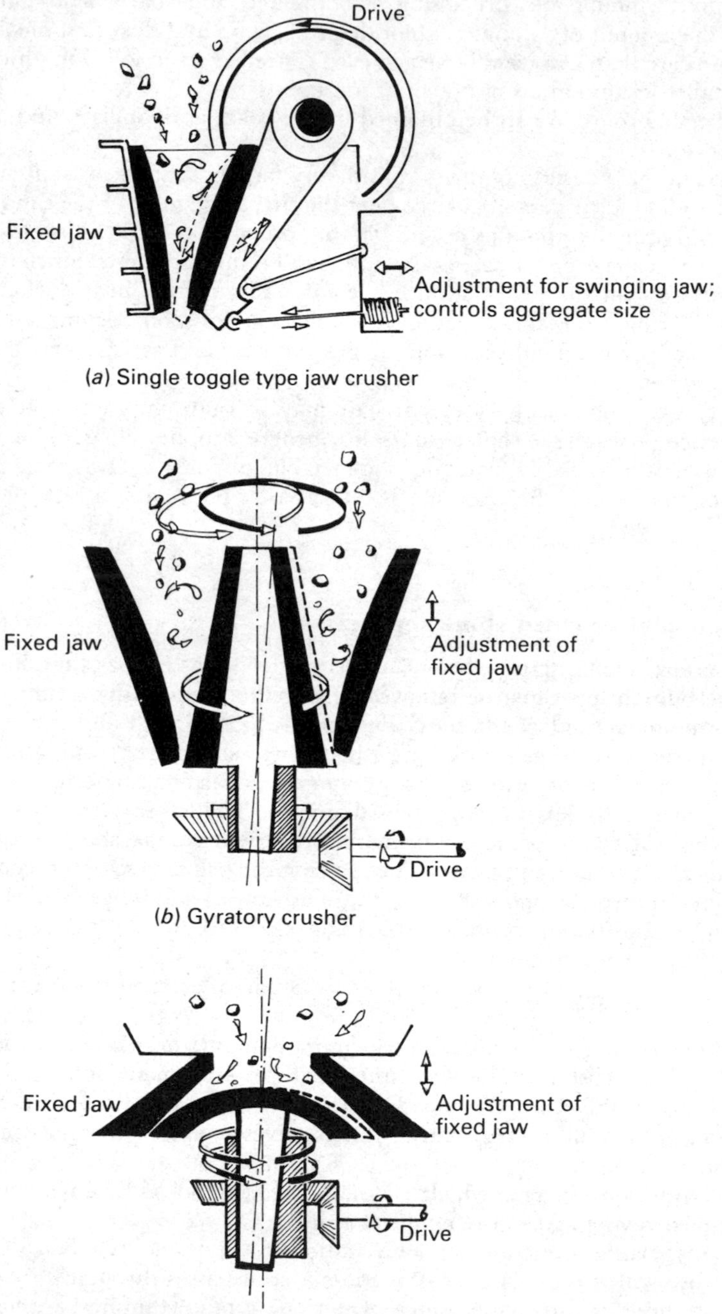

(*a*) Single toggle type jaw crusher

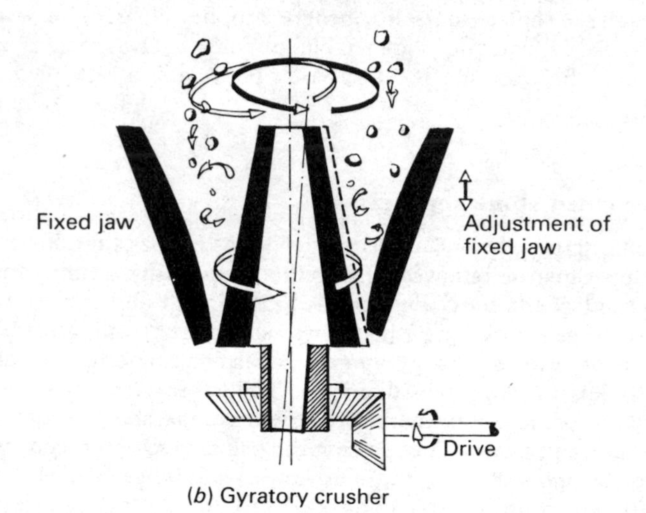

(*b*) Gyratory crusher

Fixed jaw
Adjustment of
fixed jaw
Drive

(*c*) Disc or gyrosphere crusher

Fig. 3·4 Action of jaw, gyratory and disc crushers

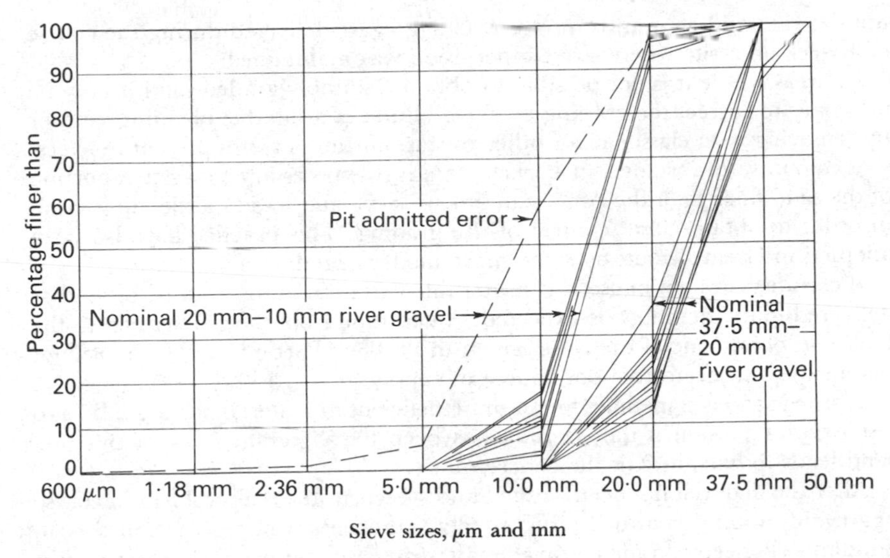

Fig. 3.5 Gradings of a typical day's delivery of 37·5 mm–20 mm and 20 mm–10 mm river gravel from a pit with a large washing and screening plant

or smooth surface of the other. Uneven wear of the rollers causes difficulties in maintaining an even size.

(6) Rod mills are replacing roll crushers for the reduction of fines, being more economical, and giving a more uniform product.

The action of jaw, gyratory and disc crushers is shown diagrammatically in Fig. 3.4. Hammer and gyratory crushers are usually regarded as the most satisfactory since the aggregate produced is more cubical in form.

Further information on factors governing the grading and shape of crushed rock is given in a survey of the literature by Shergold and Greysmith [3].

The screens may be of the revolving type used for sand and gravel, or of one of several types operated by vibration.

The commercial grading of aggregates

Aggregates cannot be screened commercially to the same degree of accuracy as that attained in the laboratory. Several factors are responsible for the grading variations which normally occur, even with modern screening plants. Perhaps the most important factor is the speed at which the material is passed through the screens, any increase in speed resulting in a greater proportion of undersized material being retained. On the other hand, the presence of oversize material in an aggregate usually results from an increase in the size of the screen meshes due to wear.

Aggregates from properly equipped washing and screening plants may be expected to be of reasonably uniform grading, but it is advisable to inspect each load as it is delivered, and to make check tests on the grading from time to time. An example of the sort of grading variation which may be expected on the site is given in Fig. 3.5. The sieve analyses were made on samples of 37·5 mm–20 mm and 20 mm–10 mm gravel aggregate delivered during the course of a week on a site where strict supervision was maintained.

In areas where it is not possible to obtain a suitably graded sand it may be necessary to correct the grading by the addition of a suitable blending sand or by removing with classifiers, or other means, portions of sizes present in excessive amounts; for instance, at Parker Dam it was necessary to waste a portion of the sand finer than the American No. 16 sieve and to add a blending sand, in order to obtain closer control of the grading. This practice has also been adopted in Great Britain by some more modern producers.

If classifiers are not in use, a considerable variation in the grading of sand as delivered to the site must be expected. A variation from the coarser end to the finer end of any one of the zones given in BS 882 : Part 2 : 1973 [(4)] should not be unexpected, even from one source of supply.

Washed sand is sometimes found to be deficient in material passing a BS 410 test sieve of 300 μm if the washing has been too vigorous. Lack of this fine sand leads to harshness in the concrete mix.

After the material has been washed and screened, the real problem in coarse aggregate grading control begins; namely, prevention of segregation due to handling. Perfectly graded gravel may segregate during a single incorrect stockpiling operation to such an extent that the workability of the concrete in which it is used will be unsatisfactory. Tests [(5)] made at Hoover Dam showed that when the fine gravel 20 mm to 5 mm was dropped from a belt into a conical pile and recovered through a gate beneath the centre of the pile, the proportion of 20 mm to 10 mm material varied from 55 per cent to 74 per cent, and the 10 mm to 5 mm material from 42 per cent to 16 per cent. Correct and incorrect methods of stockpiling and handling aggregates are shown in Figs. 3.6. and 3.7.

Sampling of aggregates for testing

Before aggregates are accepted from any particular pit or other source of supply, it is very desirable that a visit should be made in order to inspect the methods of removing overburden, the variations in strata, pockets of unsuitable material, and methods of digging. Furthermore, the final choice of the source of supply of any aggregate should be based not only on the results of tests in the laboratory, but also on the inspection of the method of production and organization at the pit. Only in this way is it possible to eliminate sources of supply likely to give trouble by large fluctuations in grading, occasional dirty loads, breakdown in continuity of supplies due to failure of worn-out plant at the pit, and other similar causes.

Representative samples of the material as stockpiled or in the bins ready for delivery should be obtained at the same time. The method of taking samples will have an important influence on the accuracy of the results, and reference should be made to BS 812 : Part 1 : 1975 [(6)], for the correct procedure.

Important points to note are:

(1) Under favourable conditions at least ten portions (increments) should be drawn from different parts of the bulk. All the increments should be combined to form the main sample to be sent to the laboratory. The quantity sent to the laboratory should not be less than that shown in Table 3.2.

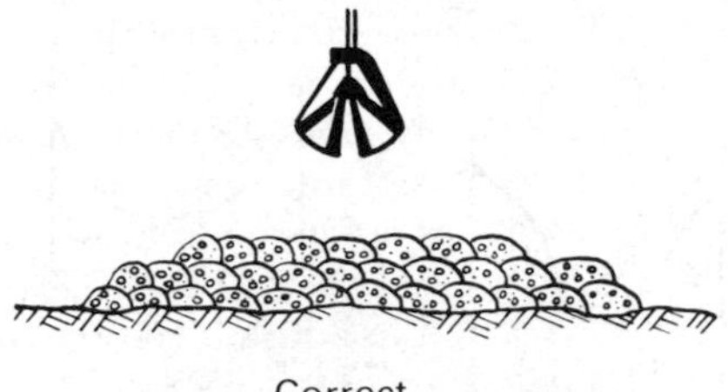

Correct

Methods which place material in the pile in individual units not larger than a lorry load and which do not permit the aggregate to run down the slopes at the edge of the pile

Incorrect

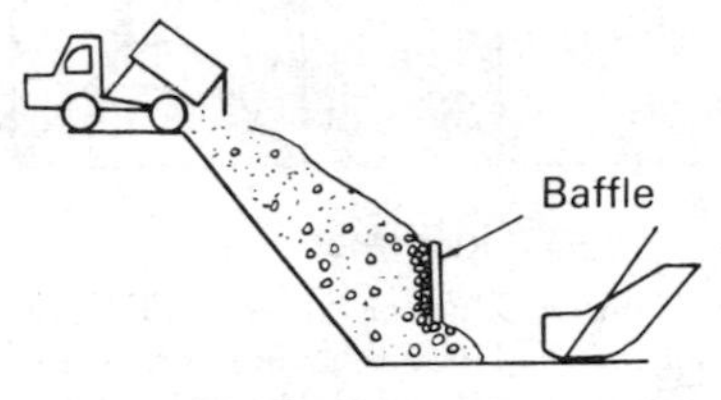

Correct

Use of baffle to trap larger particles which run down slope

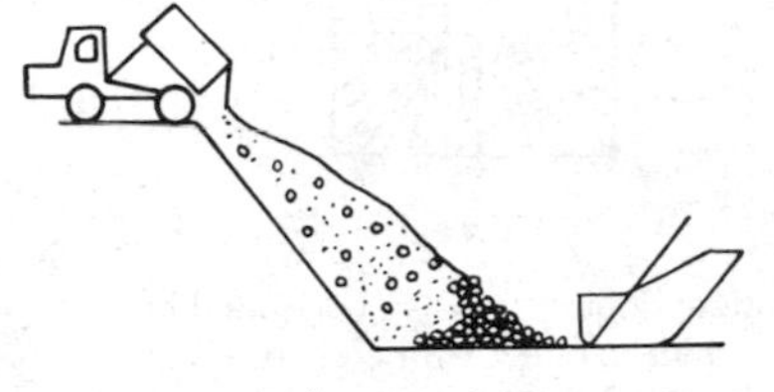

Incorrect

Methods which permit the aggregate to roll down the slopes as it is added to the pile

(a) **Handling of coarse aggregate**

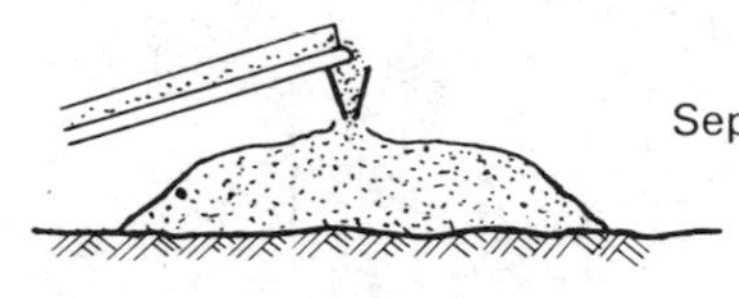

Separation

Wind

Correct

Use mobile conveyor with drop chute kept as low as possible and light side screens to prevent wind blowing smaller particles off conveyor.

Incorrect

Free fall of material from high end of stacker permitting wind to separate fine from coarse material.

(b) **Handling of fine aggregate**

Fig. 3.6

(2) In difficult circumstances the quantities shown in Table 3.2 may not be practicable, and a greater number of increments will be required in order that the main sample may be larger and therefore more truly representative of the bulk. It is difficult to sample accurately from stockpiles, where there may be segregation of the coarser material towards the base and sides. Similarly, 'sandwich loaded' vehicles, into which different sizes of aggregate have been loaded one after another, cannot be sampled satisfactorily and a proper sample can only be taken after the load is mixed.

(3) Sampling is best carried out when aggregate is being loaded into or

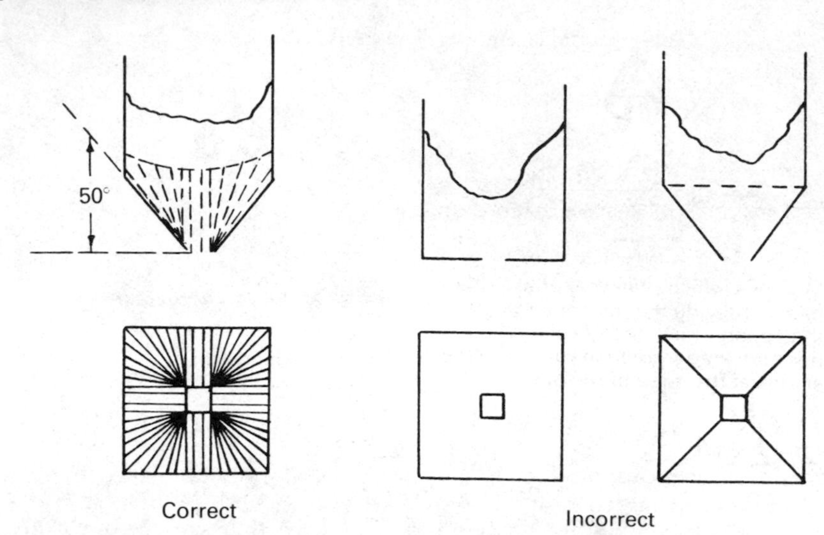

Bin bottom sloping 50° from horizontal in all directions to outlet with corners of bin properly rounded.

Flat bottom bins or those with any arrangement of slopes having corners or areas such that all material in bins will not flow readily through outlet without shovelling.

Slope of aggregate bin bottoms

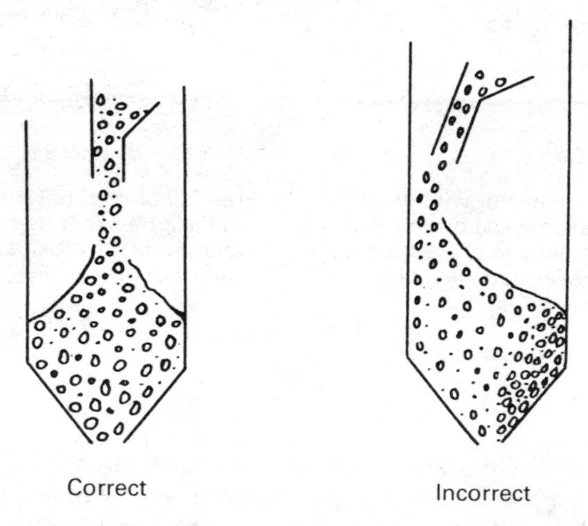

Material drops vertically into bin directly over the discharge opening permitting the discharge of more generally uniform material.

Chuting material into bin at an angle. Material falling other than directly over opening not always uniform as discharged.

Filling of aggregate bins

Fig. 3.7 Correct and incorrect methods of handling aggregate

Table 3.2 Minimum Weights for Sampling of Aggregates

Nominal size	Minimum weight of main sample kg
28 mm and larger	50
Smaller than 28 mm and larger than 5 mm	25
5 mm or smaller	13

Note. Larger samples than those shown above may be required when it is known that certain tests will require larger quantities. In addition larger samples will usually be required if trial mixes are to be made.

unloaded from vehicles, or when it is being discharged from a conveyor belt.

(4) When sampling rock direct from a natural outcrop or from a pit or quarry face, special care should be taken to ensure that samples are truly representative of the range of variations over the whole of the quarry face, vein or bed, later to become the source of aggregate.

Sample reduction

At the laboratory, the main sample should be reduced to the quantity required for testing, using a sample divider which extracts representative parts of the whole, or by quartering, Fig. 3.8. A suitable form of sample divider is the riffle box shown in Fig. 3.9. For reduction on the sample divider, the

Fig. 3.8 Quartering an aggregate

Fig. 3.9 Riffle box for dividing a sample of aggregate

material should be surface dry. When reduction is made by quartering, fine and all-in aggregates are most accurately mixed and quartered when in a damp condition, if necessary achieved by sprinkling with clean water before mixing. For full details of the method of quartering reference should be made to BS 812 : Part 1 : 1975.

Sieve analysis (grading)

The following definitions have been accepted as Standard in Great Britain:

(a) *Fine aggregate.* The term fine aggregate shall mean aggregate mainly passing a 5 mm BS 410 test sieve and containing only so much coarse aggregate as is permitted for the various grading zones described in BS 882 : Part 2 : 1973.

(b) *Coarse aggregate.* The term coarse aggregate shall mean an aggregate mainly retained on a 5 mm BS 410 test sieve and containing only so much

Table 3.3 British and Equivalent American Sieves

BS 410 test sieve nominal aperture size		Equivalent ASTM E11–70 sieve designation	
Metric	Imperial equivalent	Standard width of aperture	ASTM sieve No.
37·5 mm	$1\frac{1}{2}$ in	38·1 mm	($1\frac{1}{2}$ in)
20·0 mm	$\frac{3}{4}$ in	19·0 mm	($\frac{3}{4}$ in)
10·0 mm	$\frac{3}{8}$ in	9·5 mm	($\frac{3}{8}$ in)
5·0 mm	$\frac{3}{16}$ in	4·76 mm	4
2·36 mm	No. 7	2·38 mm	8
1·18 mm	No. 14	1·19 mm	16
600 μm	No. 25	595 μm	30
300 μm	No. 52	297 μm	50
150 μm	No. 100	149 μm	100
75 μm	No. 200	74 μm	200

finer material as is permitted for the various sizes described in BS 882 : Part 2 : 1973. Coarse aggregate may be described as uncrushed gravel, crushed gravel, crushed stone, or partially crushed gravel when it is a combination of uncrushed and crushed gravels.

(c) *All-in aggregate.* An aggregate composed of a mixture of fine aggregate and coarse aggregate.

Sieve analyses are made by passing the dried aggregate through a series of test sieves beginning with one sufficiently coarse to pass all the material. The diameters of the sieves and the sizes of the mesh openings are given in BS 410:1969[7]. The sizes of mesh in general use for the sieve analysis of concrete aggregate are given in Table 3.3.

American sieves have slightly different sizes of aperture from British sieves, a point to be remembered when studying the literature. American sieves should comply with ASTM E11 and the sieve sizes used for grading analysis of aggregates have been included in Table 3.3.

Sieving by hand is a laborious procedure, and in any laboratory where a considerable amount of sieving is to be done, it is advisable both in order to avoid wastage of labour, and to ensure greater accuracy, to use one of the sieving machines which can deal with a stack of several sieves at a time. A machine of this type is illustrated in Fig. 3.10. Sieves either 200 mm or 300 mm diameter can be used.

An example of the steps in the calculations for the grading of an aggregate is given in Table 3.4. The original sample is assumed to have been 5 kg in weight after drying. Having completed the sieving, the weight of aggregate retained on each sieve in turn is recorded, giving the figures in column 2. The weight passing each sieve given in column 3 is obtained by the cumulative addition of the weights given in column 2, starting with that material passing 150 μm; thus, the weight of material passing the 600 μm sieve is the sum of the weights retained on sieves 300 and 150 μm, together with that passing 150 μm. It is

Fig. 3.10 Sieving machine adaptable for 300 mm or 200 mm diameter sieves

Table 3.4 Example of Calculation for Grading of Aggregate

BS sieve size	Weight of material retained on each sieve kg	Weight of material passing each sieve kg	Percentage passing each sieve
37·5 mm	0	5·00	100
20·0 mm	0·50	4·50	90
10·0 mm	0·25	4·25	85
5·0 mm	0·50	3·75	75
2·36 mm	0·25	3·50	70
1·18 mm	0·50	3·00	60
600 μm	0·50	2·50	50
300 μm	0·75	1·75	35
150 μm	1·25	0·50	10
Passing 150 μm	0·50		
Total	5·00		

Table 3.5 BS 882: Part 2: 1973 Grading Limits for Coarse Aggregates

BS 410 test sieve mm	Percentage by weight passing BS sieves							
	Nominal size of graded aggregate			Nominal size of single-sized aggregates				
	40 mm to 5 mm	20 mm to 5 mm	14 mm to 5 mm	63 mm	40 mm	20 mm	14 mm	10 mm
75·0	100	—	—	100	—	—	—	—
63·0	—	—	—	85–100	100	—	—	—
37·5	95–100	100	—	0–30	85–100	100	—	—
20·0	35–70	95–100	100	0–5	0–25	85–100	100	—
14·0	—	—	90–100	—	—	—	85–100	100
10·0	10–40	30–60	50–85	—	0–5	0–25	0–50	85–100
5·00	0–5	0–10	0–10	—	—	0–5	0–10	0–25
2·36	—	—	—	—	—	—	—	0–5

Table 3.6 BS 882: Part 2: 1973 Grading Limits for Fine Aggregates

BS 410 test sieve	Percentage by weight passing BS sieves			
	Grading Zone 1	Grading Zone 2	Grading Zone 3	Grading Zone 4
10·0 mm	**100**	**100**	**100**	**100**
5·00 mm	**90–100**	**90–100**	**90–100**	95–**100**
2·36 mm	**60**–95	75–**100**	85–**100**	95–**100**
1·18 mm	**30**–70	55–90	75–**100**	90–**100**
600 μm	**15–34**	**35–59**	**60–79**	**80–100**
300 μm	**5**–20	8–30	12–40	15–**50**
150 μm	**0**–10*	**0**–10*	**0**–10*	**0–15***

* For crushed stone sands, the permissible limit is increased to 20%. A tolerance of up to 5% may be applied to the percentages given in light type.

then an easy step to convert weight passing into percentage passing.

Grading requirements for aggregates from natural sources, are given in BS 882 : Part 2 : 1973, and have been summarized in Tables 3.5 and 3.6, reproduced by permission of the British Standards Institution.

In the case of fine aggregate, it is required that the grading should lie within one of four zones, subject to certain tolerances. In a note in the Specification it is pointed out that fine aggregate complying with the requirements of any one of the four grading zones is suitable for making concrete, but the quality of the concrete will depend upon a number of factors, including the mix proportions. Generally, the ratio of fine to coarse aggregate in a mix should decrease as the fine aggregate becomes finer, i.e. passing from Grading Zone 1 to Grading

Table 3·7 ASTM C33–74a Grading Limits for Coarse Aggregates

Size number	Nominal size (sieves with square openings)	Percentage by weight passing each sieve size (square openings)												
		100 mm (4 in)	90 mm (3½ in)	75 mm (3 in)	63 mm (2½ in)	50 mm (2 in)	37·5 mm (1½ in)	25 mm (1 in)	19 mm (¾ in)	12·5 mm (½ in)	9·5 mm (⅜ in)	4·75 mm (No. 4)	2·36 mm (No. 8)	1·18 mm (No. 16)
1	90–37·5 mm (3½–1½ in)	100	90–100		25–60		0–15		0–5					
2	63–37·5 mm (2½–1½ in)			100	90–100	35–70	0–15		0–5					
357	50–4·75 mm (2 in–No. 4)				100	95–100		35–70		10–30		0–5		
467	37·5–4·75 mm (1½ in–No. 4)					100	95–100		35–70		10–30	0–5		
57	25–4·75 mm (1 in–No. 4)						100	95–100		25–60		0–10	0–5	
67	19–4·75 mm (¾ in–No. 4)							100	90–100		20–55	0–10	0–5	
7	12·5–4·75 mm (½ in–No. 4)								100	90–100	40–70	0–15	0–5	
8	9·5–2·36 mm (⅜ in–No. 8)									100	85–100	10–30	0–10	0–5
3	50–25 mm (2–1 in)				100	90–100	35–70	0–15		0–5				
4	37·5–19 mm (1½–¾ in)					100	90–100	20–55	0–15		0–5			

Table 3.8 ASTM C33-74a Grading Limits for Fine Aggregates

ASTM E11-70 sieve size	Percentage by weight passing each sieve size
9·5 mm (3/8 in)	100
4·75 mm (No. 4)	95–100
2·36 mm (No. 8)	80–100
1·18 mm (No. 16)	50–85
500 μm (No. 30)	25–60
300 μm (No. 50)	10–30
150 μm (No. 100)	2–10

Note 1. The minimum percentages shown above for material passing the No. 50 and No. 100 sieves may be reduced to 5 and 0, respectively, if the aggregate is to be used in air-entrained concrete containing more than 251 kg/m³ (4½ American bags of cement per yd³), or in non air-entrained concrete containing more than 307 kg/m³ (5½ American bags of cement per yd³).

Note 2. It is also required that the fine aggregate shall have not more than 45 per cent retained between any two consecutive sieves of those shown and its fineness modulus shall be not less than 2·3 nor more than 3·1.

Zone 4. This is dealt with more fully in Chapter 7 on design of concrete mixes, p. 107.

The zoning is so arranged that, within the British Isles, the great majority of fine aggregates should fall in one or other of the zones. The user should, therefore, be able to obtain deliveries having gradings lying within the specified zone, so giving reasonable uniformity.

ASTM C33-74a Standard Specification for Concrete Aggregates (8) contains similar grading requirements to those in BS 882, and these are summarized in Tables 3.7 and 3.8. It will be noted that the Americans specify limits for the fineness modulus of fine aggregates. This is obtained by adding the total percentage shown by the sieve analysis to be retained on each of the following ASTM sieves and dividing the sum by 100: 150 μm, 300 μm, 500 μm, 1·18 mm, 2·36 mm, 4·75 mm, 9·5 mm, 19 mm, 37·5 mm, 75 mm and 150 mm.

Combined gradings

If the design of concrete mixes is based on the grading of the aggregates, it is necessary to consider the combined grading of the fine aggregate and one or more coarse aggregates. Given the individual gradings, the theoretical proportions necessary to approximate any desired grading can be determined by trial and error, by arithmetical or by graphical methods. One graphical method (9) applicable to two or more sizes of aggregate is illustrated by the following

Table 3.9 Gradings, Percentage Passing

Aggregate	BS 410 test sieve								
	μm			mm					
	150	300	600	1·18	2·36	5·0	10·0	20·0	40·0
Sand	2	16	51	75	90	99	100	100	100
10 mm–5 mm					3	9	94	100	100
20 mm–10 mm						2	18	92	100
38% of sand	1	6	19	28	34	38	38	38	38
22% of 10–5 mm					1	2	21	22	22
40% of 20–10 mm						1	7	37	40
Combined grading	1	6	19	28	35	41	66	97	100
Required grading	0	5	21	28	35	42	65	100	100

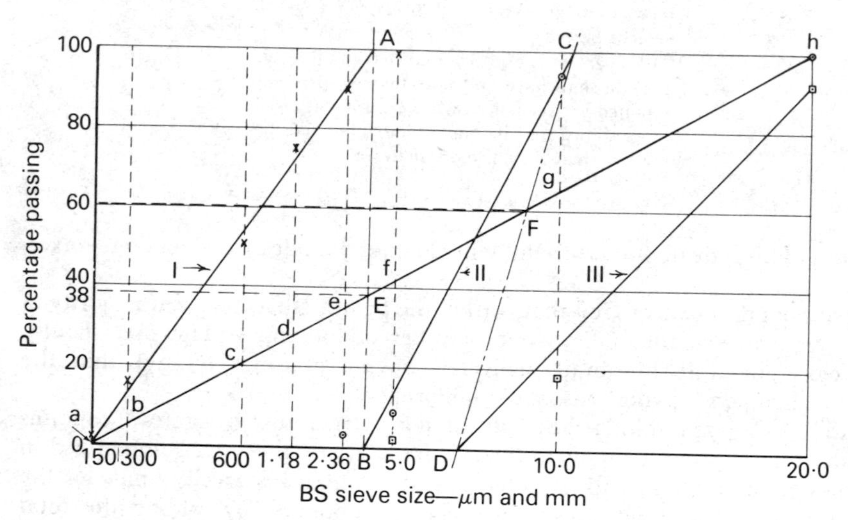

Fig. 3.11 Graphical method of proportioning aggregates approximating to a required grading

example, in which the aggregate gradings shown in Table 3.9 are to be combined to give a grading approximating to a required grading.

Rectangular coordinates are set up (Fig. 3.11) with a scale of percentage passing (0 to 100%) vertically, and a straight line, a-h, is drawn through the origin at any convenient slope.

Vertical lines are drawn through the points between a and h, where the line a-h intersects the horizontal lines, corresponding to the values of the percentage of material required to pass the various sieves—according to the required grading. These lines are the ordinates for the sieve sizes, and are marked accordingly.

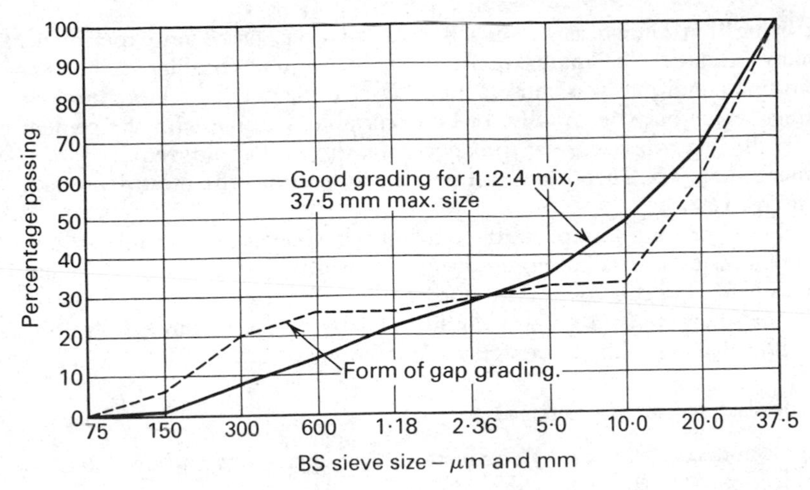

Fig. 3.12 Plotting of grading curves for comparative purposes

By this method the required grading curve has been transformed into a straight line by plotting against an irregular scale.

The values of actual percentage passing each sieve are then plotted for the separate aggregates and the steepest reasonable straight line (I, II and III) is drawn through each set of points. A straight line is then drawn between the point A, where the mean line for the finest material (I) cuts the 100 per cent passing line, and the point B where the mean line for the next finest material (II) cuts the 0 per cent passing line. Similarly, a straight line is drawn through the corresponding points C and D for the coarse aggregates.

The values of percentage passing, corresponding to the intersections E and F of the straight lines AB and CD with the initial sloping line a-h, are recorded and indicate the proportions in which the aggregates are to be combined.

In the example, the values are 38 per cent and 60 per cent; the proportions are therefore 38 per cent of sand, (60 – 38) = 22 per cent of 10–5 mm, and (100 – 60) = 40 per cent of 20–10 mm by weight. These values are checked in Table 3.9.

Combined gradings, however, can rarely be made to coincide at all sizes with the desired grading.

Comparison of gradings can be made very conveniently by plotting them on a chart as shown in Fig. 3.12. The full curve is a good continuous grading for a 1:2:4 mix using 37·5 mm maximum size aggregate and should give a workable concrete at a slump of about 50 mm with a reasonably low water/cement ratio. The dotted curve is typical of the combined grading of 37·5 mm to 10 mm aggregate and the fine sands common in many parts of Scotland. It is virtually a gap grading.

Grading curves

Mention has been made of a 'good grading curve' (Fig. 3.12). Good gradings or 'type gradings' have been suggested by various investigators, one of the earliest being that known as Fuller's Best Mix Curve, based on experiments with many gradations of material, from which the one considered the best under normal conditions was chosen. In practice there may be a large number of gradings which may be equally, or more suitable, varying with the cement content of the mix, the nature of the aggregate, the workability required, etc., and Fuller's Best Mix Curve does not necessarily produce the most dense and strongest concrete.

Glanville[10] made an important contribution concerning grading of aggregates in a series of tests linking grading with workability and water/cement ratio. This work and subsequent advances in knowledge concerning the effect of aggregate characteristcs on workability will be discussed more fully in a later chapter dealing with the design of concrete mixes.

Objectionable minerals and salts

Minerals and salts which may occur in natural aggregates and are likely to prove objectionable are:

(1) Coal and coal residues. Coal is objectionable because of its low strength and its undesirable effect on the appearance of concrete surfaces. Coal residues, as represented by breeze and clinker, are used in certain types of concrete, and give satisfactory results provided the proportion of unburnt coal is small. These materials as well as fly ash from modern electric power stations are discussed in Chapters 5 and 24.

(2) Certain sulphide minerals such as pyrites readily oxidize through atmospheric attack, causing unsightly rust stains such as that shown in Fig. 3.13, and loss of strength if present in sufficient quantity.

(3) Mica is a contaminating substance sometimes found in aggregates and is undesirable because of its soft, laminated, and absorptive character and its susceptibility to disintegration along cleavage planes. The presence of mica in sand has been found[11] to reduce the compressive strength of concrete at constant water/cement ratio, by about 3 per cent for a 1 per cent increase in the mica content.

(4) Soluble sulphates, including gypsum and other similar material. For further reference to the effect of sulphates in the aggregate see p. 395.

(5) Sea-shore and sea-dredged aggregates contain salts which, while not seriously affecting the strength, are liable to cause unsightly efflorescences on concrete structures and corrosion of embedded reinforcement.

The presence of sea shells, either flat or hollow, in the coarse aggregate, if present in sufficient quantities, may reduce the workability of the mix and a higher cement content may be required. It is considered a wise precaution not to exceed the limits as given in Table 3.1.

In addition to the presence of sea shells, sea-dredged aggregate is likely to contains salts, mainly sodium chloride. Reference has been made to this on p. 30.

Alkali-aggregate reactivity

Serious failures of concrete have been attributed in some parts of Denmark and the United States to expansion arising when the cement and certain

Fig. 3. 13 Pyrites stain on a bush-hammered gravel concrete surface. The dampness at the two construction joints is due to seepage of water from the back of the wall

minerals in the aggregate react chemically. This reaction occurs with many aggregates and, in fact, there is some aggregate–cement interaction with many aggregates commonly used, although, fortunately, in Great Britain the interaction is insufficient to cause damage to the concrete.

With aggregates in other countries the risk is a serious one, in that concrete affected may expand and completely disintegrate in a few years necessitating major repairs. Unfortunately, there are no means of identifying such aggregates from their appearance and aggregates which may be satisfactory with some cements may be unsatisfactory with others. The reactivity is connected with the alkali content of the cement, and one precaution when making concrete with aggregates not previously employed is to choose a cement with a low alkali content.

Elimination of unsatisfactory aggregates by laboratory tests is difficult, especially as the more reliable tests involving measurement of expansion take a very long time, say, six months to one year. Some indication can, however, be obtained by petrographic examination and other tests which give sufficient indication to eliminate the worst aggregates. Such an examination is, therefore, well worth while, providing it is carried out by an experienced investigator.

In view of the difficulty in identifying other than seriously reactive aggregates by rapid methods of testing, it is necessary to obtain all the evidence possible on the durability of any concrete made in the locality with the particular aggregate to be used. However, even this evidence is not completely reliable, since the aggregate may have been used with a low alkali cement, and it might cause failure with another cement with a higher alkali content; hence the recommendation given above to use low alkali cements when preparing concrete from unknown aggregates abroad.

The subject is dealt with in more detail in National Building Studies, *Reactions between Aggregates and Cement*[12]. A chemical method of test for determining the potential reactivity of aggregates is given in ASTM C289.

Clay, silt and dust

Clays and silts are commonly present in many gravel and sand deposits. In large quantities they cause a serious reduction in strength, and unsoundness due to their retarding action on the hydration of the cement. They are most dangerous when they form a coating on the particles of aggregate and thus prevent the adherence of the cement. Clays are also deleterious in concrete owing to their swelling and shrinkage which results from alternate wetting and drying; some clays when wet may contain as much as their own weight in water. When clays occur as constituents of other rocks, such as limestones, this absorptive character greatly increases the rock's susceptibility to disruption by weathering.

Stone dust is commonly present in crushed aggregates, but except in the case of certain dolerites, small quantities are not likely to have any deleterious effect on concrete.

Three methods are available for the determination of the clay, silt and dust content, the first of which is suitable for rough estimation in the field.

(1) *Field settling test* (*approximate*). The method as given in BS 812 : Part 1 : 1975 is as follows:

About 50 ml of a solution of common salt in water (approximately two teaspoonfuls to one litre of water, or a 1 per cent solution) is placed in a 250 ml measuring cylinder. Sand in the condition as received shall be added gradually until the volume of sand is about 100 ml. The volume shall then be made up to 150 ml by the addition of more salt solution.

The mixture is then shaken vigorously until adherent clayey particles have been dispersed, and the cylinder then placed on a level bench and gently tapped until the surface of the sand is level.

After three hours standing the height of the silt visible above the sand shall be expressed as a percentage of the height of the sand below.

This is an approximate volumetric method and is intended as no more than a guide to the approximate percentage of silt, loam, clay, etc., since the result will vary according to the fineness and type of clay.

The method is not generally applicable to crushed stone sands or coarse aggregates.

If the amount of clay and fine silt determined by this test is greater than 8 per cent by volume, then the test in BS 812 for the determination of the amount of aggregate passing 75 μm BS test sieve should be made.

(2) *Laboratory decantation method.* A method for the determination of material finer than 75 μm BS test sieve, that is, clay and silt, which has been in use for some time, is described in BS 812 : Part 1 : 1975. Very briefly, the method consists of washing the silt and clay through a 75 μm BS test sieve, and finding

the difference in weight between the original sample and that retained on the sieve. The difference is assumed to be clay and silt. An objection to this method is that any coarse material lost during the test counts as silt and clay. However, the test is simple and, since the sample used in the test is larger than that used in the sedimentation test described below, is less liable to sampling error.

The requirements of BS 882 : Part 2 : 1973 are that unless found satisfactory either as the result of such further tests as may be specified by the purchaser or his representative, or because evidence of general performance is offered which is satisfactory to him, the maximum quantities of clay, silt and fine dust determined in accordance with this test, shall not exceed the following:

(i)	Coarse aggregate	1 per cent by weight
(ii)	Sand or crushed gravel sand	3 per cent by weight
(iii)	Crushed stone sand	15 per cent by weight

ASTM C33 limits the material finer than No. 200 sieve (American) to 3 per cent by weight for concrete subject to abrasion and 5 per cent for all other concrete except for crushed stone sands when the limits may be increased to 5 and 7 per cent respectively.

As the particles in crushed stone dust are not usually as fine as those in natural sand, a somewhat higher quantity is usually tolerable. A figure of 6 per cent is suggested by the author as a reasonable limit. With some rocks, notably limestone, inclusions of clayey material in the rock as quarried may so increase the proportion of very fine particles as to make the limit of 3 per cent advisable.

(3) *Sedimentation method.* This method of separation of clay and silt is described in BS 812 : Part 1 : 1975. The test is somewhat involved and lengthy, involving pretreatment with sodium oxalate, through mixing of the suspension of fine material, and extraction of a small portion by means of a pipette. This portion is dried and weighed.

The test is open to the objection that the pretreatment may radically alter the character of the material being tested, and it is not recommended for general use.

Organic substances

Washing will often remove organic matter which is light in weight, or soluble in water, but will not remove heavy matter such as particles of coal, unless these occur in fine grains. Other organic substances, such as vegetable matter and humus, contain organic acids which inhibit the hydration of the cement, thereby delaying setting and reducing strength.

The colorimetric test for organic impurities which used to be included in previous editions of BS 812 was replaced in the 1967 edition by a test which measured the pH value of standard sand cement mortars under standard conditions. Use of this method has revealed difficulties which cannot readily be resolved and the 1975 edition of BS 812 does not include a test method for organic impurities.

It is therefore suggested that an unknown or suspect sand should be tested by a comparative method in which carefully proportioned 1 : 3 cement sand

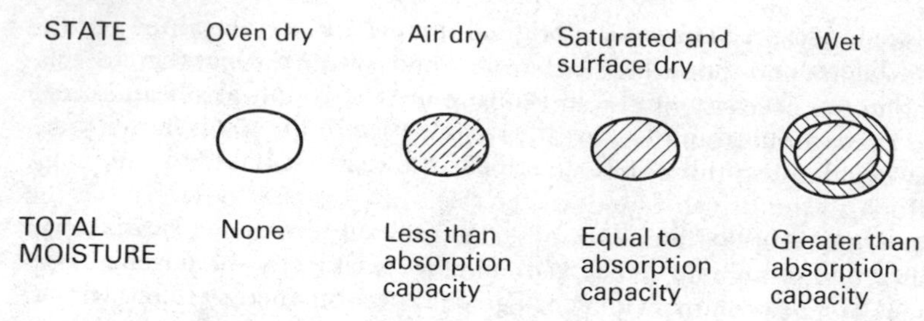

Fig. 3.14 Range of moisture contents as they affect aggregates

mortars are made, one with the sand in question and one with a known satisfactory sand of similar grading. Placing and compacting the mortars into cube moulds will permit examination at appropriate intervals for any gross retardation, and testing in compression at say 3 days would distinguish any other effects.

The colorimetric test is still retained in ASTM C40.

Absorption and surface moisture

The following explanations of certain terms used should be clearly understood:

(1) *Oven dry*. Completely dry for practical purposes.

(2) *Air dry*. Dry at the surface though containing some interior moisture, but less than the amount required to saturate the particles. In this condition an aggregate can absorb more water into itself, and may still appear dry on the surface. Absorbed moisture plays no part in the lubrication of mixed concrete.

(3) *Saturated and surface dry*. This is an idealized condition; the aggregate can absorb no more water without a film of water forming on the surface. In this condition the aggregate neither contributes to nor absorbs water from the concrete. If a wet stone is gradually dried or is wiped quickly with an absorbent duster or blotting paper, a stage is reached when the surface moisture disappears, but the stone is still saturated with absorbed water. It is then in a saturated and surface dry condition. Even in the laboratory great care is necessary, particularly with fine aggregate, in achieving the saturated surface dry condition.

(4) *Wet*. The aggregate is itself saturated and carries an excess of moisture forming a film on the surface of the particles.

The above conditions are illustrated in Fig. 3.14, which is similar to the representation included in the *ACI Manual of Concrete Inspection*[13].

The absorption of different particle sizes of the same aggregate may vary, as in the example given in Table 3.10, p. 52, so that tests on a single-sized sample should not necessarily be taken as representative of the aggregate through the whole range of gradings.

The absorption of an aggregate is often a useful guide to the permeability and frost resistance of concrete made with it.

The measurement of the moisture content of an aggregate is described in BS 812 : Part 2 : 1975. Four methods are given, the choice depending on the information required. The most useful of these tests are those giving the absorbed moisture and the surface moisture. These are the most difficult in the field owing to the necessity of drying the aggregate to a saturated surface dry condition. Even in the laboratory great care is necessary, particularly with fine aggregate, in achieving this condition.

In addition to the British Standard tests it is often useful to determine the absorption of an aggregate after only a few minutes soaking, as the rate of absorption provides a guide to the reduction in workability between mixing and placing. Further comment on this point is made in Chapter 6, p. 96.

Relative density

If a section is cut through any piece of stone and the surface magnified, it will be seen that it is to some extent honeycombed with capillaries and tiny airholes. For this reason the value obtained for the relative density of an aggregate depends on the method of test.

Methods of test and the calculations required for the determination of the relative densities are given in BS 812 : Part 2 : 1975, ASTM C127.73 and ASTM C128.73 [14].

Whichever test method is adopted the calculations are similar and are based on the following data:

A = weight of saturated surface-dry aggregate in air
$B - C$ = weight of saturated sample in water (B is the weight of container plus water plus aggregate; C is the weight of container plus water only)
D = weight of oven-dried aggregate in air

(a) Relative density on an oven-dried basis may be defined as the ratio of the weight in air of a unit volume of a permeable material (including both permeable and impermeable voids) to the weight in air of an equal volume of water.

$$\text{Relative density} - \text{oven dried} - \frac{D}{A - (B - C)}$$

(b) Relative density on a saturated and surface-dry basis may be defined as the ratio of the weight in air of a permeable material in a saturated surface-dry condition to the weight in air of an equal volume of water.

$$\text{Relative density (SSD) basis)} = \frac{A}{A - (B - C)}$$

(c) Apparent relative density may be defined as the ratio of the weight in air of a unit volume of a material to the weight in air of an equal volume of water. The volume of the material shall be that of the impermeable portion.

$$\text{Apparent relative density} = \frac{D}{D - (B - C)}$$

In concrete mix calculations for determining the solid volumes of the constituent parts, it is necessary to know the space occupied by the aggregate particles regardless of whether or not pores or voids exist within the particles. The value to be used for this is that referred to above as the relative density (SSD basis).

The term 'solid volume' is preferred to the older term 'absolute volume' as being a more apt description, since the term absolute would appear to imply that the volume contains only solid matter. This is not true since the volume taken includes voids enclosed within the cement and aggregate particles.

The most difficult part of the determination of the relative density is the decision as to when an aggregate is saturated but surface dry. In the field it is usually necessary to drive off the surplus moisture by warming on a Primus stove. Continuous stirring is required until the wet film of water just disappears from the stones, or until sand ceases to stick to the stirrer and becomes free running. In the laboratory a fan blowing over an electric fire (such as a hairdrier) is convenient. Considerable care is needed in dealing with sands.

Tests for relative density are probably best avoided in the field as far as possible, samples being sent to a central laboratory whenever such tests are required.

At least two determinations should be made, and if these do not agree, further tests should be made until agreement is reached.

Different particle sizes of the same aggregate may have different apparent relative densities. Average values from a large number of tests on a river aggregate are given in Table 3.10.

Table 3.10 Apparent Relative Density and Absorption for Different Particle Sizes of River Aggregate

Size of aggregate mm	Apparent specific gravity	Absorption % of dry weight
37·5–19	2·55	0·3
19–9·5	2·52	0·8
9·5–4·75	2·45	1·5
4·75 down	2·60	1·0

Mechanical properties of aggregates

It is required in BS 882 that the 'ten per cent fines' value when determined by the method given in BS 812 : Part 3 : 1975 shall be not less than 50 kN. For concrete wearing surfaces, the ten per cent fines value shall be not less than 100 kN. Very briefly, the method consists of placing a quantity of the aggregate, graded between stated sieves, in a cylinder, inserting a plunger, and applying a compression load so as to cause a total penetration of the plunger varying between 15 mm and 24 mm, depending on the type of aggregate. The aggregate is then resieved on the appropriate sieve and the ten per cent fines value calculated from the formula given in BS 812.

As an alternative to the ten per cent fines value, the aggregate impact test given in BS 812 : Part 3 : 1975 may be made. The aggregate impact value should not exceed 45 per cent, and for concrete wearing surfaces should not exceed 30 per cent.

For measuring the resistance of aggregates to surface wear by abrasion, an important property for aggregates used in concrete roads, a method of test is given in BS 812 : Part 3 : 1975. American tests for abrasion resistance by use of the Los Angeles machine are given in ASTM C535-69 and C131-69.

Bulk density and voids

The bulk density of a material is the weight of the material held by a container of unit volume when filled or compacted under defined conditions. It is normally expressed in kg/m³.

The bulk density of an aggregate is affected by several factors, including the amount of moisture present and the amount of compactive effort used in filling the container.

It is emphasized in BS 812 : Part 2 : 1975 that the methods specified for the determination of the bulk density and voids of an aggregate are laboratory tests intended for comparing properties of different aggregates and are not generally suitable for use in translating proportions by weight into proportions by volume on site. They give the loose weight or the 'rodded' or compacted weight per m³, depending on the method of filling the container with dry aggregate.

On the site and for conversion purposes, it is probably best to have the box filled in turn by each of the men filling the mixer hopper, and to take an average value. This average value is likely to lie somewhere between the bulk densities determined by the BS 812 methods for compacted and loose weights.

If the aggregate is damp and contains an appreciable amount of material below a 5 mm sieve, it is necessary not only to find the weight per m³ dry, but also to make tests for bulking. The weight per m³ at any moisture content can be calculated from the dry density and the percentage bulking, not forgetting to allow for the weight of moisture. Normally, the weight per m³ of sand or all-in ballast decreases from the dry weight to a minimum value at a certain moisture content, and then increases again until the material becomes saturated.

A method sometimes used in proportioning concrete is based on the void content of the aggregates, and a measurement of voids may be useful in certain cases. BS 812 gives a method of measurement for use in the laboratory. As with measurements of the weight per m³, the void content will vary according to the method of filling the gauge-box.

Bulking

When sand is moistened, films of water form on the particles and the surface tension tends to hold them apart, causing an increase in volume or bulking. Fine sand bulks more than coarse sand; aggregate retained on a 5 mm sieve is scarcely affected.

In general, sand bulks rapidly to the extent of 20 to 30 per cent, as the moisture content rises to about 4 to 6 per cent. Further increases in moisture content result in a decrease in bulking, until, when the sand is completely

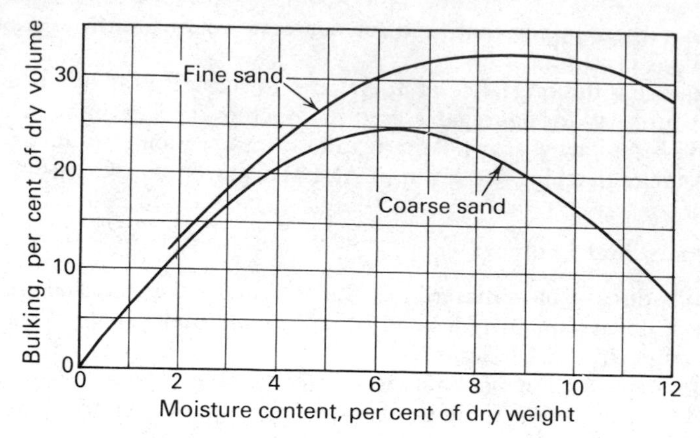

Fig. 3.15 Typical form of bulking curves

saturated, its volume is practically the same as it was in a dry condition. Curves showing the difference between the bulking of coarse and fine sand plotted against moisture content are given in Fig. 3.15. In practice it is usual to assume an average value of 20 per cent for the bulking of sand as delivered in a damp condition, unless tests are made to obtain more exact information.

All-in ballast may contain anything up to 60 or 70 per cent of sand, and it is found that bulking occurs roughly in proportion to the amount of sand present.

Bulking can be determined by filling a gauge-box or other container of known volume (A) with damp sand in the manner in which the mixer hopper will be filled. This involves the same difficulties as have been described in the last section. The sand is then either dried and filled back into the gauge-box, or the container is filled with water and the damp sand is poured in to displace the water. Whichever method is adopted, the new depth of aggregate in the container gives the unbulked volume (B).

Then percentage bulking expressed as a percentage of the dry volume

$$= \frac{A - B}{B} \times 100$$

When measuring aggregates by volume it is necessary to make an allowance for bulking in order to obtain the correct proportion of sand particles required to produce good workable concrete. If the cement content is specified as a proportion of the aggregate, omission of an allowance for bulking will also result in the use of too much cement per m³ of concrete produced.

Particle Shape and Surface Texture

The particle shape is important since it affects the workability, as is shown in Chapter 7, Design of Concrete Mixes. It may conveniently be assessed by a test for angularity suggested by Shergold [15], which is based on the percentage voids in the aggregate after compaction in a specified manner. The test gives a value termed the angularity number; the method of determination as follows briefly is given in BS 812 : Part 1 : 1975.

A quantity of single-sized aggregate is filled into a metal cylinder of about 0·003 m³ capacity having approximate internal dimensions 150 mm diameter and 150 mm high, in three layers, allowing the aggregate to slip in gently from a scoop. Single-sized aggregate is used; in the case of coarse aggregate the size limits specified are 20·0 mm–14·0 mm, 14·0 mm–10·0 mm, 10·0 mm–6·3 mm, and 6·3 mm–5·0 mm. Each layer of aggregate is then tamped 100 times by allowing a metal rod 16 mm diameter and 600 mm long, with rounded end, to fall freely from a height of 50 mm above the surface of the aggregate. Further aggregate is then added to fill the cylinder and 'rolled off' with the tamping rod. The volume of the cylinder is determined from the weight of water required to fill it (1 cm³ = 1 g).

Then the angularity number of the aggregate is:

$$f_H = 67 - \frac{100M}{CGa} \tag{1}$$

where C = volume of cylinder in cm³ given by the weight in grams of water required to fill the cylinder,

G_a = relative density of the aggregate calculated on an oven-dried basis,

M = weight of aggregate in cylinder in grams.

For estimations of workability the author has suggested[16] that a different method of expressing the result is more suitable. Thus:

$$\text{Angularity index } f_A = \frac{3f_H}{20} + 1 \tag{2}$$

where f_H is the angularity number as expressed by Shergold.

Combining (1) and (2),

$$f_A = 11 - \frac{15M}{CG_a} \tag{3}$$

Flaky materials are not satisfactory for concrete work. The surface texture of the aggregate has some bearing on the strength of the concrete produced, in that aggregates with polished surfaces do not produce such strong concrete as those with matt or slightly rougher surfaces. Evidence is accumulating to suggest that the influence of the type of aggregate on the compressive strength, and, more particularly, on the tensile strength or modulus of rupture, is greater than has hitherto been supposed. An analysis of several series of tests made by the author has indicated that limestone and granite aggregates may give strengths up to 20 per cent higher than river gravel at the same water/cement ratio. Other investigators have reported similar experiences both with limestone and other crushed stones, as compared with river gravels. It is believed that these differences are due mainly to the effects of surface texture and particle shape.

Group classification of rocks

Rocks are classified in BS 812 : Part 1 : 1975, Clause 6. The grouping is a useful system for distinguishing the various rocks such as granite, limestone, quartzite, sandstone, etc.

REFERENCES

1 BS 1047 : Part 2 : 1974, *Specification for Air-cooled Blastfurnace Slag Coarse Aggregate for Concrete*. British Standards Institution.

2 FOOKES, P. G., and COLLIS, L., Aggregates and the Middle East, *Concrete*, **9**, No. 11, 1975.

3 SHERGOLD, F. A. and GREYSMITH, MARY G., Factors governing the grading and shape of crushed rock—a survey of the literature, *Quarry Managers' Journal*, June 1947, pp. 703–712.

4 BS 882 : Part 2 : 1973, *Specification for Aggregates from Natural Sources for Concrete (including Granolithic)*. British Standards Institution.

5 *Concrete Manual*, 1st ed. 1938, 8th ed. 1975. US Bureau of Reclamation.

6 BS 812 : 1975, *Methods for Sampling and Testing of Mineral Aggregates, Sands and Filters*. British Standards Institution.

7 BS 410 : 1969, *Test Sieves. British Standards Institution.*

8 ASTM C33–74a, *Standard Specification for Concrete Aggregates*. 1976 Annual Book of ASTM Standards, Part 14. American Society for Testing Materials.

9 *The Determination of the Proportions of Aggregates approximating to any Required Grading*, Cement and Concrete Association Advisory Note No. 9.

10 GLANVILLE, W. H., COLLINS, A. R. and MATHEWS, D. D., *The Grading of Aggregates and Workability of Concrete*, Road Research Technical Paper No. 5, 2nd ed., 1947. HMSO.

11 DEWAR, J. D., *Effect of Mica in the Fine Aggregate on the Water Requirements and Strength of Concrete*, Cement and Concrete Association TRA/370, April 1963.

12 *Reactions between Aggregates and Cement*. National Building Studies, Parts I, II and III, Research Papers 14, 15 and 17. HMSO.

13 *ACI Manual of Concrete Inspection*, 6th ed., 1975. American Concrete Institute.

14 ASTM C127–73, *Specific Gravity and Absorption of Coarse Aggregate*, C128–73, *Specific Gravity and Absorption of Fine Aggregate*, 1976 Annual Book of ASTM Standards, Part 14. American Society for Testing Materials.

15 SHERGOLD, F. A., The percentage voids in compacted gravel as a measure of its angularity, *Mag. Concr. Res.*, **5**, No. 13, Aug. 1953. Cement and Concrete Association.

16 MURDOCK, L. J., The workability of concrete, *Mag. Concr. Res.*, **12**, No. 36, Nov. 1960. Previously printed in *Research and Application*, No. 11, Jan. 1960 (George Wimpey & Co. Ltd.).

4

Cement

PORTLAND cement was first produced in 1824 by Joseph Aspdin, who heated a mixture of finely divided clay and limestone or chalk in a furnace at a temperature sufficiently high to drive off the carbonic acid gas. It was not until 1845 that Isaac Johnson burnt the same materials together in a furnace or kiln until they clinkered, thus producing a cement very similar in its essential chemical constituents to modern Portland cement. Since then many refinements, changes in the scale and type of plant, and improvements in methods of control and testing have been made.

The basic materials for the manufacture of Portland cement are usually forms of limestone and clay. Chalk or marl commonly provides the calcareous material in the form of calcium carbonate, while clay or shale provide the argillaceous material, which is mainly alumina and silica. The sites of cement works are normally chosen at places where both these types of material occur adjacent to one another. The method of obtaining the raw material is shown in Fig. 4.1.

Two processes of manufacture are employed, the dry process and the wet process, the latter being the more common in this country. In the dry process the materials are crushed, dried, and then ground in ball mills to a powder which is burnt in its dry condition. In the wet process the materials are first crushed and then ground to form a slurry in wash mills. After passing through the wash mills and the slurry silos, the slurry passes to the slurry tanks which are shown in Fig. 4.2. Samples of the slurry are taken from these tanks from time to time for testing, and any correction in the chemical composition is made by changing the proportions of the calcareous and argillaceous constituents. The slurry is next pumped to a kiln. Cement kilns (shown in Fig. 4.2.) may be as large as 5·7 m (18·7 ft) in diameter and about 200 m (650 ft) in length, and with an output of as much as 76 tonnes per hour. Pulverized coal, oil, or natural gas is used as fuel and is injected into the kiln at the opposite end to that at which the slurry enters. Thus, the slurry enters the kiln at the cooler end, and by the rotation of the kiln in conjunction with its slope, passes down and gradually increases in temperature until it clinkers. The cement clinker then passes through clinker coolers as illustrated in Fig. 4.3. Having cooled sufficiently, the clinker is ground to the required degree of fineness in ball mills shown in Fig. 4.4. During grinding a 'retarder' is incorporated, gypsum being the material generally used. An addition of 2 or 3 per cent is usually necessary to slow down and control the time of setting. The friction arising from the grinding operation causes a considerable increase in temperature, and is the cause of the hot cement sometimes delivered on to a site. The temperature of the cement as it comes out of the grinding mill can be as high as 70°C.

From the mills the cement is conveyed or blown to storage silos, from which

Fig. 4.1 Digging limestone for cement manufacture

Fig. 4.2 Aerial view of cement works at Northfleet. The rotary kilns can be seen in front of the chimneys. The slurry tanks can be seen to the left of the chimneys

Fig. 4·3 Clinker cooling tubes below rotary kilns

Fig. 4·4 Air-swept cement ball mills. The steel grinding balls of various sizes are seen in the foreground

Table 4.1 Percentage Composition and Compound Content of Portland Cements

	Ordinary	Rapid-hardening	Low-heat	'Sulphate resisting'
Analysis				
Lime	63·1	64·5	60·0	64·0
Silica	20·6	20·7	22·5	24·4
Alumina	6·3	5·2	5·2	3·7
Iron oxide	3·6	2·9	4·6	3·0
Compounds				
Tricalcium silicate (C_3S)	40	50	25	40
Dicalcium silicate (C_2S)	30	21	45	40
Tricalcium aluminate (C_3A)	11	9	6	2
Iron compound (C_4AF)	11	9	14	9

* The abbreviated chemical notation given for the above compounds represent chemical formulae: C = CaO, S = SiO_2, A = Al_2O_3, F = Fe_2O_3

the cement is discharged as required, either being fed to a packing plant or direct to bulk-cement lorries. The packing plant automatically fills paper bags with cement up to a nominal weight of 50 kg (110 lb). In America, where imperial units are still used, bags of cement, referred to as sacks, weigh 94 lb.

Other ancillary plant at the cement works includes dust-extracting equipment, air compressors, and the power generators.

Chemistry of Portland cement

The chemistry of Portland cement is very complicated, and not yet fully understood. It is sufficient to recognize the main constituents and to have some idea of the influence of the four main compounds on the process of setting and hardening.

The approximate composition of Portland cements is given in Table 4.1 [1, 2]. Lime constitutes nearly two-thirds of the cement and its proportion has an important effect on the properties. An excess of lime causes unsoundness and disintegration of the cement after setting. A high lime content, but not enough to be excessive, tends to slow down the setting, but produces a high early strength. Too little lime may result in a weak cement and, if underburnt, in one which has a quick set.

Silica forms about one-fifth, while alumina is only present to the extent of about one-twentieth of the cement. A high silica content, which is usually accompanied by a low alumina content, produces a slow-setting cement of high strength, and improves resistance to chemical attack. If the reverse is the case, that is, a high alumina and a low silica content, the cement is quick setting and of high strength.

Iron oxide imparts the grey colour to cement, and acts in the same manner as the alumina.

Magnesia is limited by BS 12 : 1971 [3] to 4 per cent. A greater quantity is liable to cause unsoundness.

The sulphur content is usually calculated as sulphuric anhydride (SO_3) and is limited to 2·5 or 3·0 per cent, depending on whether the tricalcium aluminate content is greater or less than 7 per cent. Excessive quantities cause unsoundness.

The alkalis, soda and potash, are normally carried away in the flue gases when the cement is burnt, and are only present in the finished product in small quantities. If, for any reason, they are present in the cement in excessive quantities, they cause efflorescence, and increased risk of failure due to alkali–aggregate reactivity.

When cement is mixed with water, a chemical action begins between the various compounds and the water. In the initial stages, the small quantity of retarder (gypsum) quickly goes into solution, and is thus able to exert its influence on the other chemical reactions which are starting. These reactions result in the formation of the various compounds which cause setting and hardening, the four most important being:

(1) Tricalcium aluminate (three molecules of lime to one of alumina) C_3A.
This compound hydrates very rapidly and produces a considerable amount of heat; it causes the initial stiffening, but contributes least to the ultimate strength; is less resistant to chemical attack, being particularly vulnerable to the disintegrating action of sulphates in ground water; and has the greatest tendency to cause cracking, due to volume changes.

(2) Tricalcium silicate (three molecules of lime to one of silica) C_3S.
This compound jellifies within a few hours, generating considerable heat. The quantity formed in the setting reaction has a marked effect on the strength of concrete at early ages, mainly in the first 14 days.

(3) Dicalcium silicate (two molecules of lime to one of silica) C_2S.
The formation of this compound proceeds slowly with a slow rate of heat evolution. It is mainly responsible for the progressive increase in strength which occurs from 14 to 28 days, and onwards. Cements in which a high proportion of dicalcium silicate is formed, have a relatively high resistance to chemical attack, a relatively low drying shrinkage, and hence, are the most durable of the Portland cements.

(4) Tetracalcium aluminoferrite (four molecules of lime to one of alumina, and one of iron oxide) C_4AF.
The presence of the aluminoferrite compound is of little importance, since it has no marked effect on the strength or other properties of the hardened cement.

The chemical reactions outlined above, result in the formation of a mixture of gels and crystals from the solution of cement and water, which, by their adherence and physical attraction to one another and to the aggregate present, gradually set and harden to produce concrete.

It is particularly important to note that the setting and hardening is a chemical reaction in which water plays an important part, and is not just a matter of drying out. In fact, setting and hardening stop as soon as the concrete becomes dry.

Fineness of grinding

After the chemical composition, the other dominant factor affecting cement properties is the fineness of grinding. Finer grinding increases the speed with which the various constituents react with water, but does not alter their inherent properties. The tendency during recent years has been to grind cements to greater finenesses in order to produce high strengths at early ages. Rapid-hardening Portland cement is an example of this development.

Fineness of grinding is of some importance in relation to the workability of concrete mixes. While claims that an increase in the fineness of cement results in a reduction in the water/cement ratio required to produce a given consistence are open to doubt, it has been established that greater fineness improves the cohesiveness of a concrete mix. Further, the quantity of water rising to the surface of the concrete, known as 'bleeding', is reduced.

Cracking due to early thermal contraction is related to the rate of development of strength of a concrete and, in general, cements which gain strength rapidly are more prone to cracking. Increasing the fineness of any particular cement raises its rate of development of strength, and so indirectly increases the risk of crack formation. The contraction is mainly due to the fall in temperature to the ambient value, following the rise due to the evolution of the heat of hydration.

Types of cement

The brief analysis which has been given of the composition of Portland cement and the influence of the various compounds present on its properties, has indicated that a large range in characteristics is possible by varying the proportions of the raw materials, by differences in the fineness of grinding, and by other changes in technique. Since for different purposes it is advantageous to stress one characteristic more than others, various classes of Portland cement have been developed.

Analyses showing the main constituents and compounds of typical ordinary, rapid-hardening, low-heat and sulphate-resisting cements are given in Table 4.1. These are average values giving only a broad picture of the make-up of the various types of Portland cement.

It is interesting to note that the ultimate compressive strengths obtained with all the cements included in Table 4.1, are broadly about the same, and that it is only the rate of hardening which varies to a marked extent.

Rapid-hardening Portland cement generally has a high tricalcium silicate content which, combined with the fine grinding, contributes towards a high early strength. As with ordinary Portland cement there is a variation in the properties of the rapid-hardening types, and it has been found that the 1 and 3 day compressive strength requirements of BS 12 : 1971 are not always achieved.

It should be noted that the requirements in respect of setting times are the same as those for ordinary Portland cement.

American cement Type III complying with ASTM C150-74 is similar in physical and chemical properties.

The British Standard and ASTM requirements for this and various other types of cement are included in Table 4.2.

Ultra-high early strength cement is produced by separating the fines from rapid-hardening Portland cement during manufacture in a cyclone air elutriator. The result is a cement with a very high specific surface (above 700 m^2/kg). It complies with BS 12 : 1971 except that it contains a rather higher proportion

Table 4.2 Physical Requirements for Cements

Type of cement	Mortar cube strength N/mm² (lb/in²)				Concrete cube strength N/mm² (lb/in²)			Fineness m²/kg		Setting time (Vicat)	
	1 day	3 day	7 day	28 day	3 day	7 day	28 day	Air permeability	Turbidimeter	Initial not less than	Final not more than
										minutes	hours
(a) Ordinary Portland BS 12 : 1971	—	15 (2200)	23 (3400)	—	8 (1200)	14 (2000)	—	225	—	45	10
ASTM Type 1	—	12·4 (1800)	19·3 (2800)	—	—	—	—	280	160	45	8
(b) Rapid-hardening Portland BS 1222 : 1971	—	21 (3000)	28 (4000)	—	12 (1700)	17 (2500)	—	325	—	45	10
ASTM Type III	12·4 (1800)	24·1 (3500)	—	—	—	—	—	—	—	45	8
(c) Low-heat Portland BS 1370 : 1974	—	8 (1100)	14 (2000)	28 (4000)	3 (500)	7 (1000)	14 (2000)	320	—	60	10
ASTM Type IV	—	—	6·9 (1000)	17·2 (2500)	—	—	—	280	160	45	8
ASTM Type II	—	10·3 (1500)	17·2 (2500)	—	—	—	—	280	160	45	8
(d) Sulphate resisting Portland BS 4027 : 1972	—	15 (2200)	23 (3400)	—	8 (1200)	14 (2000)	—	250	—	45	10
ASTM Type V	—	8·3 (1200)	15·2 (2200)	20·7 (3000)	—	—	—	280	160	45	8
(e) Portland blastfurnace BS 146 : 1973	—	15 (2200)	23 (3400)	34 (5000)	8 (1200)	14 (2000)	22 (3200)	225	—	45	10
(f) High alumina BS 915 : 1972	42 (6000)	49 (7000)	—	—	—	—	—	225	—	2 hours and not more than 6 hours	2 hours after the initial set

of gypsum than the maximum permitted value of 3 per cent.

This cement is useful for producing very high early strength concrete. It does however have a rather high water demand and tends to produce a 'sticky' concrete.

Low-heat cement has a high proportion of dicalcium silicate mainly at the expense of the tricalcium silicate and is, therefore, slow in hardening and produces less heat than the other cements. This cement was produced mainly as a result of problems which resulted from the temperature rise in mass concrete dams.

Its use is mainly confined to gravity retaining walls, large dams, and similar massive concrete structures where the temperature rise in the mass of concrete, and the subsequent thermal shrinkage on cooling, can be reduced.

It is usually only manufactured in Britain for large contracts.

Moderate low-heat cement is manufactured in America in accordance with Type II according to ASTM C150-74.

Sulphate-resisting Portland cement has a better performance in resisting sulphate attack than ordinary Portland cement, due to the reduction in the tricalcium aluminate content. Specification requirements for this type of cement are shown in Table 4.2. It is not resistant to acids. Calcium chloride should not be added as it reduces its sulphate-resisting properties.

As with all concrete, the prime resistance to chemical attack is afforded by the production of a dense well-compacted concrete, and lean mixes that are difficult to compact should be avoided; the concrete should not be leaner than 1 to 6 by weight (cement content of 310 kg/m^3 (525 lb/yd^3) of concrete).

White Portland cement has the same properties as ordinary Portland cement but is manufactured from raw materials containing less than 1 per cent iron. Its cost is about three to four times that of ordinary Portland cement.

Coloured Portland cements are made by adding suitable pigments to ordinary Portland cement in the case of deep colours and to white Portland cement when pale shades are required.

Air-entraining cements. These are ordinary Portland cements to which an air-entraining agent has been interground during the manufacturing process. They are available in USA for Types I, II and III cement, and are required to satisfy the requirements of ASTM C175-74. They are not available in Britain where it is found more convenient and satisfactory to vary the amount of air-entraining agent to meet the particular aggregate types, gradings and sizes.

Portland-blastfurnace cement, complying with the requirements of BS 146 : 1973, is manufactured mainly in Scotland by grinding a mixture of Portland cement clinker and blastfurnace slag in the ratio of about 35 : 65. On the Continent cements are made with a greater range, 30–85 per cent in the proportion of slag used.

The properties of slag cements vary depending on the amount and characteristics of the slag. They are slower hardening and have reduced 3 and 7 day strength requirements compared with ordinary Portland cement. They are also more affected by low temperatures during the curing period and by inadequate curing[(4)] than is normal cement.

They are particularly useful for mass concrete structures because they have a somewhat reduced heat of hydration.

Portland blastfurnace cement concrete is more resistant to chemical attack, especially in sea water, than ordinary Portland cement.

More recently, use has been made in Britain of dry-ground granulated slag as an addition at the concrete mixer to produce an on-site Portland blastfurnace type concrete. Up to 54 per cent of the cement has been replaced by ground granulated slag in structural concrete.

Pozzolanic cements are produced by grinding together a mixture of 85–60 per cent Portland cement and 15–40 per cent pozzolana, which may be a naturally active material such as volcanic ash or pumice, or an artificial product such as pulverized fuel ash, burnt clay or shale. The rate of strength development is slower than that of normal Portland cement, especially at low temperatures. Pozzolanic cements have a higher resistance to chemical disintegration than the base Portland cement which they contain. Their resistance to sulphate attack is similar to that of sulphate-resisting Portland cement (see p. 64). Portland-pozzolanic cement is manufactured in America according to ASTM C595-74.

Supersulphated cements are made in Belgium and France by grinding, usually together, but sometimes separately, blastfurnace slag, calcium sulphate and an activator, usually ordinary Portland cement [5].

They have good properties of resistance to attack by sulphates and are stated to withstand attack from acids having pH values as low as 3·5.

They are used in the same way as ordinary Portland cement but it is preferable to increase the time of mixing to about five minutes.

The surface of the concrete may be rather friable and dusty, unless the exposed surfaces are coated with limewash and kept damp during the hardening period after striking the formwork. These precautions need not, of course, be taken when the concrete is subsequently buried or otherwise protected.

They have been found in the past to have rather variable setting and hardening properties, possibly due to temperature effects, so that particular care should be exercised during cold weather.

Supersulphated cement combines chemically with more water than is required for the hydration of Portland cement, and evaporation during curing must be avoided.

As with sulphate-resisting cement, mixes leaner than 1 to 6 by weight (cement content of 310 kg/m^3 (525 lb/yd^3) of concrete) are not recommended.

High alumina cement is produced by fusing together in a furnace a mixture of limestone and bauxite (aluminium ore).

This cement has an exceptionally high proportion of aluminate (35–44 per cent), resulting in a rapid development in strength; in fact, high alumina cement is as strong, or even stronger, at 24 hours, as ordinary Portland cement is at 28 days. Setting and hardening result chiefly from the formation of monocalcium aluminate and are accompanied by a rapid and high evolution of heat, a factor of considerable value when concreting at abnormally low temperatures, but a distinct disadvantage at normal temperatures in all but thin sections from which the heat can escape quickly.

A serious disadvantage of high alumina cement is that the compressive strength of concrete made with it is reduced by a conversion reaction in which the metastable calcium aluminate hydrated compounds change to more stable compounds. This reaction is generally accompanied by an increase in porosity and it may be recognized sometimes by a change from the characteristic dark grey colour to a chocolate brown colour.

Conversion takes place very rapidly under conditions of high temperature combined with a high humidity, but it will also occur at relatively normal temperatures and humidity. This latter characteristic was highlighted by the much published structural failures in 1973 and 1974; in particular by the collapse in February 1974 of two beams, made with high alumina cement, forming part of a roof over a swimming pool at a Stepney school. The conclusions of the investigation by the Building Research Establishment[6] were that the failure of the beams was caused by loss of strength due to conversion of the high alumina cement concrete, followed by sulphate attack, in conditions of high temperature and humidity.

Subsequent research by the Building Research Establishment[7] showed the following:

(a) Most high alumina cement concrete in prestressed beams in buildings more than a few years old is highly converted or is likely to become so.

(b) The strength of highly converted concrete is very variable and is substantially less than its initial strength.

(c) Once a high level of conversion has been reached, no further loss of strength due to conversion will occur.

(d) The long-term strength at a temperature of 18°C is very sensitive to water/cement ratio and there is an increasing reduction in strength below the 24 hour value for increasing water/cement ratios above 0·4. The reduction becomes more pronounced at higher temperatures.

Since, therefore, the dangers of using high alumina cement had become very obvious, the Building Regulations were amended in 1975 to give enforcing authorities the right to prohibit the use of high alumina cement for structural purposes. British Standard CP 110 'The structural use of concrete' was also amended to remove all mention of high alumina cement.

However, the amendment to the Building Regulations recognizes that high alumina cement may be used in those few situations where its sulphate resisting properties make it the only suitable material. Similarly its use for flue linings, which are subject to sulphate and acid attack, is also permitted.

Great care should be taken to keep high alumina cement apart from Portland cement, as small quantities of either may cause a flash set in case of mixture. High alumina cement can be bonded to hardened Portland cement concrete that is at least 7 days old. Conversely Portland cement concrete can be bonded to hardened high alumina cement concrete which is at least 24 hours old.

Hydrophobic cement. This is a Portland cement to which an additive has been introduced, giving the cement particles a protective coating which inhibits hydration of the cement. This enables the cement to be stored in damp conditions for a much longer period than is possible with ordinary Portland cement. The water-repellent film is broken down by abrasion during the concrete mixing process, and normal hydration then takes place. The time of mixing should be increased by at least one minute longer than the normal period; hand mixing is unsatisfactory.

Hydrophobic cement is unlikely to provide waterproofing or water-repellent properties in the finished concrete. Hydrophobic cements are produced at

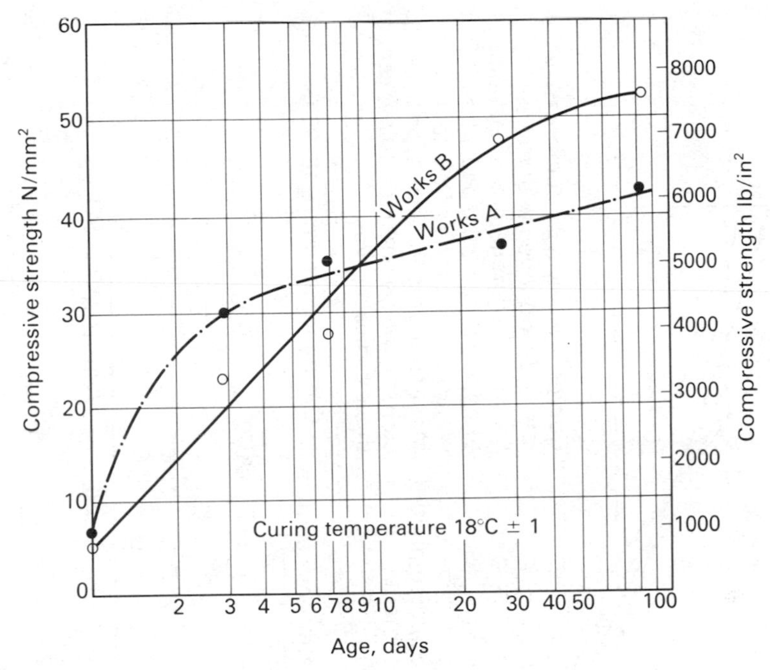

Fig. 4·5 Increase in mortar cube strength with age of two samples of ordinary Portland cement

present only at two or three places in the world and are likely to be useful and economically worthwhile only in humid and tropical climates.

The properties of hydrophobic cement are stated to comply with the requirements of BS 12 : 1971 for ordinary Portland cement.

The increase of strength with age

Reference has been made to the rate of development of strength of cement and concrete, and it has been shown that this is dependent on the compounds present. The rate of strength increase is greatest during the initial period of hardening and gradually decreases as time goes on. In the early stages of hydration it is only the outer layers of the particles of cement which take part in the chemical reaction, and if a piece of concrete is examined through a microscope, it is seen that particles of unhydrated cement still exist in the hardened paste. These unhydrated particles continue to absorb moisture from the air even after the mixing water has dried out and, by continuing the chemical reaction, gradually increase the strength and density of the concrete, a process which may continue for several years. It is the presence of unhydrated particles of cement which cause the phenomenon of autogenous healing, that is the healing together of concrete which has cracked. The effect of autogenous healing may be demonstrated by partly crushing a concrete cube, leaving it to weather for a few weeks, and then testing it again. A large proportion of the original strength is recovered.

Cements do not all harden at the same rate; Fig. 4·5[8] compares mortar

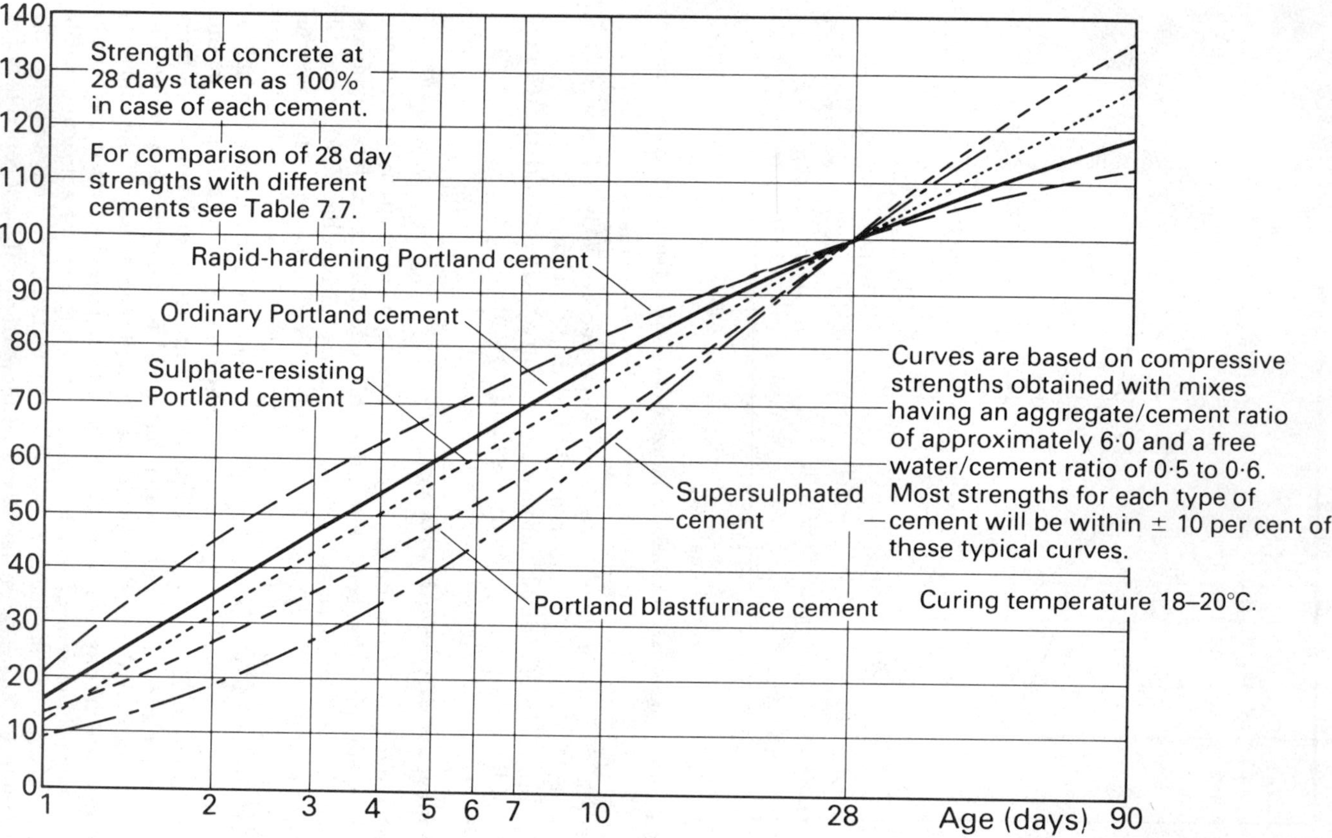

Fig. 4.6 Typical rates of strength development of different cements

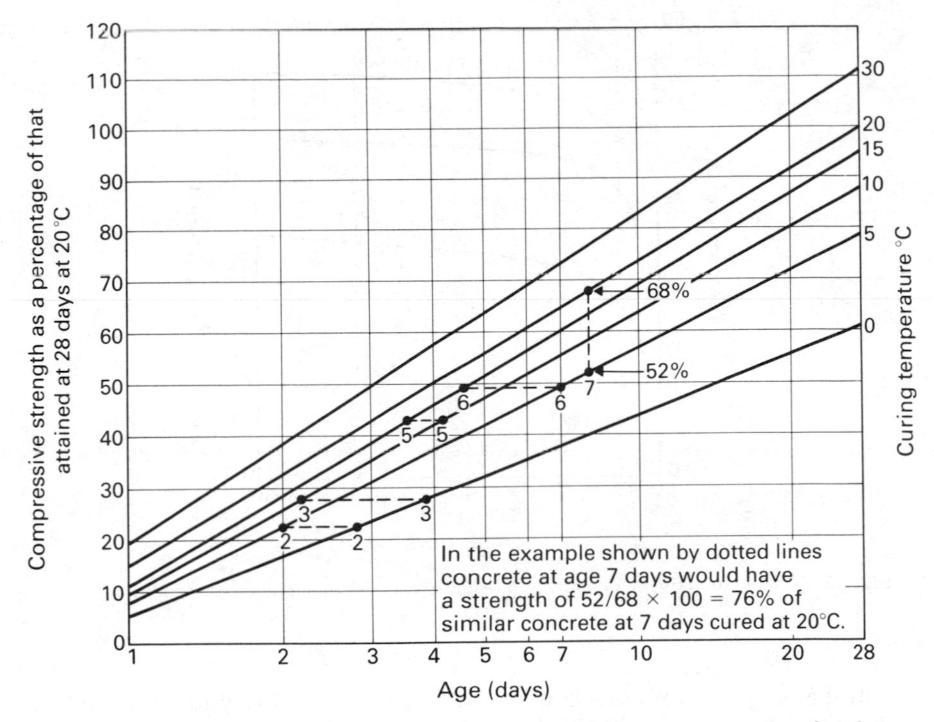

Fig. 4.7 The approximate relationship between curing temperature, compressive strength and age.
NOTE: The relationship applies only to: (1) ordinary Portland cement; (2) rapid-hardening Portland cement; (3) sulphate-resisting Portland cement

cube strengths at ages up to 90 days, for two samples of ordinary Portland cement. Some authorities suggest that at ages of a year and upwards, the variation in compressive strength, even between the different classes of Portland cement, is small compared with that at early ages.

The rates of increase in strength of concrete test cubes made with various cements and cured in water at 18°C (64°F), are given in Fig. 4.6.

Effect of temperature

The speed at which the chemical reaction proceeds during the setting and hardening of cement varies with the curing temperature. In small masses of concrete where the heat evolved by the cement is dissipated quickly, the rate of hardening varies with the surrounding temperature in the manner shown in Fig. 4.7. This diagram gives the compressive strength at ages up to 28 days as a percentage of that attained at 28 days at 20°C, for a range of temperatures. Thus, concrete cured for 7 days at 0°C only reaches 38/65, or 58 per cent of the compressive strength of similar concrete cured at 20°C, also for 7 days. The example given in Fig. 4.7 is fully described on p. 409. The relationship should be treated only as a guide since different cements harden at different rates, and an accurate assessment is not possible.

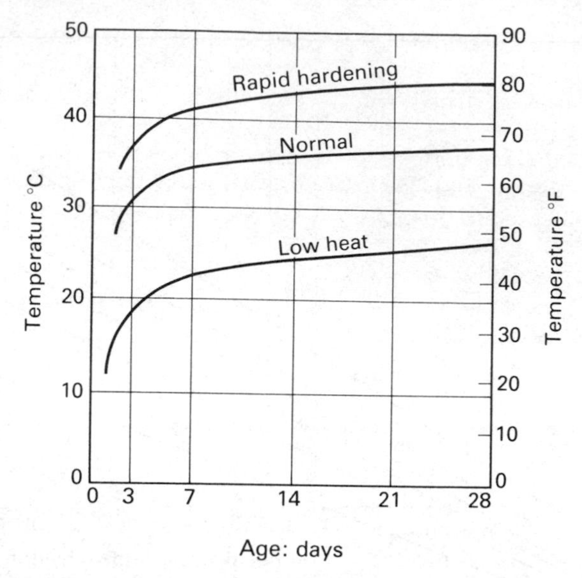

Fig. 4.8 Temperature rise in a typical 1 : 9 (weight) concrete under adiabatic conditions

If the temperature is much above 40°C, the paste is likely to set so rapidly that the concrete may stiffen before it is compacted, especially if, in addition, there is an opportunity for water to evaporate from the surface. In large sections, or in sections which are well insulated, the heat evolution of the cement may more than offset the heat losses, and the resultant temperature rise will increase the rate of hardening.

Heat of hydration

The chemical reaction when cement sets and hardens is accompanied by the evolution of heat and, as has already been stated, this is dependent on the relative quantities of the various compounds present in the cement.

For British ordinary Portland cement the heat of hydration averages 294 J/g at 7 days, rising to about 336 J/g at 28 days. A variation of ± 42 J/g on these figures may be expected for individual samples. The temperature rise in concrete cured under adiabatic conditions, i.e. when no heat loss can occur, is largest during the first few days of hardening, as is shown in Fig. 4.8 [9], gradually becoming less as the concrete ages.

The evolution of heat causes an increase in the temperature of the concrete. In actual practice this is dependent, not only on the heat evolution from the cement, but also on:

(a) volume of concrete being placed in any one operation;
(b) rate of placing of the concrete;
(c) the type of formwork;
(d) external atmospheric conditions, particularly the ambient temperature;

(e) the temperature of the concrete at the time of placing;
(f) the thermal conductivity of the concrete.

The temperature rise will obviously be greatest in a large concrete mass; in dams a maximum rise of as much as 44°C has been recorded, and in most modern dams in which normal Portland cement has been used it has been at least 22°C (see also p. 334). Dam concrete is usually fairly lean, and since the quantity of heat produced is proportional to the quantity of cement present, richer concretes in less massive structures are often subject to temperature rises of the same order. As an example, Fig. 12.5 shows the average temperature recorded in a part of a large concrete foundation where the concrete had a cement content of 326 kg/m³ (550 lb/yd³).

Properties of concrete as related to cement content

It is generally known that the strength of concrete increases with the cement content.

It is often argued that an increase in the proportion of cement increases the shrinkage. While this is correct for mixes of the same water/cement ratio, it is not necessarily important in practice, since it is usual to work to a given workability on the site, as measured roughly by the slump test. However, an excess of water combined with a high cement content may cause undue shrinkage with an increased tendency toward cracking.

If the water/cement ratio is maintained constant, an increase in the cement content improves the workability of the mix without affecting the strength.

Transport and storage

Cement for small jobs is usually packed in 50 kg bags and transported to the site by lorries. Where large quantities are used and cement silos are installed on the site, transport in bulk is more economical.

The storage of cement is entirely a matter of keeping it dry, and it is necessary to stack the bags in a shed or under whatever cover is available. Every effort should be made to prevent moisture from coming into contact with the bags at any point, and it is advisable to provide a raised floor covered with waterproof material such as tarpaulin.

Even when stored under good conditions bagged cement may lose 20 per cent of its strength after 2 months' storage, and 40 per cent after 6 months' storage. Cement can be stored in airtight bins indefinitely without deteriorating in any way, but this is not generally practicable for site use.

Air-set cement results from storage in a damp atmosphere. This is due to the moisture present in the air being absorbed by the cement and causing a partial set. The absorption of moisture from a damp atmosphere is reduced by preventing the movement of air into the store as far as practicable. As a rough guide, lumpy cement which cannot be easily crumbled in the fingers is unsatisfactory for general use.

Cement should be stacked in such a way that the cement first delivered can be used first.

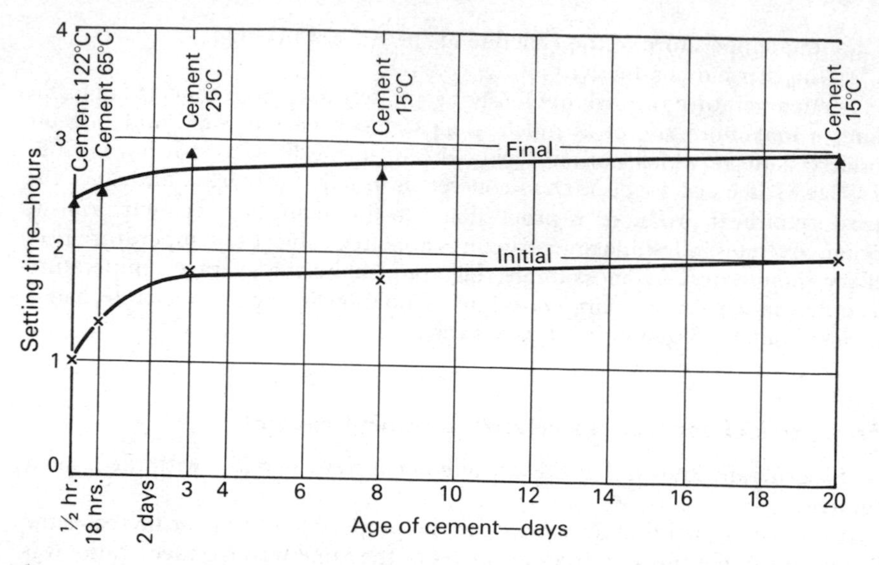

Fig. 4·9 Setting time of hot cement

Hot cement

The arrival on sites of cement in a warm or hot condition is often the subject of controversy. The high temperature is due to the friction which occurs during grinding, the heat evolved not having been dissipated during storage.

Tests at the Building Research Station (10) have indicated that hot cement has a somewhat shorter setting time than cool cement, as is shown in Fig. 4.9. Since this test is made on neat cement, the effect of the hot cement is probably exaggerated compared with the effect in mixed concrete. Other tests (11) have shown that the variation of the temperature of cement over the range 18°–71°C has a negligible effect on the workability and strength of concrete.

Cement tests

Details of the methods of testing are given in the relevant British and ASTM Standards and they should be implicitly followed if reliable results are to be obtained. In general, it is advisable to leave cement testing to an experienced operator in a properly equipped laboratory which has controlled temperature and humidity.

Cements, like other manufactured goods, are subject to some variation in quality, and it is advisable when comparing cement from two sources to obtain at least a dozen samples from different consignments and on different days. The comparison should then be based not only on the standard of quality, but also on the variability or otherwise shown by the tests. An important point which should be remembered in connection with cement tests is that strength and setting times are also affected by the relative humidity of the atmosphere. For both these tests it is essential to maintain the specified temperatures and curing conditions. The tests usually made on cement are:

(1) *Fineness.* BS 12 : Part 2 : 1971 describes a method of test for the determination of the specific surface of cement expressed as the total surface area in m^2/kg based on the Lea-Nurse permeability apparatus.

ASTM C204-73 gives a similar but modified method of test using the Blaine air permeability apparatus, whereas ASTM C115-74 gives a method of test using the Wagner turbidimeter apparatus.

The Blaine and Lea and Nurse methods give results of specific surface in close agreement with one another, but very much higher values than those obtained from the Wagner method. As a general indication the air permeability methods give results about 1·8 times higher than those from the turbidimeter method.

The relevant specification requirements for the various British and American cements are shown in Table 4.2.

(2) *Chemical composition.* The analysis of cement is complicated and should be left to an experienced chemist. Limits for the proportions of certain of the constituents are given in the relevant standards.

(3) *Strength.* The strength of cement is determined from compressive tests on mortar cubes or, alternatively, in British Standard tests by 100 mm (4 in.) concrete cubes. Provision is sometimes made for a tensile test when specially desired and provides a guide to the early strength when the speed of demoulding is important.

(*a*) *Compressive strength of mortar cubes.* In Britain, the first of two alternative methods given in BS 12 : 1971 for determining the strength of cement is based on making and testing mortar cubes. This method will, therefore, be described first, but it should be noted that this method is now little used since most testing laboratories use the second alternative method based on concrete cubes.

The mortar is made of a mixture of cement, standard Leighton Buzzard sand and water in specified proportions, and then fed into 70·7 mm (2·78 in.) cube moulds. Each filled mould is then vibrated in a standard machine which has a frequency of 12 000 vibrations per minute.

The cube mould with its base-plate is firmly held on the table of the vibrating machine by screwing down the hopper and by a thumb-bolt on the side. The hopper is filled with mortar and vibrated for 2 minutes at full speed. The mould is then removed and the whole placed in an atmosphere of at least 90 per cent relative humidity and at a temperature of 19° ± 1°C for 24 hours, after which the cubes are removed and immediately submerged in clean water until the time of testing.

This test is not affected by personal error as much as the tensile test, since the compaction is controlled by the machine. Thorough mixing of the materials is important, and a skilled operator should be employed.

Three cubes are tested in compression at each of the specified ages, generally 3 and 7 days; the minimum strengths which should be obtained are shown in Table 4.2.

For American cements the compressive strength is determined on 2 in (50·8 mm) cubes of a mixture of cement, standard Ottawa sand and water in certain proportions. The specimens are hand-compacted in a specified manner instead of using the British vibrating machine.

These slight differences in the sand quality and grading, water/cement ratio and method of compaction will result in variations in the results from the two methods and strict comparison between the specification requirements is not possible.

Mortar cube strengths correlate reasonably well with those obtained with the same cement used in a 1 : 2 : 4 concrete mix, and a water/cement ratio of 0·6. Tests made on 220 samples of cement at the Building Research Station [12] indicate that the concrete strength using the above mix may be expected to be about 0·7 times the mortar cube strength at the same age. The range in the ratio was about 0·6 to 0·8 in this test series. The mortar cube test, therefore, provides an indication of the compressive strength likely to be obtained in practice with cement similar to that used in the test.

(*b*) *Compressive strength of concrete cubes.* The second method given in BS 12 requires the preparation of 100 mm (4 in) concrete cubes with certain specified types of aggregate, 20 mm ($\frac{3}{4}$ in) maximum size, and sand. It is this method which is now commonly used for testing cement in Britain.

The combined grading of the aggregate is adjusted until a mix, in which 310 ± 1 g cement and 186 ± 1 g water is used to give sufficient concrete for each 100 mm cube, has a slump of 15 mm–50 mm ($\frac{1}{2}$ in–2 in). These specified quantities give a concrete with an aggregate/cement ratio of approximately 6 and a water/cement ratio of 0·6. The concrete is transferred to each mould in two approximately equal layers, each of which is given at least 35 strokes of a standard tamping bar. After demoulding, the cubes are stored in water at a temperature of 19° ± 1°C until tested in compression at 3 and 7 days. Three cubes are tested at each age. The required strengths are given in the relevant British Standards and are included in Table 4.2.

The requirements in BS 12 (as stated in Appendix D) permit the test to be carried out using any coarse and fine aggregate meeting certain specified requirements. However, it is now considered desirable to restrict the choice of the aggregate type to

(*a*) Mountsorrel granite in two size fractions for the coarse aggregate.

(*b*) Dried silica sand in five size factors for the fine aggregate. The characteristics and sources of this sand are specified in BS 4551 : 1970 [13].

Using these specified aggregates it becomes possible to standardize the batch quantities and thereby obtain a greater measure of agreement between the test results on cement samples from different test laboratories.

(*c*) *Variations in cement strengths.* Cements usually have strengths well in excess of the specification, but a large variation is often observed. Table 4.4 summarizes the results from a large number of mortar cube tests, at ages from 7 to 90 days, on cements delivered to sites in the United Kingdom. A further illustration of variations both at 7 and 28 days is given in Fig. 4.10. In this case the results given in the histograms relate to concrete cubes and are derived from tests made by Ready Mixed Concrete Limited on batches of ordinary Portland cement from 35 sources in the United Kingdom. The variations in strength are the result of changes in the characteristics of the basic materials, quality of the fuel used, plant inefficiencies and other factors. A small percentage of the variation may be ascribed to errors in the making and testing of the cubes. As indicated previously when discussing the increase of strength with age, cements harden at different rates and this accounts for some variation of strength at early stages; the variation in the strength of later ages may be smaller, although no indication of a smaller standard deviation for British cements at later ages was apparent in the tests quoted in Table 4.4.

Table 4·4 Summary of BS Vibrated Mortar Cube Tests on a Number of Cements

Country of manufacture	Type of cement	Compressive strength N/mm²					Standard deviation N/mm²				
		1-day	3-day	7-day	28-day	90-day	1-day	3-day	7-day	28-day	90-day
England	Ordinary Portland	7·0 (14)	25·0 (15)	33·5 (15)	47·5 (14)	53·0 (15)	1·8 (14)	2·6 (15)	2·4 (15)	4·5 (14)	4·8 (15)
Belgium	Ordinary Portland	5·5 (3)	18·0 (11)	30·5 (11)	45·0 (3)	48·0 (3)	*	2·9 (11)	4·0 (11)	*	*
Germany	Ordinary Portland	9·5 (3)	22·5 (8)	30·5 (8)	46·5 (3)	49·5 (3)	*	*	*	*	*
Sweden	Ordinary Portland	13·5 (2)	22·5 (3)	33·0 (3)	44·0 (2)	52·0 (2)	*	*	*	*	*
England	Sulphate-resisting	9·0 (2)	21·5 (6)	30·5 (6)	45·0 (4)	—	*	*	*	*	—
Belgium	Super-sulphate-resisting	7·5 (4)	29·0 (10)	39·5 (10)	54·0 (6)	—	*	7·5 (10)	6·5 (10)	*	—

(1) The figures in brackets indicate the number of batches tested.
(2) The compressive strengths quoted are the average of the results of tests on the number of batches indicated, three cubes being tested at each age from each batch.
(3) Standard deviation not assessed for less than 10 batches of cement.*

(4) *Setting time.* The test for setting time involves both a determination of the quantity of water necessary to produce a paste of normal consistency, and the test for setting time itself. Figure 4.11 shows the Vicat needle used for the test. The normal consistency is that at which the plunger penetrates to within 5 mm to 7 mm from the bottom of the Vicat mould, $4 \pm \frac{1}{4}$ minutes from the moment of adding water to mix the cement paste. The initial setting time is the period which elapses between the time when the water is added to the cement, and the time at which the thin needle does not penetrate beyond a point 5 mm from the bottom of the mould. The cement is considered as finally set when, on applying the needle with the angular attachment, it makes an impression on the block while the attachment fails to do so. It should be noted that the initial and final setting times are a gauge of the period which elapses before a certain consistency of paste is reached; setting and hardening actually commence before the initial setting time and proceed long after the final setting time.

(5) *Soundness.* The Le Chatelier method of test for soundness is a simple little test to determine whether undue expansion is likely to occur. Providing a bath controlled at 20 ± 1°C is available, this test can easily be done on the site. British cements rarely give a measurable expansion, however, and the test is for that reason not often made.

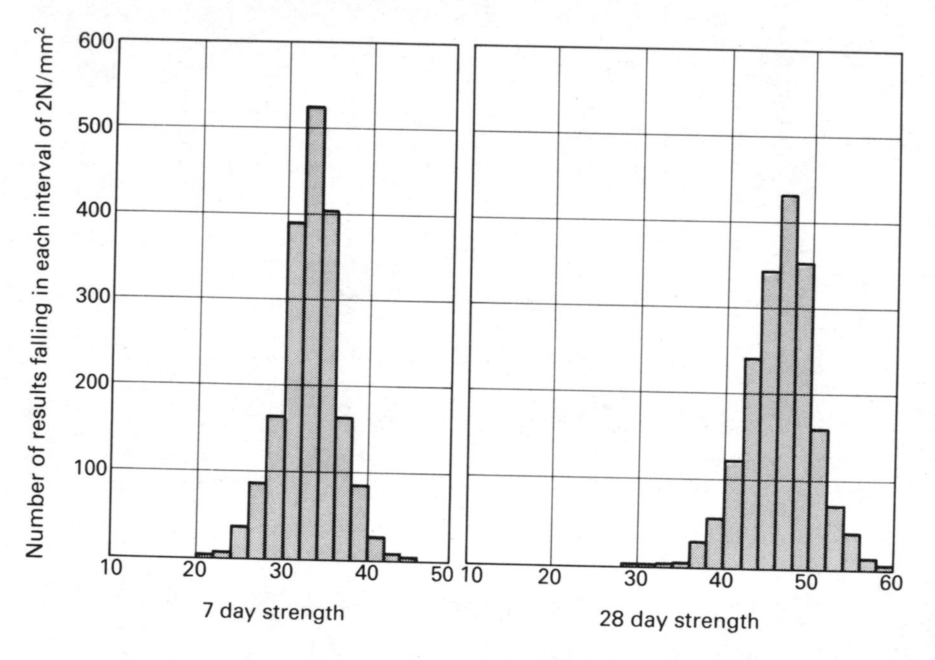

Fig. 4.10 Histograms for compressive strengths of samples of ordinary Portland cement taken in the United Kingdom

Fig. 4.11 Measurement of the initial setting time of cement by means of the Vicat needle

REFERENCES

1 LEA, F. M., Modern developments in cements in relation to concrete practice, *J. Inst. Civ. Engrs*, Feb. 1943.

2 LEA, F. M., *The Chemistry of Cement and Concrete*, 3rd ed., 1970. Edward Arnold (Publishers) Ltd., London.

3 BS 12 : Part 2 : 1971, *Portland Cement (Ordinary and Rapid-hardening)*. British Standards Institution.

4 FULTON, F. S., *Concrete Technology*, 1964. The Portland Cement Institute, Johannesburg.

5 BAXTER, D. J. and BOARDMAN, J. M., The manufacture, properties and use of super-sulphated cement in civil engineering works, *Civ. Eng. Publ. Wks Rev.*, April 1962.

6 BATE, S. C. C., *Report on the Failure of Roof Beams at Sir John Cass's Foundation and Red Coat Church of England Secondary School, Stepney*. Building Research Establishment Current Paper CP 58/74, June 1974.

7 BUILDING RESEARCH ESTABLISHMENT, *High Alumina Cement Concrete in Buildings*. Building Research Establishment Current Paper CP 34/75, April 1975.

8 SNOW, F. S., LEA, F. M., MURDOCK, L. J. and BURKE, E., Would the strength grading of ordinary Portland cement be a contribution to structural economy? *Introductory Notes, Proc. Inst. Civ. Engrs*, Part III, Dec. 1953.

9 DAVEY, N., *Temperature Rise in Hydrating Concrete*, Building Research Technical Paper, No. 15. HMSO.

10 DAVEY, N., *Hot Cement*, Building Research Bulletin No. 7, 1930. HMSO.

11 KEENE, P. W., *An Investigation of the Effect of Cement Temperatures and Ambient Temperatures on the Workability and Strength of Concrete*, Cement and Concrete Association DN/17, April 1962.

12 NEWMAN, A. J., Materials for concrete, *Proceedings of a Symposium on Concrete Quality*, Nov. 1964. Cement and Concrete Association.

13 BS 4551 : 1970, *Methods of Testing Mortars and Specification for Mortar Testing Sand*. British Standards Institution.

5
Admixtures

Introduction

ADMIXTURES, ranging from additions of chemicals to the use of waste materials for which an outlet is being sought, have been advocated for use in concrete almost from the time when cement was first invented.

The use of an admixture should only be considered where it is desired to modify for a particular reason the properties of either fresh or hardened concrete, or both, which cannot be modified or obtained by changes in the composition or proportions of the normal concrete mix. For instance, a harsh mix may be made more plastic and cohesive by the addition of a plasticizer, a pore filler, an air-entraining agent, a change in the proportion of sand to coarse aggregate, a change in the sand grading, or by the use of additional cement.

Because an admixture is generally introduced into a concrete mix in a relatively minute quantity, the degree of control must be higher than is usual for ordinary concrete work to ensure that overdosages are unlikely to occur. There is always a temptation for the operative to add that little bit extra, especially if conditions appear to him to be more adverse than usual. The trouble and expense involved in achieving the required control can often outweigh any advantages obtainable by using admixtures. Excess amounts of an admixture can be highly detrimental to the strength or other properties of the concrete.

In Great Britain there are well over 100 proprietary admixtures available on the market, and the selection of a suitable one for a particular purpose can be confusing. In order to be able to assess the suitability of an admixture the active ingredients must be known, because, while some properties of the concrete may be enhanced by one ingredient, detrimental effects to other properties may be caused by other ingredients. As an example, a plasticizer was being used on site in concrete for airfield runways, the control being exercised over the concreting was good and an automatic dispenser was in use for measuring the correct dosage of admixture. Three days after placing one bay of concrete it had not set and slumped on removing the side forms. It was established that a valve in the dispenser had stuck, resulting in a ten-fold increase in the amount of admixture for each batch of concrete. This overdosage did not result in excessive workability but the small amount of retarder present became the predominating factor.

The control of the quality of admixtures has been tightened in recent years, particularly since the introduction of CP 110:1972 which states that the engineer should be provided with the following data:

(1) the typical dosage and detrimental effects of underdosage and overdosage;

Table 5.1 Admixture Acceptance Test Requirements

Category of admixture	Water reduction	Stiffening time: Time from completion of mixing to reach a resistance to penetration of: 0·5 N/mm²	3·5 N/mm²	Compressive strength: Percentage of control mix (minimum)	Age
Accelerating	—	More than 1 h	At least 1 h less than control mix	125 100	24 h 7 and 28 days
Retarding	—	At least 1 h longer than control mix	—	90 95	7 days 28 days
Normal water-reducing	At least 5%	Within ± 1 h of control mix	Within ± 1 hr of control mix	110	7 and 28 days
Accelerating water-reducing	At least 5%	More than 1 h	At least 1 h less than control mix	125 110	24 h 7 and 28 days
Retarding water-reducing	At least 5%	At least 1 h longer than control mix	—	110	7 and 28 days

(2) the chemical name(s) of the main active ingredient(s) in the admixture;

(3) whether or not the admixture contains chloride and, if so, the chloride content of the admixture expressed as a percentage of equivalent anhydrous calcium chloride by weight of the admixture;

(4) whether or not the admixture leads to the entrainment of air when used at the manufacturer's recommended dosage.

The importance of testing admixtures, particularly with the cement and aggregates to be used on a site, cannot be over-emphasized. Such tests should be designed to show that no adverse side effects are likely, such as increased drying shrinkage, reduced bond strength, reduced ultimate compressive strength, and reduced modulus of elasticity.

A concrete of fundamentally poor quality will not be converted into a good concrete by any type of admixture.

Performance standards for admixtures

Performance criteria covering the use of accelerating admixtures, retarding admixtures and water-reducing admixtures are given in BS 5075 : Part 1 : 1974. This standard specifies admixture acceptance tests which demonstrate the ability of a particular formulation to meet stipulated performance requirements; and admixture uniformity tests carried out by the vendor, or at an independent laboratory at the request of the purchaser, which demonstrate that a particular consignment of admixture is similar to material which has previously been submitted to the tests.

The requirements from the acceptance tests are given in Table 5.1. These requirements are based on tests made with comparative mixes of concrete with and without the addition of the admixture. It should be noted that the stiffening-time test, which is carried out on mortar sieved from the concrete determines the times required for a standard needle to indicate a resistance to penetration of 0·5 N/mm^2 and 3·5 N/mm^2. Work has shown that the time to reach a resistance of 0·5 N/mm^2 corresponds approximately to the extreme limit for placing and compacting concrete, and the time to reach a resistance of 3·5 N/mm^2 gives a guide to the time available for the avoidance of cold joints.

The requirements for pigments for colouring concrete are given in BS 1014:1975. Information on pigments is given on p. 350.

Accelerators

Accelerators are generally used for accelerating the strength development of concrete in cold weather. Calcium chloride has been the most widely used accelerator in the past because of its cheapness and predictable performance. Moreover, it has a very marked effect on the rate at which concrete hardens during the first few hours of its life.

However, calcium chloride suffers from the big disadvantage that its use can result in the corrosion of the reinforcement steel within the concrete, particularly if the dosage is large and if it is not properly controlled. For example, in one instance, corrosion of mild steel bar reinforcement, resulting in spalling of the concrete within twelve months, was found to be due to a twelvefold dosage of calcium chloride.

Because of this danger, calcium chloride should not be added to any dense reinforced concrete. The danger is recognized by several authorities in the United Kingdom and in other countries who have imposed a positive ban on the use of calcium chloride as a concrete admixture. In addition, a recent amendment to British Standard CP 110 recommends that calcium chloride should never be added in prestressed concrete, reinforced concrete and concrete containing embedded metal, because of the risk of corrosion.

The beneficial effects of calcium chloride may, however, be utilized in plain unreinforced concrete when early rapid hardening is required. Its use in these circumstances can permit early striking of the formwork and a consequent saving in costs due to faster production.

Shacklock[(1)] has recommended that concrete containing calcium chloride should be placed within half an hour of mixing. If the time taken is longer, the mix may stiffen to such an extent that site operatives may be tempted to adopt the bad practice of adding more water, causing a reduction in the compressive strength. Shacklock has also given information on the increase in drying shrinkage which occurs when calcium chloride is added to concrete; he quotes increases of about 70 per cent at 14 days and about 15 per cent after three months when calcium chloride is added at 2 per cent by weight of cement.

Another disadvantage of calcium chloride is that it is liable to reduce the resistance of the concrete to sulphate attack and its use is, therefore, not recommended when it is necessary to use sulphate-resisting Portland cement in the concrete.

Due to the corrosion problem associated with calcium chloride, there is a demand for a low priced effective chloride-free accelerator. Many chemicals may act as accelerators when added to concrete but they are liable to suffer from erratic performance since they can be sensitive to the chemical composition of the cement and the ultimate strength of the concrete may be impaired.

Certain chloride-free proprietary admixtures are based on calcium formate which has been found to be an effective accelerator. However, there is insufficient information to enable it to be regarded as being completely trouble-free as far as corrosion of the steel reinforcement is concerned.

Another category of accelerator covers the various chemicals used to make the concrete or mortar set within a few minutes. These set accelerators are useful in various situations such as, for example, the sealing of a hole in order to stop water seepage under pressure. Aluminium chloride is a typical example of this kind of accelerator but other chemicals based on carbonates, silicates and silicofluorides are also used. Since the strength of the concrete or mortar is liable to be reduced by the inclusion of these chemicals, they must be used with discretion.

Retarders

Retarders, as this name implies, delay the beginning of the setting and hardening of concrete, and under normal conditions in the United Kingdom their use is seldom necessary. They are most useful in hot countries where concrete has to be transported for some distance, and for concreting in large quantities in very hot weather, when it is important to avoid the formation of 'cold joints'.

Overdosage is likely to affect seriously the setting of concrete. If deep pours are envisaged the delayed setting is likely to increase formwork pressures considerably.

The chemical composition of retarders is similar to that of water-reducing admixtures. Many proprietary admixtures, in fact, exhibit the combined characteristics of both retardation and water-reducing properties.

Water-reducing admixtures (plasticizers)

A number of chemicals have been marketed which, added in very small percentages to the mix, give some increase in workability. Alternatively they can be regarded as a means by which the water content of the concrete can be reduced without suffering a loss in workability. Hence they are generally classified as water-reducing admixtures, but they are also sometimes referred to as plasticizers, wetting agents and dispersing agents.

The extent to which the water content of the concrete may be reduced by the use of these admixtures depends on the characteristics of the mix, but it is generally between 5 and 10 per cent. The increase in strength may be up to 10 per cent as a result of any reduction in the water content.

Water-reducing admixtures are mainly of benefit when added to concrete made with 'difficult' aggregates and so liable to harshness or segregation of the mix, or to special and unusually rich mixes where very high strengths are needed in conjunction with good workability.

Water-reducing admixtures are based on lignosulphonates (a by-product of the wood pulp industry) or salts of hydroxylated carboxylic acid. A typical dosage is 0·2 per cent by weight of cement.

Shacklock[1] has reported that whereas admixtures in the lignosulphonate group are likely to produce a rather more cohesive concrete with a reduced tendency to bleed, admixtures in the hydroxy-acid group have less influence on cohesiveness and may tend to increase bleeding. These differences are partly due to the fact that the lignosulphonate admixtures are likely to entrain up to 2 per cent of air into the concrete, whereas the hydroxy-acid type of admixture does not normally produce any air entrainment. Both types have a retarding effect on the setting of the concrete.

Little has been published on the long-term effects of any of these materials.

Pore Fillers

A number of materials are available in powder form which can be used to increase the cohesive characteristics of a concrete and thereby improve the resistance to bleeding. These materials include quick lime, slaked lime, diatomaceous earth, bentonite, kaolin, and rock flour. In addition pulverized fuel ash (fly ash) and ground blastfurnace slag may also be considered as pore fillers although these materials are normally used for their pozzolanic properties (see below).

In their role as pore fillers, these materials are of benefit for harsh lean concrete mixes which suffer from a shortage of fine particles. In general, however, they are of very doubtful value since, whilst they provide increased workability and cohesiveness, they usually necessitate an increased water/cement ratio in the concrete with a corresponding loss in strength. Moreover, the drying shrinkage of the concrete is increased by these materials, some of which are, in effect, little different from the clay, silt and stone dust so carefully excluded from concrete aggregates.

In many cases, there seems no advantage in adding a pore filler rather than using a richer mix.

Pozzolanas

Pozzolanas are those materials which react with any lime set free during the setting of cement, and so improve the durability of concrete, including its resistance to attack by sulphates, peaty waters, etc., Callet[5], however, instances a failure of pozzolanic cement at Le Havre (see also p. 384). Pozzolanas, which in themselves possess little or no cementitious properties, are used either in addition to, or to replace up to 70 per cent of the cement. They reduce the rate of hardening of concrete, and this is one of the principal objections to their use. The evidence available suggests that the ultimate strength for replacements of up to at least 20 per cent of the cement by pozzolanas is not very different from that when cement alone is used.

Pozzolanas can occur in a natural form, as for example, volcanic ash, scoria and pumice. Of greater significance however is the artificially produced form of pozzolana known as pulverized fuel ash (pfa) or 'fly ash'. Since this is a waste product which is produced in large quantities, it is considered separately.

Pulverized fuel ash (pfa). Pulverized fuel ash is produced during the burning of powdered coal, the chief source of supply being power stations. The particles of the ash are predominantly spherical in shape and their fineness resembles that of cement varying between 250 and 500 m^2/kg. Its pozzolanic properties are derived from the presence of about 50 per cent finely ground silica. It contains some unburnt coal, the proportion of which should not exceed 7 per cent if the fly ash is to be used as an admixture in concrete. The proportion of this unburnt material is apt to vary from time to time, being highest when the burners are working inefficiently immediately after being lit, or when the boilers are not working under full load, and frequent testing is important if the occasional use of material with a high carbon content is to be avoided. Pulverized fuel ash varies in properties and it is necessary to determine independently the quality of material from any particular source.

The properties of pulverized fuel ash for use in concrete are covered by BS 3892 : 1965 which states that the loss-on-ignition value should not exceed 7 per cent. It also states that the sulphate content of the ash, expressed as SO_3, should be below 2·5 per cent, where the quantity of ash to be used is less than the quantity of cement, and below 1·5 per cent if it exceeds it.

In general pulverized fuel ash has a somewhat similar effect to natural pozzolanas, in that the initial setting and hardening may be delayed, but the ultimate strength may be as high as that of concrete made with cement alone. This applies only for proportions of pulverized fuel ash up to about 20 per cent used as a replacement for cement.

There have also been suggestions that pulverized fuel ash might be used to replace a proportion of the sand in the mix. Tests have indicated that replacement of up to 20 per cent of sand by pulverized fuel ash results in an increase in workability in harsh mixes, but not in mixes which already contain an adequate proportion of fines. Some increase in strength occurs under the same conditions, providing the workability is maintained constant. The drying shrinkage is not affected by its use in the proportions stated above and the temperature rise in mass concrete may be reduced. Pulverized fuel ash has been used in the concrete of several dams and other structures involving the placing of large pours of concrete.

Since one of the main problems with pulverized fuel ash is the variability in quality, there is a need for controlled processing to improve the uniformity. To cater for this need, plant has been installed in recent years at selected power stations in Great Britain, by an independent company, to process and grade pulverized fuel ash to within specified limits for fineness and carbon content. The resulting product, known by the trade name of 'Pozzolan' is sold with a guaranteed pozzolanic quality.

Pozzolanic Portland cements, made from blending cement and pulverized fuel ash, are produced in several countries and there are standards relating to these cements. In Great Britian the production of pozzolanic Portland cement is uneconomic because of the costs of transporting the pulverized fuel ash to the cement works (or alternatively the cement to the power stations) plus the costs of the handling facilities. Production is limited to only one source where the output is relatively small.

Ground blastfurnace slag

Ground blastfurnace slag is used to produce Portland blastfurnace cement which has been in production for many years and which is described briefly on p. 64. More recently ground blastfurnace slag has also become available for mixing on site, under the trade name 'Cemsave'. This material is produced by grinding the granulated blastfurnace slag to a powder at the steelworks. The powder produced in this way has a fineness similar to that of ordinary Portland cement but is distinctly lighter in colour. It is transported to site, for blending with ordinary Portland cement, in the mixer along with the other concrete constituents.

Concrete containing ground blastfurnace slag has similar properties to concrete made with pulverized fuel ash. Ground blastfurnace slag is not, however, a pozzolana and it does not set if it is mixed with water. It has to be activated by means of Portland cement and the resulting hydration reactions then form specific hydrated compounds.

The replacement of some of the cement by ground granulated slag reduces the rate of strength development, particularly at low curing temperatures, but the same ultimate strength may still be achieved. When 30 per cent replacement of cement by ground blastfurnace slag is adopted, the 28 day strength of the reference mix is achieved within two months. For replacements in excess of 50 per cent the time required to reach the 28 day reference strength might well be one year or more.

The use of ground blastfurnace slag is beneficial in situations where a reduced rate of heat of hydration is desirable, as, for example, in large pours. It may also be beneficial in giving the concrete improved resistance to sulphate attack.

Atwell had described [6] the large scale use of ground blastfurnace slag on a site where cement replacements of up to 54 per cent were used. Substantial savings in cost were reported because of the close proximity of the source of production of the slag, but they were offset by extra costs during the winter due to the slow rate of heat development in the concrete. Because of the slow rate of heat development, it was found necessary during winter conditions to heat the concrete to at least 10°C and provide covers and occasionally heaters during curing. Also in order that the contractor could maintain his production schedule on the striking of soffit formwork, it is reported that the slag content in the concrete had to be reduced from 30 per cent to 13 per cent and eventually to zero in the case of the 25 N/mm² grade concrete.

Water-repellent admixtures

A large number of water repellent admixtures have been placed on the market, their function being to prevent the absorption of rain into the concrete. They are of no use against water under pressure. The water-repellent material, i.e. that which causes the water to run off, as 'off a duck's back', is usually a metallic soap or vegetable or mineral oil. Gradual leaching by rainwater reduces their effectiveness in time.

The value of those admixtures which are claimed to waterproof concrete is doubtful, even if they do possess some waterproofing effect, since they are completely ineffective against water penetration through cracks, joints, honeycombed areas, etc., where most leakage is likely to occur. Good concrete is practically impermeable in itself without any admixture.

Air entrainment

Most air-entraining agents for concrete consist of a neutralized vinsol resin which is mixed into the concrete at a dosage which is typically about 0·05 per cent by weight of cement. They are used widely in the United States where it is generally accepted that the durability of concrete, particularly as regards frost resistance and resistance to de-icing salts, is considerably improved by the entrainment of a small quantity of air in the form of very fine bubbles.

Air-entraining agents have not been used to the same extent in Great Britain, mainly because the incidence of frost damage has been so much lower than in the United States. The customary use of lower water contents in Great Britain may be an important factor. They are, however, being used more extensively in concrete for road and runway constructions, particularly to provide increased resistance to the combined effects of frost and the salt used for de-icing.

The Department of Transport Specification for Road and Bridge Works[2] issued in 1976 requires that the concrete used in the construction of at least the top 50 mm (2 in) of paving to be used as a wearing course shall contain an air-entraining agent. The quantity of agent required is that which will give a total quantity of air of $4\frac{1}{2} \pm 1\frac{1}{2}$ per cent by volume of the mix.

Guidance on the design of air-entrained concrete is given in the 'Design of normal concrete mixes' issued by the Department of the Environment[3]. It is stated that, in general, the strength of concrete is reduced by the addition of entrained air and that it may be assumed that a loss of 5·5 per cent in compressive strength will result for each 1 per cent by volume of air entrained in the mix (see also Fig. 13.1).

However, the workability is increased by an amount roughly equivalent to an increase of 0·2 in the water/cement ratio for each 1 per cent of entrained air. If the reduction in strength is acceptable, greater workability is therefore obtained without the use of an unduly high water/cement ratio, which would give rise to risk of frost failure (see also p. 379).

If, however, account is taken of the fact that air-entrained concrete needs less mixing water to obtain a given workability, it is found that, in general, an air-entrained concrete has a strength which is about 10 per cent less than a non-air-entrained concrete of similar cement content and workability.

The air which is entrained is in the form of small round bubbles having a diameter ranging from about 20 to 2000 μm (0·0008 to 0·08 in). To obtain the maximum protection from damage due to freezing and thawing, it has been recommended[1] that the quantity of air in the matrix of the concrete should be about 13 per cent. On this basis the average air content of the fresh concrete, for different maximum sizes of aggregate, should be as follows:

Maximum size of aggregate mm (in)	Average air content per cent
40 (1½)	4
20 (¾)	5
14 (½)	6
10 (⅜)	7

These values are given in the BS Code of Practice CP 110 for concrete which is required to provide resistance to the effects of salt used for de-icing.

A reduction in the tendency of concrete to bleed is also usually obtained with concrete which is air-entrained. As bleeding results in the formation of fine cracks below the larger particles of aggregate, which provide paths for the penetration of water, a reduction in bleeding is beneficial from the point of view of permeability and durability.

Air entrainment should not be used unless site control is good, since the amount of air entrainment may vary considerably with changes in sand grading, errors in proportioning, workability of the mix, and temperatures. Checks on the air void content should be made at frequent intervals using apparatus such as that shown in Fig. 5.1 [4], since each one per cent increase in air voids is likely to result in a loss in strength of between 5 and 6 per cent. Providing a good quality air-entraining agent is used, the air bubbles produced are fairly stable and it is unlikely that there will be any significant loss of entrained air during normal handling and compaction.

Careful consideration should be given to the relative economies of using an air-entraining agent, instead of the addition of extra cement; changes in the resulting strength, and the desirability of avoiding complication especially on small sites, should be borne in mind.

Air-entraining agents are sometimes added to lightweight concrete in large dosages to give air content in the concrete of up to 30 per cent. By this means, the density of the concrete is considerably reduced and thermal insulation properties improved.

Superplasticizers

Concrete admixtures known as superplasticizers have recently been developed and have been used to a considerable extent in Germany and Japan. There use in Great Britain, however, has been very limited. When added to concrete, superplasticizers have the effect of increasing the workability of the concrete very considerably. They are, in fact, regarded as a means of producing 'flowing' concrete without the undesirable segregation that generally occurs with very wet concrete. Alternatively, they may be used to increase the strength of the concrete because they allow the water content to be reduced whilst still maintaining the same workability.

Superplasticizers, which are supplied as liquids, can be thought of as chemicals which extend the performance of normal plasticizers. Superplasticizers are, however, quite different chemicals and the majority of the brands available fall into two main categories, namely

(a) sulphonated melamine formaldehyde condensates

(b) sulphonated naphthalene formaldehyde condensates.

Because of the flowing properties which superplasticizers impart to con-

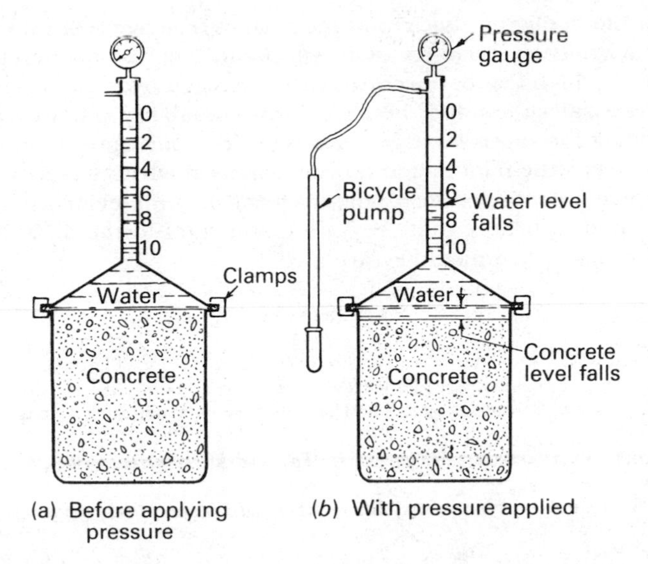

Fig. 5.1 Diagram of pressure-type apparatus for determining the air content of air-entrained concrete
When the air pressure is increased, the air in the concrete is reduced in volume; this slightly lowers the surface of the concrete and causes the level of the water in the tube to fall

crete, they may be useful where it is necessary to place concrete in difficult locations as, for example, where the reinforcement is heavily congested. They also enable concrete to be placed quickly in appropriate situations such as floor slabs where the self-levelling tendency of the concrete reduces the effort required for placing and finishing. It follows that superplasticizers should not be used where the concrete is to be placed on a slope exceeding about 3° to the horizontal. However, these advantages must be weighed against the fact that superplasticizers are very much more expensive than normal plasticizers and substantially increase the cost of the concrete.

Superplasticizers will not fluidify all concrete mixes successfully and the mix must be designed to suit. An important influence is the amount of combined fines (cement and sand having a particle size of less than 300 μm). Useful information on this aspect, as well as many others, is given in the report of a Joint Working Party of the Cement Admixtures Association and the Cement and Concrete Association on 'Superplasticizing Admixtures in Concrete'[7]. One particular requirement, which is recommended in this report, is that the mix should contain 450 kg/m³ (760 lb/yd³) of combined fines when the maximum aggregate size is 20 mm (¾ in). As a guide, concrete mixes which are suitable for pumping, i.e. cohesive with a high sand content, are also suitable for the addition of a superplasticizer. Trial mixes before the work starts are essential.

Since the effect of the high workability lasts for only about 30 to 60 minutes, depending on temperature, after the addition of the superplasticizer to the concrete, it is important that the concrete is placed without undue delay.

Control of the dosage is also important since overdosage will make the concrete so fluid that severe segregation will occur. The use of superplasticizers requires a very high degree of control to be exercised over the batching of the concrete, especially the water, because if the workability is not correct at the time of adding the superplasticizer, excessive flow and segregation will result.

There is very little information on the long-term effects of superplasticizers on concrete and contained steel reinforcement, but the evidence available (8) gives no reason to believe that they are detrimental to the durability of the reinforced concrete in which they are used.

REFERENCES

1 SHACKLOCK, B. W., Admixtures as an aid to concrete construction. The science of admixtures. *Proceedings of a Symposium organized jointly by the Concrete Society and the Cement Admixtures Association, London, 6 November 1969*. The Concrete Society, Publication 54.004, 1970, pp. 7–19.

2 DEPARTMENT OF TRANSPORT, *Specification for Road and Bridge works*. 1976, pp. 194, HMSO, London.

3 DEPARTMENT OF THE ENVIRONMENT, *Design of Normal Concrete Mixes*. 1975, pp. 30, HMSO, London.

4 BS 1881 : PART 2 : 1970, *Methods of Testing Concrete. Part 2. Methods of Testing Fresh Concrete*. British Standards Institution.

5 CALLET, M. P., Concreting in sea-water. Notes from a paper published in *The Dock and Harbour Authority, Concr. Constr. Eng.*, Nov. 1949, p. 364.

6 ATWELL, J. S. F., Some properties of ground granulated slag and cement. *Proc. Inst. Civ. Engrs*, Part 2, **57**, June 1974.

7 *Superplasticizing Admixtures in Concrete*, Report of a Joint Working Party of the Cement Admixtures Association and the Cement and Concrete Association. Cement and Concrete Association, 1976.

8 QUINION, D. W., *Superplasticizers in Concrete—a Review of International Experience of Long-term Reliability*. Construction Industry Research and Information Association, Report 62. September 1976, pp. 25.

6

Water/Cement Ratio and Workability

WATER has two functions in a concrete mix, firstly, to enable the chemical reactions which cause setting and hardening to proceed, and secondly, to lubricate the mixture of gravel, sand and cement in order to facilitate placing.

As in other chemical reactions, the cement and water combine in definite proportions, and for normal Portland cement 1 part by weight of cement requires about 0·25 parts by weight of water for hydration. Concrete containing such a small proportion of water would, however, be very dry and exceedingly difficult to compact. Extra water is, therefore, added to lubricate the mix, that is to make it workable, and since all this water evaporates when the concrete dries, leaving voids, it is important that the water used to provide lubrication should be kept to a minimum.

If concrete is not fully compacted, numerous bubbles of air may be trapped, resulting in further voids. There are, then, two main sources of voids in concrete; entrapped air bubbles, and the water required for lubrication, which later evaporates. Air bubbles are more easily eliminated from wet mixes than from dry mixes; the latter require a considerable amount of work to liberate the air. It follows that for any given method of compaction representing a certain amount of work done there is an optimum water content at which the volume of entrapped air plus the volume of water voids is a minimum. Concrete with the minimum volume of total voids is the densest and strongest and, therefore, normally the most desirable.

The conclusion to be drawn from the above is that the densest and strongest concrete is obtained by using the minimum quantity of water consistent with the degree of workability required to give maximum density. The degree of workability must be considered in relation to the method of compaction and the type of construction, in order to avoid the necessity for an excessive amount of work to obtain maximum density. This statement can be illustrated by comparing the stiff concrete placed in roadwork by heavy vibratory equipment with the more fluid concrete necessary for thin reinforced concrete wall sections rammed by hand. If stiff concrete were used in the second case, honeycombing would almost certainly occur, and the density and strength would be far less than in the roadwork.

The definition of the term water/cement ratio needs clarification. The difficulty arises from the presence in a batch of concrete of water from three possible sources:

(1) Water absorbed into the aggregate, (w_a)
(2) Surface water on the aggregate, (w_s)
(3) Water added during mixing, (w_m).

Water from sources (2) and (3) together provides what might be termed the free water in the mix and the author has throughout this book adopted the basis that:

$$\text{Water/cement ratio} = \frac{w_s + w_m}{W_c} = \frac{w}{W_c} \qquad (1)$$

where W_c denotes the weight of cement.

In this equation it is assumed that the aggregate is wet, or damp but internally saturated; if it should be dry the water to be added at the mixer,

$$w_m = w_a + w \qquad (2)$$

The relationship between the water/cement ratio, the richness of the mix, the grading of the aggregate, and the workability and strength of concrete was first studied by Professor Abrams in America. The conclusions drawn from this work led to the formulation of Abrams' water/cement ratio law, as follows:

> 'With given concrete materials and conditions of test, the quantity of mixing water used determines the strength of the concrete, so long as the mix is of workable plasticity.'

This law implies that for fully compacted concrete, sound aggregates and a given cement the strength depends only on the ratio of water to cement. Research in recent years has shown that there is some variation from this law, and this is dealt with in Chapter 7, Design of Concrete Mixes. The strength: water/cement ratio relationship given in Chapter 7 is quoted for an age of 28 days only. Reference should be made to Chapter 4, Cement, for information on the effects of age and temperature.

Increases in the water/cement ratio have an adverse effect on such properties of concrete as permeability, resistance to frost action and weathering, resistance to abrasion, tensile strength, creep, modulus of rupture, and shrinkage (see Chapter 2).

It may be concluded that for nearly all purposes concrete having the least water/cement ratio necessary to give the degree of workability required for full compaction without an undue amount of work, will give the best results.

Workability

The term workability is difficult to define precisely, and Newman[1] has proposed that it should be defined in at least three separate properties:

(1) Compactibility, or the ease with which the concrete can be compacted and the air voids removed.

(2) Mobility, or the ease with which concrete can flow into moulds around steel and be remoulded.

(3) Stability, or the ability of concrete to remain a stable coherent homogeneous mass during handling and vibration without the constituents segregating.

To these might be added the ease with which a good finish is obtainable, especially for vertical surfaces cast against formwork and floor slabs which are to be power trowelled.

No one test gives complete guidance on the degree to which these properties are developed in a mix. The compacting factor, the V-B test and the slump test provide the most useful general guide at present. Of these the slump test is probably the best known in the field, both in this country and in the United States, while the V-B test and the compacting factor test have been used to an increasing extent in the laboratory. The slump test is a good measure of mobility and stability in concretes of medium or high workability. The compacting factor test was devised at the Building Research Station and has certain advantages in that it measures the degree of compaction for a certain amount of work done. It is much more sensitive to changes in the water/cement ratio of rich mixes than of lean mixes where particle interference occurs[1]. Cusens[2] concluded that the small amount of work applied in the compacting factor test did not provide an accurate measure of the work required to compact dry, harsh mixes with compacting factors less than 0·8. The V-B test is more satisfactory for rich or dry mixes which are to be vibrated to obtain high strengths.

Other methods attempt to measure the consistency of the concrete while still in the mixer. They depend on measurements of torque, or overturning forces. Although they indicate changes in the consistency of concrete mixes, their use must be supplemented by frequent slump tests.

Slump test

The slump test was devised in America nearly fifty years ago and has been universally adopted as a check of the consistency of concrete in the field. While it has been shown that with all except rich workable concretes it is liable to random variations, the slump test does provide a simple practical method for maintaining reasonable uniformity in the consistency of concrete produced in the field.

The method of making slump tests is described in BS 1881 : 1970. The test is made with a metal cone 300 mm high, having a bottom diameter of 200 mm and a top diameter of 100 mm, on a representative sample of the concrete, the procedure being as follows.

The slump cone should first be inspected to make sure the internal surface is clean, dry, and free from set cement. The cone is then placed on a smooth, flat, non-absorbent surface, preferably a steel plate, and the operator holds the mould firmly in place while it is being filled, by standing on the foot pieces. The mould is filled to about one-fourth of its height with the concrete which is then tamped, using 25 strokes of a 16 mm rod, 600 mm long, rounded at the lower end, Fig. 6.1. (*a*). The filling is completed by three further layers similar in height to the first, and the top struck off so that the mould is exactly filled. The feet should still remain on the foot pieces until surplus concrete has been cleaned from round the base of the cone. The mould is then removed by raising it vertically immediately after filling, Fig. 6.1 (*b*). The moulded concrete is finally allowed to subside and the height of the specimen measured after coming to rest, Fig. 6.1. (*c*).

The American slump test as given in ASTM C143-74 is similar except that the mould should be dampened and that it should be filled in three layers, each approximately one-third the volume of the cone.

The British Standard requires that the slump of concrete sampled in the

Fig. 6.1 (*a*) Tamping concrete in slump cone
Note that the cone is held down by keeping the feet on the small lugs at the bottom

Fig. 6.1 (*b*) Lifting the cone vertically

Fig. 6.1 (*c*) Method of measuring slump
The horizontal arm is 300 mm above the base plate. Alternatively, the slump cone is placed alongside and used as a basis for measuring the slump

field shall be measured immediately after sampling. It is often more useful to measure the slump at a fixed time after mixing, or after transportation to the point of placing. This arises from the change in slump which occurs with time.

The consistency is recorded in terms of mm of subsidence of the specimen during the test, which is known as the slump.

The random variations which occur between slumps are in some measure due to the fact that three distinct types of slump may occur. These are illustrated in Fig. 6.2[3]. The first consists of a general and uniform subsidence without breaking up, and may be termed a 'true' slump. The second is a 'shear' slump which occurs when the top half of the cone shears off and slips down an inclined plane. The third type results from a complete collapse of the cone. The second type of slump occurs frequently with normal concrete mixes, as, for instance, 1 : 2 : 4 mixes, and it is difficult to decide just how to measure the slump. In the illustration given, some operators might measure the minimum subsidence of 35 mm while others might measure the mean value of 110 mm. When this type of slump occurs, the test should be repeated, in order to obtain a shape closer to the 'true' slump. If the repeat test still gives a shear slump, the fact should be recorded on the test report.

The workability of concrete is related to the amount of applied work required to compact it to its maximum density, and is, therefore, not related to

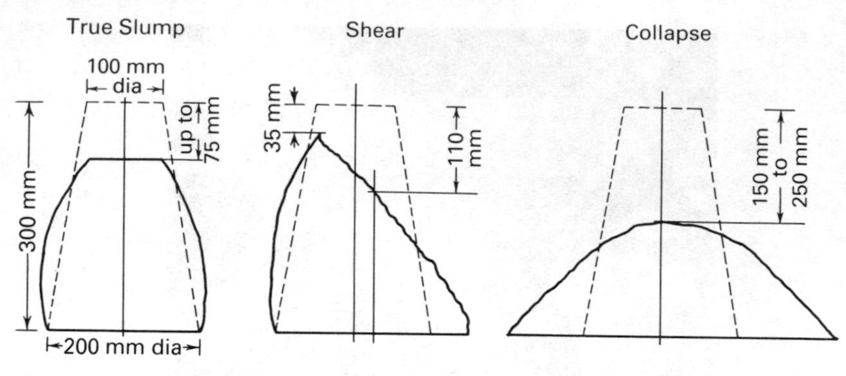

Fig. 6.2 Types of slump

the slump. It is, in fact, possible to obtain the same slump with mixes requiring very different amounts of work to compact them. This constitutes the chief disadvantage to the test.

Tests in the United States have suggested that as the temperature of the mixed concrete increases, the slump decreases. The decrease is of the order of 20 mm for a 10°C increase in temperature.

Compacting factor test

The compacting factor test was devised to measure the degree of compaction obtained by doing a standard amount of work on the concrete, and therefore bears a close relation to workability.

A full description of the development of the test is given in Road Research Technical Paper, No. 5[3]. The method of test is described fully in B.S. 1881 : 1970, to which reference should be made.

The apparatus used for the compacting factor tests is shown in Figs. 6.3 and 6.4. It consists essentially of two inverted cone-shaped hoppers and a cylindrical container. The dimensions are shown in Fig. 6.4. For the old dimensions, which will not apply after 1980, see BS 1881 : 1970.

In making the test the top hopper is filled with a representative sample of the concrete. When completely filled, a hinged door at the bottom is released and the concrete allowed to fall into the second hopper. The filling of the second hopper is thus effected by a standard method, and personal differences are minimized. The concrete is similarly released from the second hopper and falls into the cylindrical container. Surplus concrete is struck off by simultaneously working two steel floats from the outside to the centre. The cylinder contains a certain volume of concrete which has been subjected to a definite amount of work in falling from the second hopper. The contents of the cylinder are then weighed to the nearest 10 g giving the weight of partially compacted concrete (W_p).

The cylinder is then refilled from the same sample in layers approximately 50 mm deep, the layers being heavily rammed or preferably vibrated so as to obtain full compaction. The top surface is again struck off level with the top of the cylinder and the weight of concrete contained again determined. This is

Fig. 6.3 The compacting factor test. The concrete is about to be dropped into the bottom cylinder

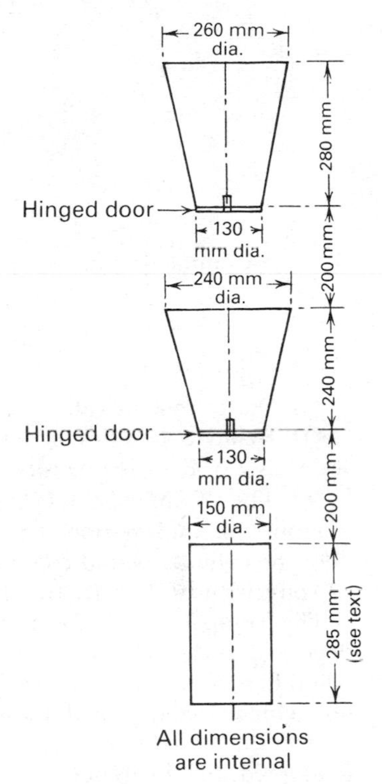

Fig. 6.4 Dimensions of compacting factor apparatus

known as the weight of fully compacted concrete (W_f).

The compacting factor is then the ratio (W_p/W_f) of the weight of partially compacted concrete to the weight of fully compacted concrete.

The weight of fully compacted concrete can also be calculated theoretically on the basis of the solid volumes occupied by the various materials, assuming no air voids. The data required are:

(1) Mix proportions by weight.

(2) Water absorbed by the aggregate, since this will also be included in the weight of the partially compacted concrete. Note that in the definition of water/cement ratio (p. 89) the water absorbed by the aggregate is not included. In Table 6.1 the absorbed water provides weight but does not occupy volume, and is therefore not calculated under M/G.

$$\text{If } W_p = 10\,995 \text{ g}$$
$$V = 0{\cdot}005 \text{ m}^3 \text{ for dimensions shown in Fig. 6.4}$$

$$W_f = \frac{\Sigma M \times V \times 10^6}{\Sigma\left(\frac{M}{G}\right)}$$

$$= \frac{7{\cdot}66 \times 0{\cdot}005 \times 10^6}{3{\cdot}17}$$

$$= 12\,082 \text{ g}$$

$$\text{Compacting factor} = \left(\frac{W_p}{W_f}\right)$$

$$= \frac{10\,995}{12\,082} = 0{\cdot}91$$

(3) Water/cement ratio.

(4) Relative densities of all the materials. Cement may be taken as 3·15 and water as 1·0. Relative density in case of aggregates. (See p. 51).

(5) The volume of the cylinder. (V m³).

From this information it is necessary to calculate the weight of concrete (W_f) the cylinder would contain if all the air voids were eliminated. This may be conveniently done by the method given in Table 6.1.

The compacting factor is the ratio of the observed weight of concrete deposited in the cylinder to the calculated weight without air voids. The result will differ from that obtained by refilling the cylinder if full compaction was not achieved, and provides a useful check.

It will be noted that, whichever method of determing the compacting factor is adopted, the difference in the two weights is due to air voids, and the closer the values, the less the air voids and the higher the compacting factor. The workability therefore increases as the compacting factor approaches unity.

The workability of concrete decreases considerably during the first few minutes after mixing, as is shown in Fig. 6.5. The change is greater when using a dry aggregate (curve A) than when using one which has been previously saturated (curve B), due to the rapid initial absorption of the dry aggregate. Since the interval between mixing and placing is likely in practice to average

Table 6.1 Method of Calculation of Compacting Factor

Material	Mix proportions by weight M	Relative density G	$\frac{M}{G}$
Coarse aggregate	4·0	2·58	1·55
Fine aggregate	2·1	2·63	0·80
Cement	1·0	3·15	0·32
Water:			
(1) effective	0·5	1·0	0·50
(2) absorbed by aggregate	0·06	—	—
Sum	7·66		3·17

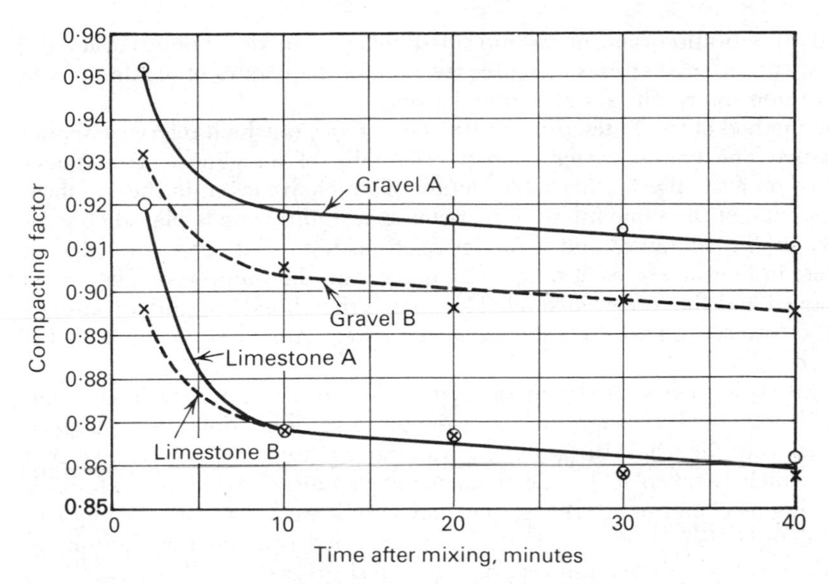

Fig. 6.5 Workability of concrete with gravel and limestone aggregates
Mix in each case was 1 : 6 (cement : aggregate) and river sand was used for all mixes.
All mixes: effective water/cement ratio 0·6.
Allowance for absorption of aggregate was made by:
A Extra water added at time of mixing (simulating the use of a dry aggregate).
B Sufficient water added to dry aggregate to give saturation ½ hour before mixing.
Mixing and curing temperature: 18°C (64°F).
Evaporation was prevented before and after mixing.

10–15 minutes, it is suggested that for test purposes a time interval of 10 minutes is allowed between the time of discharge from the mixer and the time of testing by the compacting factor method. This time interval also has the advantage that the change in compacting factor with time is relatively small compared with that in the first few minutes. This recommendation differs from that given in BS 1881 : 1970, which provides for only six minutes delay between adding the water to dry mixed materials in the laboratory and completing the workability test. Six minutes is, in the author's opinion, too short.

The test is not suitable for use generally on the site, except on larger jobs where a site laboratory is established. In the central laboratory, the compacting factor test provides a reliable guide to the workability of a concrete mix. It is used to ensure that a mix designed for use in the field has the required workability for the type of work to be done with the compacting equipment available, while containing the minimum quantity of mixing water.

A general indication of the compacting factors suitable for different types of structure and methods of compaction is given in Table 7.5. The slump which may be expected is also given.

V-B consistometer test

The V-B test is particularly suitable for the measurement in the laboratory of the workability of concrete mixes of very low workability which are compacted by vibration, and of air-entrained concrete mixes. The accuracy decreases with increasing sizes of aggregate, and for aggregates of maximum size over 20 mm the results are of doubtful value.

The method of test is described in BS 1881 : 1970, to which reference should be made. The consistometer consists essentially of a cylindrical container clamped on a small vibrating table, details of which are given in the specification. A cone of the same internal diameter as a slump cone is placed concentrically in the container and a funnel fixed on top. The cone is filled with concrete in four layers, each tamped 25 times as in the slump test. The funnel is removed and the top levelled off. The cone is then lifted out and a transparent disc lowered on to the subsided cone. The slump may be measured if required.

Vibration is then started and the remoulding of the concrete in the container is observed through the transparent disc. The time in seconds is recorded until the whole surface of the transparent disc is covered with cement grout, which is taken to be the moment when full compaction is attained. Recording mechanisms or other means may also be used to record the moment of full compaction. The workability by this method is defined as the time in seconds measured above, and reported as V-B degrees.

Cleanliness is of vital importance when using this type of apparatus, and care is necessary to avoid errors due to distortion of the central plunger.

An indication of the relation between slump, compacting factor and V-B time is given in Fig. 6.6[4]. This should only be regarded as a generalized relationship since the type of fine aggregate and other factors can have a significant effect.

Factors governing workability

The compacting factor may be shown[5] to be related mainly to the mix proportions, to the water content, and to the grading, shape and surface texture of the aggregate. The relationship is discussed under the following headings.

(*a*) *Aggregate Grading*

It has been evident for some time that workability was linked in some way with the surface area of the aggregate, and an approximate method based on research up to 1955, was published in the 2nd edition of this book. It was evident, however, that the surface area coefficient did not apply to material finer than about 150 μm sieve, nor to the cement. The new evidence now available indicates that other factors intervene; the surface area coefficient, for instance, did not provide fully for the greater compacting factor obtained by increasing the maximum size of the aggregate, e.g., from 20–37·5 mm. The reason for this lies in the fact that, in addition to having a smaller surface area requiring wetting, a more dense packing is possible. If the aggregate is assumed, for simplicity, to consist of spheres, it can be shown that the overall volume occupied by a large sphere is only some 70 per cent of the overall volume occupied by eight times the number of spheres of half the diameter, although the net volume of solids is, in both cases, the same. In the case of the

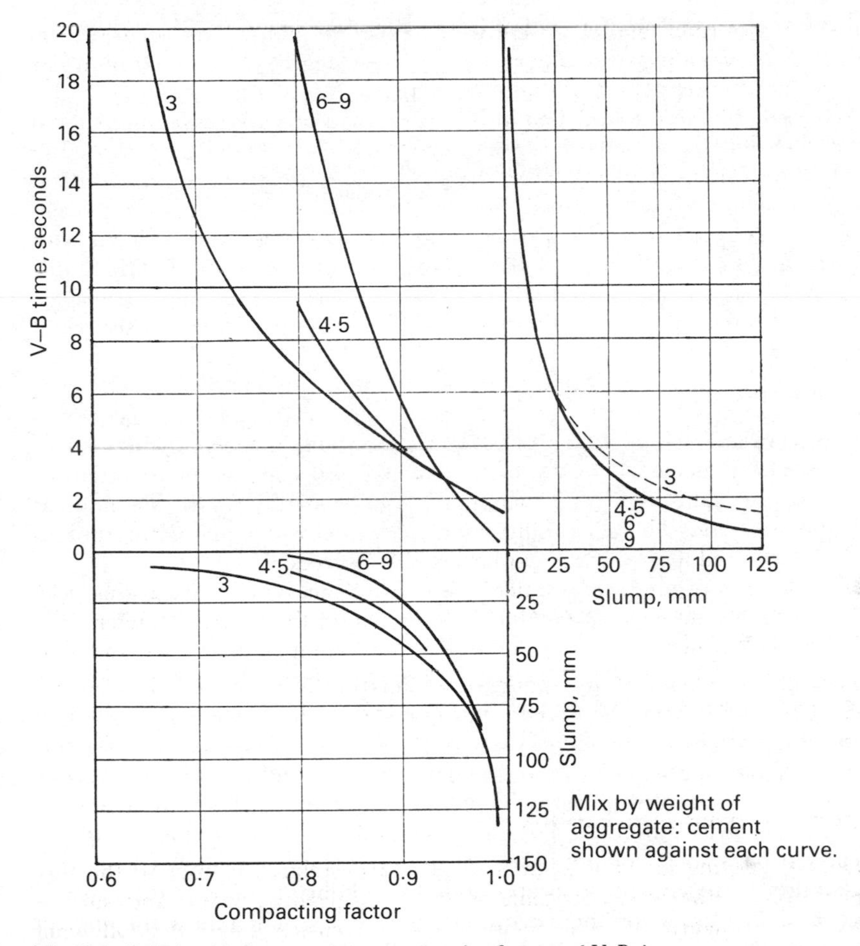

Fig. 6.6 Relationship between slump, compacting factor and V–B time

smaller spheres, the interstices must be filled with mortar, which, in the normal course of events will contain voids. There is, therefore, an advantage in using larger aggregate, although the actual increase in density which results cannot be calculated theoretically.

As already stated, the particles finer than 150 μm sieve and cement were previously neglected when considering surface area coefficients. It is now evident that, not only is the surface area coefficient of no importance for these finer particles and the cement, but that the value also requires modification for other sizes. This is because the smaller sand sizes in themselves provide some lubrication in the mix without requiring wetting to the same extent. As a result of detailed analysis[5] it is suggested that the aggregate particle size is assessed on the basis of a surface index f_s, as given in Table 6.2. It should be noted that the surface index is empirical, based on test evidence, although to some extent related to the surface area coefficient.

Table 6.2 Surface Index Calculations

Sieve size within which particles lie	Surface index for particles within sieve sizes indicated	Example of calculation of surface index for grading given	
		Percentage of particles within sieve size	f_s
75–37·5 mm	−2½	—	—
37·5–20 mm	−2	50	−100
20–10 mm	−1	14	−14
10–5 mm	1	12	12
5–2·36 mm	4	6	24
2·36–1·18 mm	7	6	42
1·18 mm–600 μm	9	5	45
600–300 μm	9	4	36
300–150 μm	7	3	21
Passing 150 μm	2	—	—
			66
	Add constant		330
			396
	× 10⁻³		0·396

The surface index for any aggregate or mixture of aggregates is then calculated by multiplying the individual figures given in Table 6.2 by the percentage of particles of the aggregate within the sieve sizes indicated, summing the results allowing for positive and negative values, and adding the constant indicated. An example is given in Table 6.2; to reduce the figure to a convenient size it is divided by 1000. For a precise result, the surface index should be calculated on a solid volume basis (see p. 101), since it is shown later that workability is dependent on proportions by volume. However, the error in using a weight basis for the sieve analysis is very small unless the specific gravity of different sizes of particle is markedly different.

(*b*) *Particle shape*

Various attempts have been made to measure the shape and angularity of aggregates. Perhaps the simplest is that suggested by Shergold [6] and termed the 'angularity number' of the aggregate. This is described fully on p. 54. For purposes of mix design it has been found more suitable to use the value obtained by calculating the angularity index, as also described. This is in effect the proportion of voids above 33 relative to the 'optimum' volume of solids, i.e. 67 per cent. Examples of angularity indices for a number of aggregates are given in Table 6.3.

Another method of expressing particle shape is the shape index [7]. This expresses the particle shape in terms of the specific surface and number of particles. The shape index may vary with the particle size, and it follows that the angularity of the sand sizes is likely to vary in a similar manner. The angularity of the sand sizes is therefore as important as that of the coarse aggregates.

Table 6.3 Examples of Angularity Indices

Type of coarse aggregate	Shape	Angularity number	Angularity index (f_A)
Quartzite gravel, natural	Rounded	1	1·15
Flint gravel, natural	Irregular	6	1·9
Limestone, crushed	Angular	7	2·05
Crushed granites and basalts, and crushed flint gravel	Angular	8 to 10	2·2 to 2·5

In the absence of test data natural uncrushed sands may be taken at $f_A = 1{\cdot}8$ to $1{\cdot}9$

(*c*) *Combining effects of grading and shape*

Analysis has shown that the angularity index, as calculated above, can be used to modify the surface index simply by multiplying them together ($f_S f_A$). If aggregates of different angularity indices are used in the same mix, then the average angularity index is calculated in proportion to the percentage of each type of material in the mix.

(*d*) *Influence of mix proportions*

It is the proportions by volume of the various constituents of a concrete mix which affect workability, so that the method of assessment of compacting factor described here is based on proportions by volume.

It needs little thought to realize that as a mix becomes richer, the grading and angularity of the aggregates become less important. An analysis of a large number of tests results[5] has shown that when the aggregate/cement ratio (A_v) is reduced to 2, the effects of grading and angularity become negligible. The combined surface and angularity indices ($f_S f_A$) therefore require modification, depending on the aggregate to cement ratio based on solid volumes. The solid volume is obtained by dividing the dry weight of each material by the respective relative density. A modified index is therefore obtained as follows:

$$f_S f_A (A_v - 2) \qquad (1)$$

where A_v = volume of aggregate to one volume of cement on the basis of solid volume of particles.

The value of A_v is calculated from the mix parts by weight, taking into account the relative density of the materials.

$$A_v = \frac{3{\cdot}15\, A_w}{G_A C_w}$$

where A_w = total weight of dry aggregate
C_w = weight of cement
G_A = average relative density of aggregate (see p. 51 for definition).

If the aggregate varies in relative density with particle size, the average relative density should be determined. London river gravel, for instance, varies in relative density from 2·55 to 2·45 between the 37·5 mm and 5 mm sieves,

Table 6.4 Values of Penetration (Vicat Apparatus) Obtained with Neat Cement Paste for Varying Water Contents

Water percentage	Depth of penetration mm	Remarks
18	Nil	
19	1	Caused subsidence but
20	3	no actual penetration
21	5	up to 9 mm
22	7	
23	9	
24	12	
25	24	
26	40+	Total penetration

and London area sand usually has a relative density of about 2·60. Many hard limestones and aggregates of metamorphic or igneous origin are in the range 2·65 to 2·70.

(*e*) *Water content*

It has been shown previously that the compacting factor is related to the water content of the concrete when expressed as a percentage by volume of the materials in the mix. Further work (5) has confirmed this concept of the water content by volume in the mix as being important, although a modification is required in assessing the quantity of water available to influence workability. An interesting experiment may be made with neat cement, which demonstrates that the consistency of a cement water paste changes quite suddenly with an increase in water/cement ratio, as measured by weight, of only about 1 per cent. The test can be done quite simply using the consistency plunger of the Vicat needle apparatus. Depths of penetration similar to those shown in Table 6.4 are obtained for ordinary Portland cement. This sudden changing of a cement paste from a stiff silt-like material almost to a fluid, with such a small change in water content, greatly simplifies the problem of determining the water content available for lubrication of the mix. It is generally found that this change in consistency occurs at water percentages of between about 23 and 28, so that excess water beyond these percentages is available for lubrication of the mix.

Then water content available for lubrication is

$$w/c - w_v \qquad (2)$$

where w_v is the water/cement ratio in the Vicat needle test at which a sudden change in consistency occurs, being generally between 0·23 and 0·28.

For most practical purposes the value of w_v may be assumed to be 0·25; at the time a mix is designed the actual cement to be used will often not be known. In the remainder of this chapter this value of 0·25 is used. Where any doubt exists, or where other types of cement are being used, the simple Vicat test may be used to obtain a more precise value for the reduction in water/cement ratio required.

Calculation of compacting factor

If the various influences discussed above and summarized on lines (1) and (2) above, are brought together, it would appear that the compacting factor is a function of:

$$\frac{w_c - 0{\cdot}25}{f_S f_A(A_v - 2)}$$

The necessary constants have been calculated from numerous test results[5], and the equation giving the compacting factor then becomes:

$$\text{cf} = 0{\cdot}74\left\{\frac{10(w/c - 0{\cdot}25)}{f_S f_A(A_v - 2)} + 0{\cdot}67\right\} \tag{3}$$

The basic principles described above have been applied particularly to the assessment of the compacting factor of concrete mixes. They are, however, applicable to workability whether measured by this method or other methods.

Segregation

Segregation is the separation of the various constituents of a concrete mix due to differences in particle size and relative density. There is a tendency for the coarser and heavier particles to settle and for the finer and lighter materials, particularly the water, to rise to the surface.

A close relation exists between segregation and many of the imperfections which occur in concrete construction such as honeycombing, weak and porous layers, surface scaling and sand streaks. Segregation is usually a result of using too much mixing water, badly graded aggregate, especially if the quantity of sand is deficient, or unsatisfactory methods of handling.

Segregation has been shown to be a serious risk where the surface index (f_S) falls below about 0·45 for 37·5 mm maximum sized aggregate and 0·5 for 20 mm maximum sized aggregate. The proportion of fines may then be insufficient adequately to fill the voids between the coarse particles. In no-fines concrete, which is made with 20–10 mm coarse aggregate only, with a surface index of 0·23, a honeycombed effect is obtained.

Control of consistency in the field

The aim in controlling consistency and workability in the field is to produce uniform concrete and to maintain a reasonably constant water/cement ratio. The slump test is the usual method of control apart from the maintenance of a constant water/cement ratio by making tests on the surface moisture content of the aggregates, and allowing a calculated quantity of water at the mixer.

The slump test is useful in two ways:

(1) Provided the grading of the aggregate, the proportioning, and the temperature of the materials are reasonably uniform, the slump indicates variations in the water/cement ratio. Even when tests are made on the moisture content of the aggregate, and an amount of water is added at the mixer calculated to give the correct water/cement ratio, the consistency and slump of the concrete must be the final guide. In practice, the moisture content of the

Table 6.5 Effect of Change of Water Content on Slump

Change in slump in mm		Percentage change in water content
From	To	
25	50	5
50	75	$4\frac{1}{2}$
75	100	3
100	125	$2\frac{1}{2}$
125	150	2

NOTE. Slump tests are somewhat unreliable and the figures given may not apply to individual slump tests.

aggregate varies throughout the stockpile, and tests can only give an average value.

On the other hand, the slump may vary by 25 mm or more when only a 1 per cent variation in the moisture content occurs in the sand used in the mix, and any change of slump should therefore be corrected by an adjustment in the water added at the mixer. It is usual when close control of the water/cement ratio is required, to allow a tolerance on the calculated quantity of water to be added at the mixer, the final adjustment depending on observations of the consistency of the concrete as it is made. Further reference will be made to this point in connection with the control of water content at the mixer.

(2) If the quantity of water added at the mixer is constant, and the moisture content of the aggregate is reasonably uniform, the slump test should indicate changes in the grading of the aggregate or errors in the proportioning of the materials.

Fig. 6.7 Mix too harsh

Fig. 6.8 Oversanded mix

Fig. 6.9 Mix of satisfactory consistency

On the site it often happens that the initial batches of concrete are a little too stiff or too plastic, and the question arises as to how much water to add or subtract to obtain the correct consistency, as measured by the slump test. Values from which the required percentage change in water content for any change in slump may be calculated are given in Table 6.5.

A further method of estimating consistency in order to maintain uniformity from batch to batch, is to watch the shape and appearance of the concrete as deposited in the transport container.

Examples of mixes of varying consistency are given in Figs. 6.7 to 6.10.

The slump may be reduced during transportation of the concrete, and when the length of haul is considerable it is advisable to make tests on the same concrete both at the mixer and at the point of deposition, in order to deter-

Fig. 6.10 A concrete mix of satisfactory consistency

mine the extent of the loss. When the distance from the mixer to the point of deposition is considerable, a loss of several inches may be recorded. This is exceptional and should not be permitted if over-wet mixes, which are liable to segregate during transport, have to be produced at the mixer to give the correct consistency at the point of deposition.

REFERENCES

1 NEWMAN, K., Properties of concrete, *Structural Concrete*, **2**, No. 11, Sept./Oct. 1965. Reinforced Concrete Association.

2 CUSENS, A. R., The measurement of the workability of dry concrete mixes, *Mag. Concr. Res.*, **8**, No. 22, March 1965. Cement and Concrete Association.

3 GLANVILLE, W. H., COLLINS, A. R. and MATHEWS, D. D., *The Grading of Aggregates and Workability of Concrete*, Road Research Technical Paper No. 5, 2nd ed., 1947. HMSO.

4 DEWAR, J. D., *Relations between Various Workability Control Tests for Ready-mixed Concrete*, Technical Report TRA/375, Feb. 1964. Cement and Concrete Association.

5 MURDOCK, L. J., The workability of concrete, *Mag. Concr. Res.*, **12**, No. 36, Nov. 1960. Previously printed in *Research and Application*, No. 11, Jan. 1960 (George Wimpey & Co. Ltd.).

6 SHERGOLD, F. A., The percentage voids in compacted gravel as a measure of its angularity, *Mag. Concr. Res.*, **5**, No. 13, Aug. 1953. Cement and Concrete Association.

7 *Variation of Specific Surface and Particle Shape with Aggregate Particle Size*, Report TRA/121, July 1952. Cement and Concrete Association.

7
Design of Concrete Mixes

THE objective in designing a concrete mix is to determine the proportions of cement, fine and coarse aggregate, and water which will satisfy the following requirements:

(1) *Compressive Strength.* The attainment of a compressive strength at 28 days (or other specified age) which meets the requirement of the designer of the structure.

(2) *Workability.* The provision of adequate workability to enable the site to transport, place, and fully compact the concrete with the available equipment. The selection of the most suitable workability will normally be the contractors responsibility and is of particular importance when the concrete will be pumped or tremied.

The mix should be cohesive in order to avoid the possibility of honeycombing, water gain (collection of water under aggregate particles resulting from bleeding) and other troubles associated with segregation. In order to satisfy this requirement, an adequate cement content is essential (see also (3) below).

(3) *Durability.* Durability is related to compressive strength; the greater the strength, the more durable the concrete. However, there are many occasions when the specified strength of the concrete can be satisfied by mixes containing higher water/cement ratios than those which will provide adequate durability in the environment to which the concrete will be subjected. In such cases the actual water/cement ratio and the density of the concrete will be the governing factors, and the strength will probably be higher than required strictly for structural purposes. The current practice in Britain is to relate durability for different types of exposure to minimum cement contents (see Chapter on durability).

(4) *Finish.* Lack of cohesion can be one of the causes of a poor finish when concrete is cast against vertical formwork, e.g. sand streaking and colour variations, and can also lead to problems in trowelling horizontal surfaces to a smooth and dense finish. In order to satisfy this requirement, and provide the dense concrete necessary for protection of the reinforcement, an adequate fines content is essential, the fines consisting of both the sand and the cement paste. The method of mix design described later should result in good cohesive concrete producing good finishes and a freedom from honeycombing and segregation.

With all mix designs a further objective is that of using the locally available materials, which will normally be cheaper than importing aggregates from further afield. Satisfactory concrete is possible with careful attention to the proportions adopted, using aggregates which do not comply with requirements of BS 882, especially for grading.

Compressive strength required

Concrete mixes are usually designed to provide a mean compressive strength at 28 days after mixing, which will provide an adequate margin over the characteristic or the minimum strength to satisfy the designer.

Cube strengths follow a normal distribution (see Appendix A) so that if a sufficient number of cubes is made, there will be a few which will be very high or very low. Provision was made for this in the concept of 'minimum strength' by recognizing that the failure of $1\frac{1}{2}$–$2\frac{1}{2}$ per cent of cube strengths was likely to occur. The newer term 'characteristic strength' is only different in allowing for a failure rate of 5 per cent. By reference to Appendix A it may be seen that the 'minimum strength' (2 per cent failure rate) is (0·32 × standard deviation) below the 'characteristic strength'.

In this chapter the 28 day strength only is considered. Reference should be made to Chapter 4 on cement for information on the effects of age and temperature on crushing strengths.

For various reasons listed in Chapter 27 dealing with quality control on the site, considerable variation occurs in the crushing resistance of concrete. It is, therefore, necessary to aim at a mean strength considerably in excess of the minimum requirement. For large civil engineering works and ready-mixed concrete the margin provided in CP 110:1972[1] is suitable. This may be taken as the smaller of the values given by (1) or (2).

(1) 1·64 times the standard deviation of cube tests on at least 100 separate batches of concrete of nominally similar proportions of similar materials and produced over a period not exceeding 12 months by the same plant under similar supervision but not less than $\frac{1}{6}$ of the characteristic strength for concrete of grade 7, 10 or 15, or 3·75 N/mm² for concrete of grade 20 or above (see Table 7.2 for grade numbers).

(2) 1·64 times the standard deviation of cube tests on at least 40 separate batches of concrete of nominally similar proportions of similar materials and produced over a period exceeding 5 days but not exceeding 6 months by the same plant under similar supervision, but not less than $\frac{1}{3}$ of the characteristic strength for concrete of grade 7, 10 or 15 or 7·5 N/mm² for concrete of grade 20 or above.

It is suggested that where there are insufficient data to satisfy (1) or (2), the margin for the initial mix design should be taken as $\frac{2}{3}$ of the characteristic strength for concrete of grade 7, 10 or 15, or 15 N/mm² for concrete of grade 20 or above.

Statistics such as those above are unlikely to become available on the majority of sites. On many structures the number of cubes which it will be reasonable to expect for any one mix produced under similar conditions will be in tens for the whole job (see also Chapter 27, Quality Control, Inspection and Testing). On these sites it will be more appropriate to exercise control using a minimum strength. A suitable margin then needs to be added to the minimum strength to give the mean strength requirement for mix design purposes. Table 7.1 gives suggested allowances, the choice of value to be used being based on a forecast or knowledge of the methods to be used in construction and the standard of quality control. For mean strengths of below 20 N/mm² the variation may be taken as

Table 7.1 Suggested Allowances for Variations in Cube Strengths for Use in Quality Control[2]

Method of proportioning, type of concrete, and control measures	Variations above or below average strength beyond which investigation is advisable (for mean strengths above 20 N/mm²)
All materials by weight. High quality concrete of uniform workability in large quantities. Competent supervision by a qualified engineer.	7
All materials by weight. High quality concrete in members of varying sections requiring slight changes in workability. Competent experienced supervision.	8·3
All materials by weight. General concreting where constant supervision is not justified. Lean mixes.	10·5
Aggregates by volume using deep bins for measurement. Cement by full bag or by weight. Good control.	8·3
Aggregates by volume. Cement by full bag or by weight. Measurement of aggregates by marks in loading hopper or by gauge boxes, without constant supervision.	10·5
Aggregates by volume. Cement by bag or part bag. Little control.	12·5
All materials by volume.*	14·5–16·5

NOTE.—Total variation is twice the value quoted. Allowances are intended to avoid more than 2 per cent of cubes failing to reach the required standard.

* Cement should never be measured other than by weight. The figures are given to indicate the extent of the variation which results if weight measurement is not adopted.

$$\frac{\text{mean strength expected}}{20} \times \text{variation shown in Table 7.1.}$$

The degree of supervision has a much more important effect than the type of equipment employed.

The alternative is to adopt one of the 'prescribed mixes' given in CP 110 : 1972[1]. These have been revised in BS 5328 : 1976, 'Methods of Specifying Concrete'; mixes for grades 20, 25 and 30 having slightly less cement. It should be noted, however, that no provision is made for control cube tests when prescribed mixes are used, a situation which many engineers will not find acceptable.

Code of practice

The BS Code of Practice CP 110 : 1972[1] and BS 5328 : 1976 make provision for nine grades of concrete the grade number corresponding with the characteristic strength. These grades and the lowest grade for compliance with appropriate use are given in Table 7.2.

Table 7.2 BS Grades of Concrete and their Use

Grade (& equivalent characteristic strength N/mm²)	Lowest grade for use shown
7	plain concrete
10	
15	reinforced concrete with lightweight aggregate
20	reinforced concrete with dense aggregate
25	
30	concrete with post-tensioned tendons
40	concrete with pretensioned tendons
50	
60	

The list of prescribed mixes does not include grades 40, 50 or 60.

Methods of designing concrete mixes

Numerous methods are available for designing concrete mixes including that given in 'Design of Normal Concrete Mixes'[3], which follows broadly similar principles to those used in Road Note No. 4 but has been metricated and expresses the mix proportions in terms of quantities of materials per unit volume of concrete rather than proportions by weight. This brings it more into line with European and American practice.

The 'Design of Normal Concrete Mixes' is based on the use of aggregates complying with the requirements of BS 882 for grading (see p. 41) and for this reason has the following limitations:

(1) Good concrete can be produced using sands not falling within one of the 4 zones in BS 882, provided that the design takes this into account. There is no provision for mix design using such aggregates.

(2) Gap graded coarse aggregates, in which the middle sizes of aggregate are absent, can produce good concrete, although extra care is necessary to avoid segregation. Again there is no provision for the use of such gradings of aggregate.

(3) Particle shape, particularly that of the fine aggregate is not adequately dealt with.

(4) The sand contents derived from the graphs tend to be lower than desirable for adequate cohesion in the fresh concrete. This has an important bearing on durability.

The mix design method given in 'Design of Normal Concrete Mixes' is similar to that recommended by the American Concrete Institute[4] and consists basically of the following steps:

(1) Selection of the water/cement ratio needed to satisfy the average (target) compressive strength at 28 days determined by providing an adequate margin over the characteristic strength, taking into account the anticipated or known variability of the materials and production.

(2) Selection of the free-water content in kg/m³ which is related to the

maximum size of the aggregate, whether or not the coarse aggregate is crushed or uncrushed, and the workability.

(3) Determination of the cement content in kg/m³ from (1) and (2), checking that the cement content is greater than any minimum which may be specified.

(4) Determination of the total aggregate content in kg/m³, based on the free-water content and the relative density of the combined aggregate.

(5) Selection of the proportion of fine aggregate, depending on the grading, water/cement ratio and the maximum aggregate size.

Provision is made for adjustments to the mix after trials. For further information reference should be made to the publications named.

Mix design

The method of mix design described in this chapter was first introduced in the 2nd edition of the book, 1955 and has become widely used. Some modification has been made as a result of experience, but the method is still based on the relationship between workability, the characteristics of the aggregates, the mix proportions, and the water content, as described in Chapter 6.

Equation (3) on p. 103 may be rearranged for the evaluation of other factors as may be convenient, For example the aggregate/cement ratio (Av) may be determined from:

$$Av = \frac{7{\cdot}4(w/c - 0{\cdot}25)}{(\text{cf} - 0{\cdot}50)f_S f_A} + 2 \tag{4}$$

where $(w/c - W_v)$ is the water content of the mix available for lubrication (as indicated in Chapter 6, W_v may be taken as 0·25 for most practical purposes)

f_S is the surface index which takes account of the grading of the aggregate

f_A is the angularity index which allows for the shape of the aggregate.

Note that the constant 2 in equation (4) arises because for mixes richer than 1 : 2 of cement : aggregate by solid volume the grading of the aggregate and its shape have little influence on workability (see definition of solid volume on p. 52).

One advantage of the use of this equation is that the effect of small changes in one or more factors on the other factors is readily calculable. The equation may be applied to all types of aggregates of widely varying grading, including gap gradings. The latter can be used satisfactorily provided extra care is taken to avoid segregation, the risk of which is increased by the absence of the middle sizes of aggregate.

Conversion to mix proportions by weight may be done by the formula

$$A_v = \frac{3{\cdot}15\, A_w}{G_v C_w} \tag{5}$$

where A_w = total weight of dry aggregate
C_w = weight of cement in same units

G_v = average relative density of the aggregate on a saturated surface-dry basis.

The above applies for all mixes in which there are sufficient fines to fill the voids between the larger particles. When the total fines becomes small, as indicated by a small surface index, segregation is liable to occur, and the compacting factor is reduced.

In assessing compressive strengths, although the strength is related to the water/cement ratio as defined on p. 89, the surface index again has an influence, causing a gradual decrease above a factor of about 0·6, and a much more rapid decrease, particularly in the case of the leaner mixes below a factor of about 0·5. This rapid decrease arises from the presence of insufficient fine materials to fill the voids between the larger particles, a point well illustrated in Fig. 7.1 by the extension of the dotted curves for mixes with a water/cement ratio of 0·4 to include no-fines concrete, which normally has a surface index of a little over 0·2. No-fines concrete contains about 30 per cent air voids.

A discussion of this and other factors which influence mix design follows.

Water/cement ratio

An inspection of Fig. 7.1 indicates that for any compressive strength at 28 days there is a comparable water/cement ratio; for instance, with ordinary Portland cement (vertical scale), a compressive strength of 43 N/mm^2 requires a water/cement ratio of 0·5, providing the surface index is between about 0·5 and 0·65. For indices below or above these limits, the strength falls away as indicated above and it is, therefore, preferable to grade the aggregate to remain within them. If a variation in surface index is expected, it is preferable that the index should exceed 0·65, rather than that it should fall much below 0·5, since the strength drops more slowly for the higher indices than for lower values and there is less chance of segregation.

The curves in Fig. 7.1 are plotted for ordinary Portland cement having a BS 12 concrete cube strength in the region of 35 N/mm^2 at 28 days. For most mix design purposes, it is sufficient to assume that British cements will attain this strength or even higher, and where evidence is available on the BS 12 concrete cube strength of the cement to be used, an adjustment should be made. The increase or decrease in concrete strength may be taken as roughly proportional to the percentage increase or decrease in the BS 12 concrete cube strength. Suggested adjustments for different types of cement, in the absence of other data, are given in Table 7.3. Note that the average compressive strength required is divided by the factor given, before using vertical scale, Fig. 7.1.

Note that for high-strength concrete (i.e. over about 45 N/mm^2 as a minimum requirement) requiring a mix rich in cement, the grading and shape of the aggregate become less important. The sand content may be reduced with advantage. In mixes as rich as 1 : 2½ or 1 : 3 of cement : total aggregate, the strength of the cement itself becomes of major importance.

The curves in Fig. 7.1 are plotted for an irregular flint gravel aggregate found in the London area. It has been found, however, that hard crushed limestone, some granites, and other aggregates give rather higher strengths

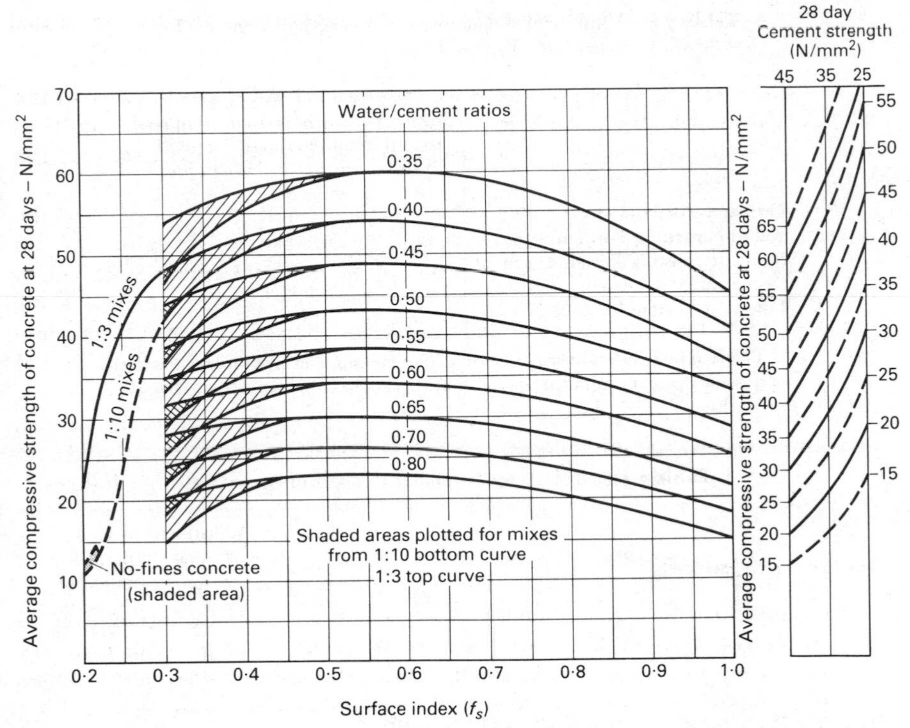

Fig. 7.1 Relationship between compressive strength, water/cement ratio and surface index

Notes.—Vertical scale for ordinary Portland cement having a standard concrete cube strength of 35 N/mm² at 28 days. When the cement is known to have a different standard cube strength an adjustment to the scale may be made, or the small diagram to the right may be used.

Aggregate—Irregular river gravel and sand (for other aggregates see Table 7.4 for adjustments).

Compaction Complete elimination of air voids should give strengths equal to that at peak of curves. If compaction is sufficiently vigorous, there will be less drop in strength above about $f_s = 0{\cdot}6$ than is indicated by the curves. Below about $f_s = 0{\cdot}45$ the strength will not be so much improved by vigorous compaction because there are insufficient fines to fill voids. The curves are based on test cube results with average site compaction, the wetter mixes by hand, the drier mixes by vibration.

than irregular flint gravel, while some rounded and polished aggregates give slightly lower strengths. The advantage in strength for a given water/cement ratio indicated for certain aggregates is largely counterbalanced in practice because, being crushed aggregates, they are angular and they require a higher water/cement ratio to give the necessary workability than irregular or rounded aggregates.

Suggested correction factors for use in determining the water/cement ratio necessary to give a required strength are given in Table 7.4. In case of doubt, it is a good plan to design two mixes, one on the basis of the curves given in Fig. 7.1, and the other on whatever increase or decrease is thought likely from Table 7.4, using the more conservative value where a range is given. The two mixes can later be checked by comparative trial mixes.

Table 7·3 Adjustment to Compressive Strength of Concrete for Type of Cement

Type of cement	Factor by which average compressive strength required should be divided in order to use vertical scale, Fig. 7.1
Ordinary Portland cement	1
Rapid-hardening Portland cement	1·1
Sulphate-resisting Portland cement	1
Supersulphated cement	1·0 to 1·25
Low-heat Portland cement	0·6 to 0·7

This Table gives an average value. Concrete made from different batches of cement will vary from this value in proportion to the BS 12 concrete cube strengths.

Table 7·4 Adjustments for Different Types of Aggregate

Aggregate		Adjustment to 28 day crushing strength before determination of water/cement ratio by use of Fig. 7.1, vertical scale. Multiply by coefficient below
Coarse	Fine	
River gravel (irregular)	River sand (irregular)	nil
River gravel (rounded and polished)	River sand (irregular)	1·05
River gravel crushed (angular)	River sand (irregular)	nil
River gravel (irregular)	River sand crushed (angular)	nil
Crushed whinstone or granite (angular)	River sand (irregular)	0·95
Crushed limestone (sub-angular)	River sand (irregular)	0·85–0·95
Crushed limestone (sub-angular)	Crushed limestone (sub-angular)	0·80–0·90
Crushed whinstone or granite (angular)	Crushed whinstone or granite	0·85–0·95*

* Crushed whinstone fines sometimes reduce strength rather than causing an increase as indicated here.

Workability

The workability required may be assessed from Table 7.5, for the conditions of placing and compaction anticipated. If concrete will be required for several different purposes from the one mixer, it is wise, in the interests of simplicity, to design the mix for the most workable concrete necessary. Frequent changes of mix on the site are confusing and often time-consuming, so that any possible saving in cement is largely nullified.

The water/cement ratio has been determined from Fig. 7.1, and the

Table 7.5 Uses of Concrete of Different Degrees of Workability

Degree of workability	Slump in mm	Compacting factor	Use for which concrete is suitable
Very low to low	0 to 25	0·80–0·87	Vibrated concrete in roads or other large sections, where powerful vibrating equipment can be used. Vibrated piles, precast beams and railway sleepers, and similar work where a high strength is required, say 40 N/mm² or over at 28 days.
Low to medium	25 to 50	0·87–0·93	Roads with ordinary forms of vibrating and finishing machines, and manually operated compactors and finishers of the vibrated beam type. Mass concrete foundations without vibration. Simple reinforced sections with vibration.
Medium to high	50 to 100	0·93–0·95	Hand-tamped roads with slump 50–75 mm. For normal reinforced work without vibration and heavily reinforced sections with vibration; pumping.
High	100 to 175	Over 0·95	For sections with congested reinforcement. Other work where placing is difficult. Pneumatic placers and tremies. Not normally suitable for vibration.

required compacting factor has been decided from Table 7.5. The next steps are to determine the mix proportions for the particular aggregate to be used, and to decide the proportions of coarse and fine aggregates.

Aggregate properties

The value of the angularity index (f_A) is determined by the type of aggregate (see p. 55 and Table 6.3).

A suitable surface index (f_S) is chosen, being based on a knowledge of the reliability of the aggregate supply, an inspection of the quarry or pit and sieve analyses. The more uniform the supply the lower the surface index acceptable, within the following limits.

Values suggested for the aggregate (or combined aggregates) used in a concrete mix are as follows:

(1) 37·5 mm max. size. Aggregates drawn from a pit or quarry efficiently maintained, $f_S = 0{\cdot}55$.

(2) 20 mm max. size. Aggregates from a similar source to that indicated under (1), $f_S = 0{\cdot}60$.

(3) 10 mm max. size. Aggregates of reasonably uniform grading $f_S = 0{\cdot}70$.

These values should be regarded as for guidance, and may be reduced by up to 0·05 if reliance can be placed on the grading of the aggregate, particularly that of the sand, being more than usually uniform. If aggregate is drawn from several sources or is likely to be variable, it may be advisable to increase the above values. The aim should be to produce a design of mix which suits the average grading to be expected, but which will still provide sufficient fines when the aggregates reach their coarser limits.

If possible, samples of aggregate representing the range in grading expected should be subjected to sieve analysis. The average grading of these aggregates should be determined, and hence their surface indices. The next step is to determine the proportions in which the aggregates need to be proportioned to give the chosen combined surface index. An example to illustrate the method of doing this is given below.

Having determined the proportions of coarse to fine aggregate for average gradings, the surface indices should be determined for the extremes of grading, using the same proportions in all cases. If the lowest value for f_S falls below 0·5 for 20 mm aggregate or 0·45 for 37·5 mm aggregate, the proportion of fines should be increased, that is, assuming that the aggregates at the extremes of grading cannot be rejected.

If the higher limit for f_S exceeds 0·9, the mix may become too cohesive for a rotating drum mixer to handle. A more satisfactory upper limit is 0·8.

Assuming then that a value for surface index has been decided, and also a value for the angularity index, $f_S f_A$ can be calculated, and hence A_v; see equation (4) (p. 111).

Average relative density

As explained in Chapter 3, the relative density of aggregates can vary with the particle size and if an accurate value is to be obtained it is necessary to make a determination for each size used in the mix. For the purposes of equation (5) (p. 111) the average relative density of the combined aggregate is then determined.

Grading and surface index calculations

The proportions of two aggregates of known individual surface indices required to give a predetermined combined surface index can readily be determined as follows:

Assume the use of one part aggregate with surface index y.

Then parts aggregate a with surface index x

$$= \frac{z - y}{x - z}$$

where z denotes surface index for combined aggregate.

A useful example to illustrate these points may be given by considering aggregates complying with the grading requirements of BS 882 : 1973. The grading requirements of this specification are summarized in Tables 3.5 and 3.6, from which it will be seen that certain tolerances are allowed in the grading of the coarse aggregate, while the sand is required to be within one of four zones.

On the site it is not unusual to find that a sand, even from the same pit, varies considerably in grading, and it is reasonable to assume that it may vary as between consecutive deliveries anywhere within a zone. As an example, the surface index may vary from 0·93 to 1·19 for sands lying within Zone 2. Note that the sands giving these values for f_S are not at the coarser and finer limits, as indicated by the left- and right-hand figures in the respective column.

Variations in the grading of the coarse aggregate have comparatively little effect on the surface index, and may usually be ignored.

On the site it is usually only practicable to cope with one mix which must prove satisfactory, both as regards strength and workability, when using any sand within the zone.

It has been suggested under the heading Workability that a suitable surface index for the combined aggregate, 20 mm max. size is 0·60.

Having decided upon f_S = 0·60 as suitable, the next step is to determine the proportions of coarse aggregate to the average of the sands (in this example the average for sands varying within Zone 2), which will give a surface index of 0·60.

Assume that for the coarse aggregate, 20–5 mm, the value of f_S = 0·30, and the value of f_S for sand lying in the middle of Zone 2 is about 1·10.

Then, using formula (6) above, the proportions required to give a combined value of f_S = 0·60 are:

$$a = \frac{0{\cdot}60 - 0{\cdot}30}{1{\cdot}10 - 0{\cdot}60} = \frac{0{\cdot}30}{0{\cdot}50} = \frac{1}{1{\cdot}67}$$

i.e. proportions required are 1 : 1·67 of average sand : coarse aggregate.

Now with these same proportions determine the combined surface index using the sand in Zone 2 having a surface index of 1·19:

$$f_S = \frac{1{\cdot}67 + 0{\cdot}30 + 1 \times 1{\cdot}19}{2{\cdot}67} = 0{\cdot}633$$

And similarly for the coarse grading having a surface index of 0·93:

$$f_S = \frac{1{\cdot}67 \times 0{\cdot}30 + 1 \times 0{\cdot}93}{2{\cdot}67} = 0{\cdot}536$$

Both values are within the acceptance limits suggested on p. 115.

Combining with Angularity

The angularity index will have been assessed roughly from the examples given in Table 6.3, or more exactly, and preferably, by tests as described on p. 55.

For this example, assume f_A = 1·9

Then $f_S f_A = 0{\cdot}6 \times 1{\cdot}9 = 1{\cdot}14$ (average)

with a range from $f_S f_A = 0{\cdot}633 \times 1{\cdot}9 = 1{\cdot}20$

to $f_S f_A = 0{\cdot}536 \times 1{\cdot}9 = 1{\cdot}02$

Deciding mix proportions by volume

In the example given on p. 112, the water/cement ratio required to give an average compressive strength at 28 days of 43 N/mm^2 was 0·5.

Substituting this value, and that of $f_S f_A$ in equation (4) it is found that for a compacting factor cf = 0·85, A_v = 6·6.

The mix proportions by weight may be determined from equation (5).

For the design example quoted above, where A_v = 6·6, the mix by weight is 1 : 5·35 for aggregate with average relative density = 2·55. With a sand: coarse aggregate ratio of 1 : ·67, the mix becomes 1 : 2 : 3·35 by weight, assuming the sand and coarse aggregate to have the same relative density.

Water/cement ratios and variations in grading

It is useful to assess the influence of the variation in grading of the sand within Zone 2:

(1) assuming the water/cement ratio is maintained constant

(2) assuming the workability is maintained constant and the water/cement ratio is varied. This is the usual occurrence on site, since a constant workability is important during placing.

Case (1)

$$\frac{w/c - 0{\cdot}25}{f_S f_A} \text{ ranges from } \frac{0{\cdot}25}{1{\cdot}20} \text{ to } \frac{0{\cdot}25}{1{\cdot}02}, \text{ i.e. from } 0{\cdot}21 \text{ to } 0{\cdot}245$$

With A_v = 6·6, the range in cf is from 0·83 to 0·89 (equation 4).

Case (2)

$$\frac{w/c - 0{\cdot}25}{f_S f_A} \text{ must remain constant} = 0{\cdot}22$$

$$\therefore \quad w/c - 0{\cdot}25 = f_S f_A \times 0{\cdot}22, \text{ i.e.} = 1{\cdot}20 \times 0{\cdot}22 \text{ to } 1{\cdot}02 \times 0{\cdot}22$$
$$= 0{\cdot}26 \text{ to } 0{\cdot}22$$

$$\therefore \quad w/c \text{ ranges from } 0{\cdot}25 + 0{\cdot}26 \text{ to } 0{\cdot}25 + 0{\cdot}22$$
$$\text{i.e. from } 0{\cdot}51 \text{ to } 0{\cdot}47.$$

The resulting range in strength for this variation in w/c ratio can be assessed from Fig. 7.1.

If either the range in workability or in strength is considered excessive for the purpose for which the concrete is required, a less variable sand must be found. The variations indicated in the example are probably as large as are desirable for good quality concrete.

Mix design charts

Mix design charts based on the surface index method are given in Figs. 7.2 and 7.3, for irregular and angular aggregates of the commonly used 20 mm maximum size. The charts may be used to choose mix proportions and to determine cement contents without knowledge of the actual gradings of the coarse and fine aggregates. When these are known the proportion of coarse to fine aggregate can be adjusted to give the appropriate surface index as described on p. 116.

As an example of the use of charts, assume the aggregate is irregular in shape and that a mean 28 day compressive strength of 40 N/mm² is required. Start on the left of Fig. 7.2 and follow the diagonal line to allow for cement strength. For a 28 day cement strength of 35 N/mm² (a safe working strength

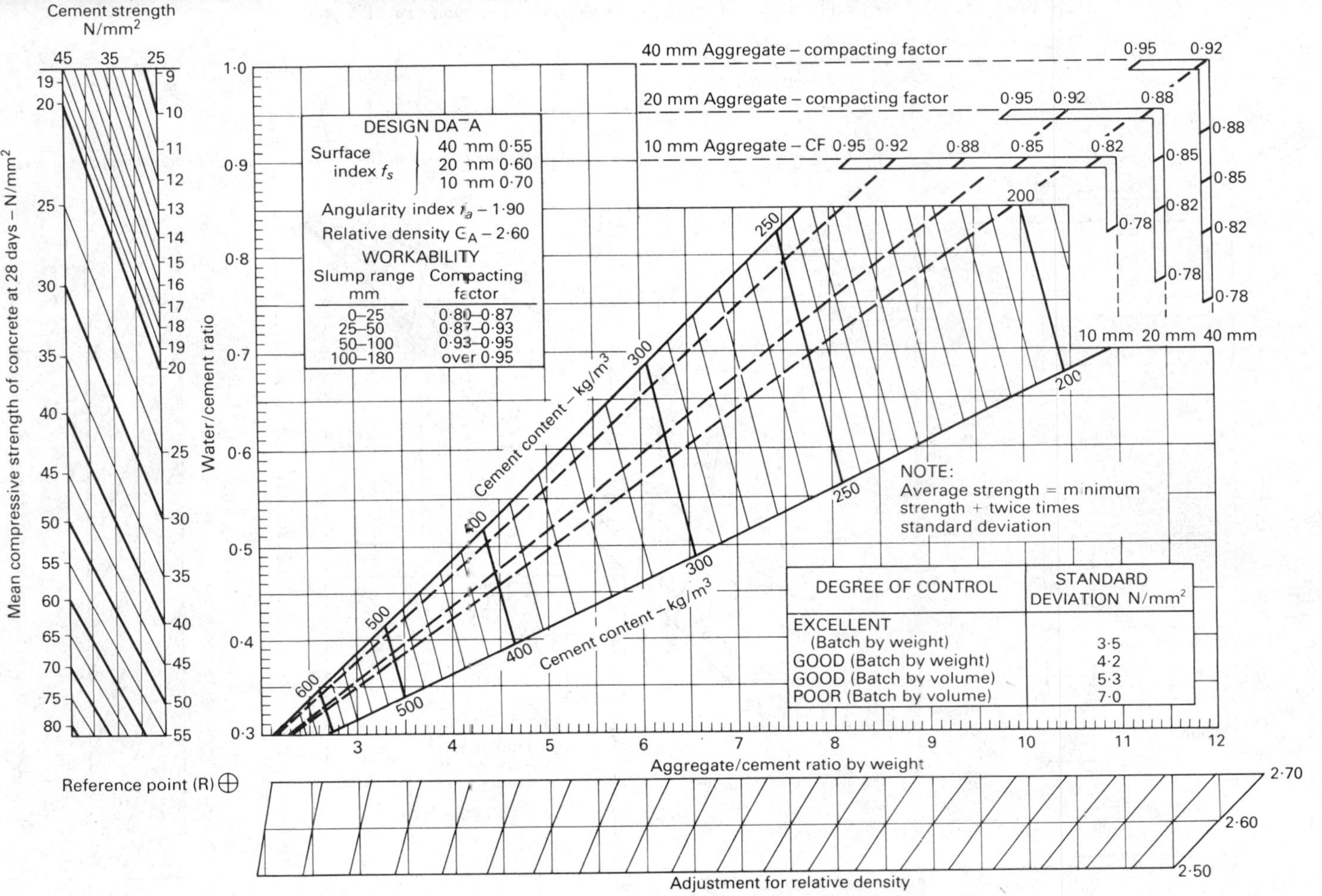

Fig. 7.2 Relationship between a/c ratio, w/c ratio, cement content and relative density for concrete mix designs for estimating purposes (20mm irregular gravel)

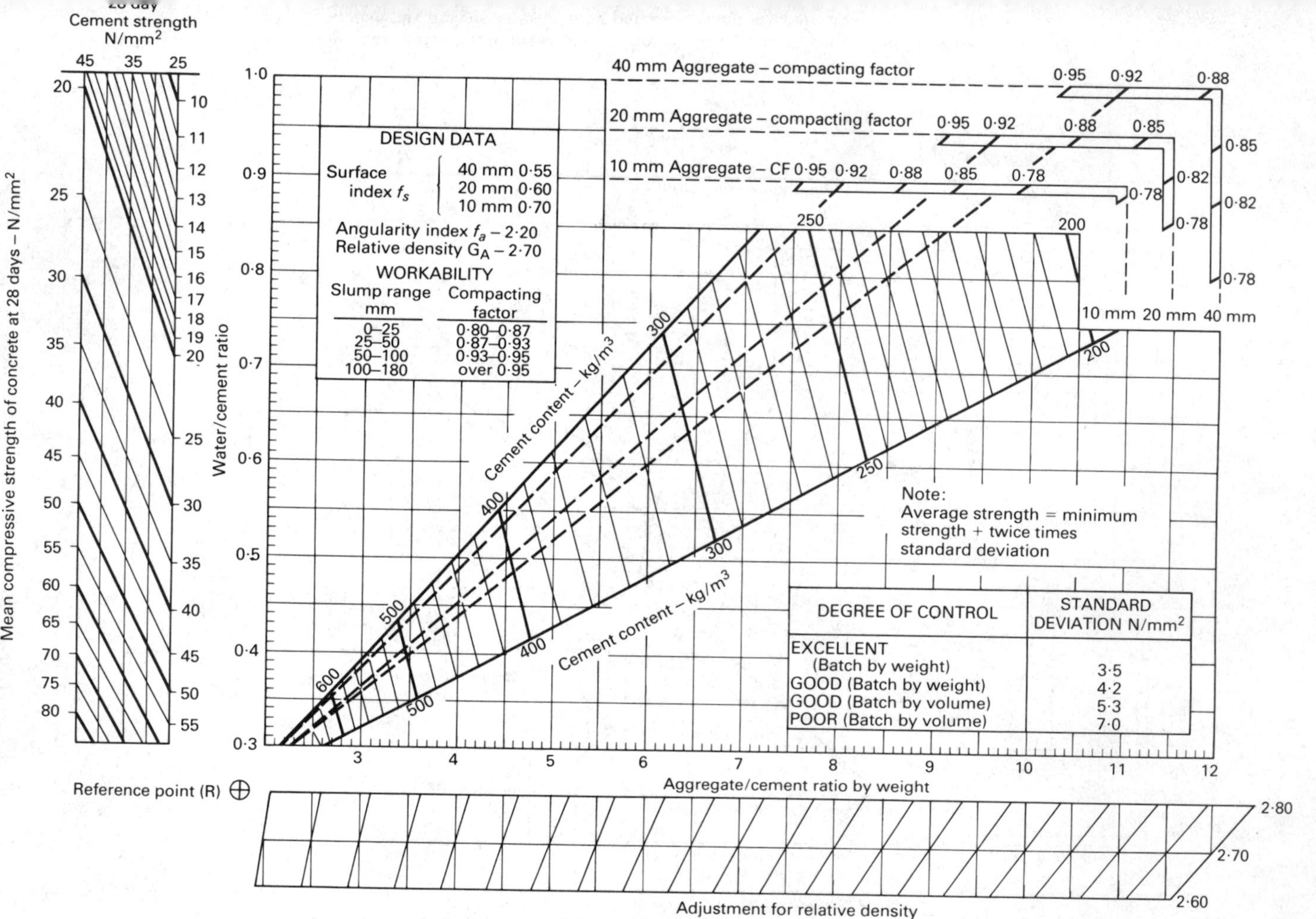

Fig. 7·3 Relationship between *a/c* ratio, *w/c* ratio, cement content and relative density for

Table 7.6 Minimum cement contents to provide durability

Exposure	Reinforced concrete				Prestressed concrete				Plain concrete			
	Nominal maximum size of aggregate (mm)				Nominal maximum size of aggregate (mm)				Nominal maximum size of aggregate (mm)			
	40	20	14	10	40	20	14	10	40	20	14	10
	kg/m³	kg/m³	kg/m³	kg/m³	kg/m³	kg/m³	kg/m³	kg/m³	kg/m³	kg/m³	kg/m³	kg/m³
MILD: e.g. completely protected against weather, or aggressive conditions, except for a brief period of exposure to normal weather conditions during construction.	220	250	270	290	300	300	300	300	200	220	250	270
MODERATE: e.g. sheltered from severe rain and against freezing whilst saturated with water. Buried concrete and concrete continuously under water.	260	290	320	340	300	300	320	340	220	250	280	300
SEVERE: e.g. exposed to sea water, moorland water, driving rain, alternate wetting and drying and to freezing whilst wet. Subject to heavy condensation or corrosive fumes.	320	360	390	410	320	360	390	410	270	310	330	360
Subject to salt used for de-icing.	260	290	320	340	300	300	320	340	240	280	310	330

for modern British ordinary Portland cements) the required water/cement ratio is then 0·53. Now move horizontally across the main chart to meet the line for compacting factor. Assuming this is 0·92 for a maximum aggregate size of 20 mm, the meeting point will be at an aggregate/cement ratio by weight of 5·25. The cross lines give the cement content. In this example the value is about 350 kg/m³. A convenient method of adjusting the mix proportions to allow for differences in the average relative density of the aggregate is given at the bottom of the chart. In the example given the aggregate/cement ratio would become 5·45 for a relative density of 2·70 or 5·0 for a relative density of 2·50.

Cement content of concrete

It is possible to design concrete mixes to satisfy strength requirements, but which will be permeable and therefore not suitable for situations where durability and watertightness are important. Specifications giving only a strength requirement are therefore not satisfactory, and a minimum cement content should be insisted upon as an essential part of any specification for concrete. The minima given in Table 7.6 are included in CP 110 : 1972. Reference may also be made to BS 5328 : 1976.

It is unwise to go below these minimum cement contents even when the quality control on site is exceptionally good. In many situations it is preferable to increase the cement content above the minimum recommended in Table 7.6, by say 20 kg/m³, in order to ensure the production of a good cohesive mix. This will be easier to place and compact, and bearing in mind variations on site or in ready-mixed supplies, will help to reduce the risk of segregation, honeycombing, etc.

The cement content per m³ of concrete may be determined quite accurately, if the mix proportions by weight, the water/cement ratio, and the relative density of the aggregates are known from the formula:

$$C = \frac{1000}{0{\cdot}317 + R/G + w/c} \tag{6}$$

where C denotes cement content in kg/m³
R denotes ratio of aggregate: cement by weight
G denotes average relative density of aggregates
w/c denotes water/cement ratio by weight.

An indication of cement content can also be obtained from the charts, Figs. 7.2 and 7.3.

It is assumed that there are no losses of water from the concrete during transport and placing, and that the concrete is fully compacted. Full compaction implies that there are no air voids, a condition never completely achieved in practice. The error is, however, small since the amount of air entrapped should, with good compaction, only amount to 1 to 2 per cent of the total volume.

In air-entrained concrete the total air content rises to about 4 to 7 per cent of the volume of the compacted concrete, and it becomes necessary to make an adjustment.

The weights of material per m³ of concrete calculated by this method do not allow for loss of materials on site, or loss of water during transit or placing.

In assessing overall quantities of materials required on site, an allowance should be made for loss of concrete either at the mixer or during transport and placing. These losses may be considerable, and due allowance should be made depending on site conditions, and the effectiveness of any measures taken to minimize losses.

Conversion from proportions by weight to proportions by volume

The conversion from weight proportions to volume proportions and vice versa involves assumptions for the weights per m³ of the fine and coarse aggregates. The values for the aggregates may be obtained by experiment, but some variation must be expected. In the following example they are taken as 1600 and 1440 kg/m³ respectively.

Once the proportions by weight have been established, equivalent proportions of aggregate by volume may be calculated as in the following example:

	Cement	Fine aggregate	Coarse aggregate	Water
(1) Weight of material: kg	50	100	170	27·5
(2) By volume dry to 50 kg cement	50	$\frac{100}{1600} = 0{\cdot}063\ m^3$	$\frac{170}{1440} = 0{\cdot}118\ m^3$	
(3) Add for bulking		$+ 20\% = 0{\cdot}013\ m^3$		
(4) Proportions of damp aggregate to 50 kg cement	50	0·076 m³	0·118 m³	27·5 kg

Line (4) gives the final proportions for use in the field under the conditions stated.

REFERENCES

1 CP 110: Part 1 : 1972, *The Structural Use of Concrete: Design, Materials and Workmanship*. British Standards Institution.

2 MURDOCK, L. J., The Control of concrete quality, *Proc. Inst. Civ. Engrs*, Part 1, July 1953.

3 DEPARTMENT OF THE ENVIRONMENT, *Design of Normal Concrete Mixes*, 1975 (2nd impression 1976). HMSO.

4 *Recommended Practice for Selecting Proportions for Normal and Heavyweight Concrete*, ACI standard (ACI 211. 1–74), May 1974. American Concrete Institute.

8

Batching Aggregates, Cement and Water

SITE control of concrete begins with an inspection of the aggregates as they are delivered. Methods of testing have already been described in Chapter 3, but a few remarks on rapid methods of checking aggregates as they are delivered may be useful. Sieve analyses can be made on coarse aggregates without previously drying them, and the results obtained are accurate enough to indicate serious errors. It is advisable to obtain the sample from different parts of the load after it has been deposited, as the grading may not be the same at the bottom of the lorry as at the top.

Sands can be sieved quickly if they are washed through the sieves with running water. The sand separated on each sieve is weighed in the sieve after allowing a few minutes for the water to drain away. While this method tends to give high values for the smaller particle sizes it does provide an indication of serious changes in grading. It is preferable, however, to have the sand in a dry condition before sieving, and it is usually possible to arrange a calor gas ring or primus stove and metal trays for rapid drying of the sand. The same equipment can be used for rapid moisture content determinations for fine and coarse aggregates.

Clay lumps can be picked out by a visual inspection. Silt and clay are found by shaking up the aggregates with water and allowing them to settle. After a little experience with the particular aggregate in use, it is possible to differentiate between clean and dirty samples by the muddiness of the water immediately after the larger particles have settled.

A first indication of the cleanliness of a sand can be obtained by picking up a little sand and rubbing it between the palms of the hands. If the palms remain clean the sand may be considered clean. If the palms are stained then it may be an indication that excessive amounts of clay and silt are present and that a more accurate determination of the amount is required. Details of tests suitable for use on site are given in Chapter 27.

In regions where dust or sand is carried in the wind, it may be necessary to erect protective screens to prevent contamination. Some form of shelter is also of advantage in hot desert regions to protect the aggregate from the heat of the sun's rays, so reducing the temperature of the concrete at the time of mixing as much as possible.

The deposition of the aggregates should be carefully regulated to prevent different sizes from getting in the same heap, and to make certain that different sizes are separated sufficiently to preclude any intermingling (see Fig. 8.1).

Occasionally the method of handling the aggregates enables stockpiles of the different aggregate sizes to be formed some distance apart. Generally the stockpiles are close together and it is therefore essential to provide some form of partition to separate the different sizes. Suitable partitions can be made by

Fig. 8.1 Typical arrangement for small batching and mixing set-up

driving old rails into the ground and laying railway sleepers or other timber between them, by building brick or block walls, or by making rough concrete walls. Where there is more than one size of coarse aggregate it is advisable to display prominent notices, giving the sizes in bold figures, such that the drivers of delivery lorries can see at a glance where each size should be tipped.

If the mixer is to stay in one position for any length of time, the area over which the aggregate is deposited should be covered with a layer of concrete 75–100 mm (3–4 in) thick, otherwise it is necessary to regard the first 150 mm (6 in) or so of aggregate as wasted in the mud.

The concrete should be laid to fall away from the mixer in order to facilitate free drainage of water from the aggregate. Variations in the moisture content of coarse aggregate as delivered are generally not sufficient to have much effect on the control of the water/cement ratio compared to the effect of the variations which commonly occur in the moisture content of the fine aggregate, i.e. the sand. Therefore, only in special circumstances need the coarse aggregate be stockpiled for some hours to allow for drainage before use. Given reasonable drainage it can be expected that the moisture content of fine aggregate will be reasonably uniform throughout the heap, after having been deposited for at least 14 hours, except for the bottom 300 mm or so, which will remain saturated. To allow for drainage, it is desirable to have two separate stockpiles of fine aggregate, so that fresh deliveries can be tipped into one while the other stockpile is being drawn on. If two stockpiles are provided and if care is taken not to draw from the bottom of the heap a reasonably uniform water/cement ratio can be maintained in the concrete produced, without having to change the amount of added water from one batch to the next. Duplicate stockpiles of fine aggregate are, however, seldom practicable or economical where only moderate quantities of concrete are required. This is one reason why it is generally more difficult to achieve good control on small works than on large.

Methods of handling aggregates to prevent segregation have already been discussed. The stockpiles at the back of the mixer should be so arranged that the aggregates are deposited and fed to the mixer economically and in the correct manner. When coarse aggregate is delivered graded from the maximum size to 5 mm, it sometimes happens that the aggregate producer fills the lorry with, say, two grabs of 20 mm to 10 mm and one grab of 10 mm to 5 mm. The aggregate as delivered is unmixed and on small jobs some difficulty may be experienced in obtaining a uniform mix. The best solution is to insist that the aggregate producer mixes the sizes efficiently at the pit. If, however, it is necessary to mix the aggregate on the site, this should be done by taking portions from different parts of the heap.

In some cases it is possible to obtain from suppliers coarse and fine aggregates already combined. These premixed aggregates (generally referred to as all-in-aggregates) are seldom satisfactory for good quality concrete because of segregation during transportation and tipping.

Some of the points mentioned above are illustrated diagrammatically in Fig. 8.2. The diagrams are mainly concerned with smaller works; for larger works the aggregate would probably be lifted by dragline, grab, or conveyor system to overhead storage hoppers discharging directly into weigh batchers. The slope of hopper bottoms should not be less than 50° to the horizontal if a free flow of aggregate is to be maintained.

The possibility of contamination of the aggregate behind the mixer by extraneous matter should not be overlooked. As an example, serious and extensive spalling of concrete at the entrance to a coalmine was traced to the deposition of spent carbide on the edge of the heap of aggregate, by the miners going off duty. Leaves, particularly pine needles, can delay the setting of cement, and cause local weakness. A broken bag of plaster and inclusion of a small quantity in the mix can cause a serious expansive failure. In food factories, aggregate may become contamined with sugar, very small quantities of which are likely to delay or prevent setting of the cement.

Methods of proportioning concrete mixes depend, to some extent, on the quantity of concrete required. On most works it is now generally accepted practice for all to be batched by weight. For small contracts batching may be done by weight or volume, depending upon the nature of the work.

Volume batching of aggregate

The accuracy of any method of measurement in which a gauge box or hopper is filled to a certain level depends on:

(*a*) bulking of the sand. The extent of bulking is dependent on the moisture content of the sand and is discussed in detail on p. 53;

(*b*) the closeness with which the material packs. If the material packs closely with few air voids, the solid volume of material is greater than if the material is packed loosely.

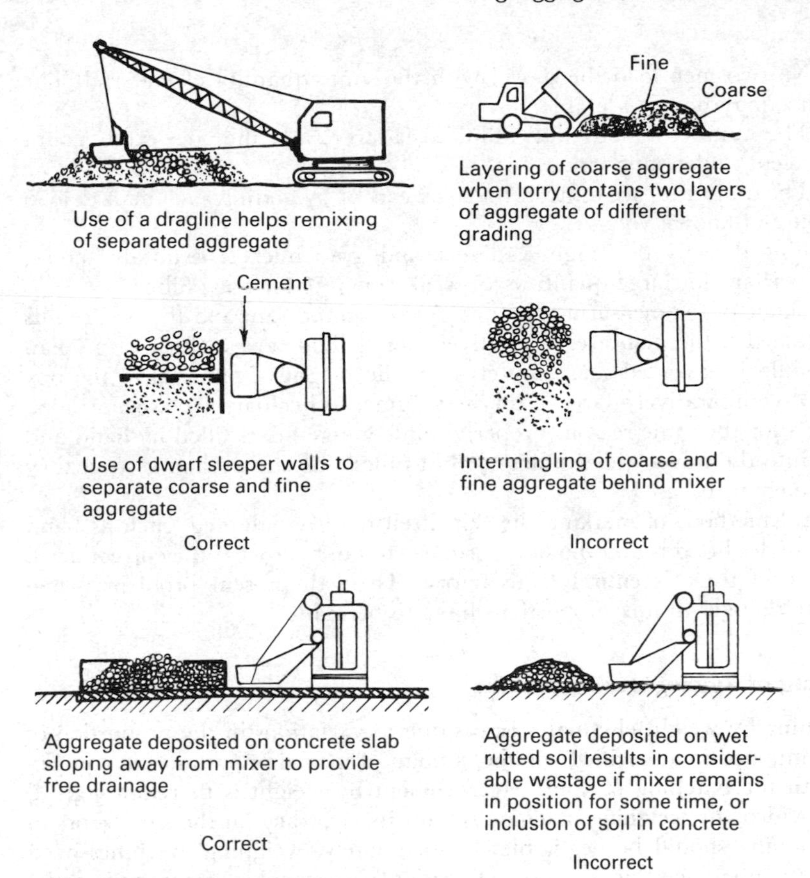

Fig. 8.2 Handling of aggregate at the mixer

In the volume batching of aggregates there are therefore two possible sources of error: variations in the solid volume of aggregate contained in a specified measured volume; and errors in the measured volume. The first variation can be minimized although it is difficult to eliminate it, but errors in measurement can be reduced by careful attention to the type of measure.

Although variations occur in volume batching this method is probably as good as weigh batching on small works where supervision is not good, and where proper maintenance of weigh batching equipment is not possible. Volume batching need not be greatly inferior to weigh batching; experience shows that the standard of supervision on the site is of far more importance than the type of equipment used. In the case of lightweight concrete production, volume batching of the lightweight aggregate may often be preferable to weigh batching. This is because of the large variations which can occur in the absorbed moisture content of the lightweight aggregates. Also small inaccuracies in weighing cause large variations in volume.

Occasionally the practice of batching by filling the mixer skip with so many shovelfuls of each aggregate is adopted. This is bad practice for the following reasons:

(1) No two men load the shovel with the same quantity of material. The difference may be considerable.

(2) The number of shovelfuls required is large, and the men rarely count them except under supervision.

(3) The quantities are difficult to check except by getting each man to load a measure of known volume.

Batching directly by dragline, digger and grab buckets is unsatisfactory, except perhaps for large quantities of weak concrete for mass filling.

The shape of the measuring-box or hopper is important and the best results are obtained if the plan area is relatively small. Flat boxes with a large plan area should be avoided since any error in the height of material in the box causes a comparatively large error in volume. Wheelbarrows are not good measures for the same reason. Properly made gauge-boxes filled by hand and tipped into the mixer are satisfactory, but unless the supervision is strict they are unlikely to be used.

Several methods of marking the skip itself may be adopted, such as band marks, angles fitted round the sides, or bars running across at the correct level, but none of these is entirely satisfactory. They all present problems when frequent changes in mix proportions have to be made.

Batching of aggregates by weight

Batching by weight eliminates errors due to variations in the proportion of voids contained in a specified volume, a point of special importance in connection with the batching of sand. Measurement by weight is therefore logical, and provided the weighing machine retains its accuracy on the site, errors in proportioning should be negligible. Unfortunately, weighing machines need careful maintenance and regular calibration if reasonable accuracy is to be maintained. This is not always easy to arrange on small or isolated sites.

However, wherever good and regular maintenance is practicable, weigh batching is to be preferred to volume batching. An important advantage with weigh batching is the greater uniformity between successive batches of concrete.

Different types of weigh batching plant are available for use on large and small contracts. On large sites weigh hoppers fed from large overhead storage bins are commonly used. These weigh hoppers may discharge by gravity straight into the mixer or on to a belt conveyor feeding the mixer (Fig. 8.3).

Weigh buckets with balanced mechanical lever systems are widely used on large batching plants, and the more modern equipment incorporates a remote indication of weight by electrical means. Electrically operated slave dials may, however, suffer from the disadvantage of being dust-prone and this can lead to inaccuracy.

On smaller sites, the materials are often weigh batched in a loading hopper which is integral with the mixer (Figs. 8.1 and 8.4). In its lowered position, the hopper rests on a sealed hydraulic capsule connected to a pressure gauge inside a weighing dial on the mixer. This weighing mechanism operates hydraulically and gives a direct progressive reading of the total weight of

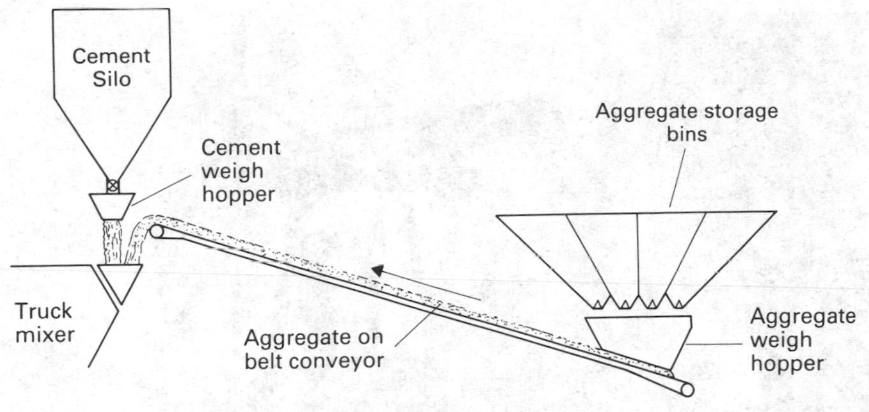

Fig. 8.3 General arrangement of typical dry batching plant

Fig. 8.4 A 30 tonne overhead mounted cement silo which automatically weighs and delivers cement direct into a mixer having a loading hopper fitted with an integral weighing mechanism for batching the aggregates

Fig. 8.5 Integral weigh hopper in elevated position discharging directly into the mixer. Note the load capsule exposed under the raised hopper

material in the hopper as it is being filled. When the hopper has been filled with the required batch weights of each material, it is raised to allow the contents to fall out through the discharge point directly into the mixer (Fig. 8.5). Such arrangements are generally referred to as integral weigh batchers.

Another device used for weigh batching is the load cell, which basically consists of a steel cylinder fitted with electrical strain gauges. The elastic deformation of the steel cylinder under load applied from a weigh hopper is measured by the strain gauges on the Wheatstone bridge principle. Since the deformation is related to the applied load, the weight of material in the hopper can be readily indicated by an electrical signal. Load cells are completely sealed and require very little maintenance.

A useful auxiliary which is fitted to most of the smaller mixers is the power scraper shovel which is used for feeding aggregates into the loading hopper as

Fig. 8.6 View of a power scraper shovel fitted to a 350 litre (12 ft^3) mixer for loading aggregate into the mixer hopper

shown in Figs. 8.1 and 8.6. This saves manpower but it needs to be used with care to avoid batching more than the quantity required. A particular factor to guard against is the situation when an operator finds, through repetition, that a certain weight of aggregate requires a given number of scraper bucketfuls and he subsequently works on this basis. This is equivalent to an approximate volume batching procedure and it can clearly introduce errors.

On large batching and mixing plants, a fully mechanized version of the scraper shovel is used. A scraper bucket is slung from a slewing boom in the manner of a drag-line machine and is operated by an operator in a steel cab (see Fig. 9.5). With some automatic versions, the whole operation may be controlled by the mixer driver who merely presses a button to slew the boom over the selected aggregate bay and then starts the scraping cycle.

BS 5328 : 1976 'Methods for Specifying Concrete' requires that the accuracy of measuring equipment for batching purposes shall be within ± 3 per cent of the quantity of total aggregates, cement or water being measured.

Aggregate weighing machines require regular attention if they are to maintain their accuracy. Check calibrations should always be made by adding weights in the hopper equal to the full weight of aggregate in the batch. Errors may occur for a number of reasons such as:

(1) The knife-edges and other moving parts become clogged or blunt, causing stiffness of movement.

(2) The dashpot damping device on beam and jockey weight types may become detached or be filled with the wrong type of oil or water.

(3) Again on the beam and jockey weight types, the beam gets out of balance due to movement of the counter-balances, either through ignorance or by loosening of the locking device.

(4) Levers or rods may be bent and bind on some part of the framework.

(5) With dial types the interior mechanism may become worn, or the springs may become weak or broken. If the weighing bucket is not locked in position during transit, sufficient wear of these parts can occur during a single journey to cause substantial errors.

(6) Perspect covers warp and may impede the free movement of pointers causing quite considerable errors.

(7) If the weigh batcher has two weighing buckets, and both are in use, the one for sand and the other for gravel, confusion may easily arise as to which is to be filled with sand and gravel, resulting in the relative proportions being reversed. It is therefore essential to label the buckets clearly to indicate the material with which they should be loaded.

(8) Spectators sometimes add their weight to the bucket.

(9) Spillage of aggregate sometimes accumulates and gives local support to the weigh bucket; this can easily occur with most integral weigh batches as they seldom have much clearance above the ground. The regular clearing of aggregates from under the weight bucket is, therefore, important.

The method of delivering the materials into the weigh hopper can have some influence on the accuracy of the batching. Delivery from overhead hoppers with servo-operated gates is likely to be more accurate than by other methods.

In the case of large batching plants, there has been considerable development in recent years in the use of power-assisted and remotely activated controls, compressed air being used as the power source. These controls allow the batching operator to be accommodated away from the weigh hoppers in a comfortable control cabin, equipped with a console and weigh dials. This cabin can be located at ground level if so desired. Fig. 9.5 shows a general arrangement of a typical batching plant, based on the weigh hopper system, which utilizes these developments.

Many large plants are now available with automatic weigh-batching. This generally means that the operator has only to press one or two buttons to put into motion the weighing of all the different materials, the flow of each being cut off when the correct weight is reached.

In their most advanced forms, automatic plants are electronically operated on a punched card system; this type of plant is particularly suitable for the production of ready-mixed concrete, in which very frequent changes in mix proportions have to be made to meet the varying requirements of different customers.

With most large plants, particularly those made in the USA, recorders can be fitted which note automatically and graphically the weight of each material delivered to each batch; that is, they are meant to record the actual weight as distinct from the intended weight. In practice their accuracy generally leaves something to be desired.

Continuous weighing of aggregates

In addition to the batch weighing systems described above, there is also a system of proportioning which is generally referred to as continuous batching. This system is sometimes used on large batching plants in conjunction with continuous mixers.

The principle of continuous weighing is based on the aggregate being fed from an overhead hopper on to a short conveyor belt at a controlled rate. The conveyor belt is provided with a weigh roller which is counter balanced by a pivoted lever arrangement to suit the feed rate. If the weigh roller detects a change in weight from that selected, the feed rate through the radial gate from the hopper is adjusted automatically by electric switches and linked mechanisms until a balance is achieved.

A separate belt weigher must be provided for each aggregate size and each of these feeds a main conveyor belt taking the combined flows of aggregates to the mixer. In this way, the mixer receives a stream of material of layered and uniform cross-section representing the aggregates proportioned to the required mix design. A belt weigher must also be provided for the cement being fed from the overhead silo to the mixer.

A good standard of control is possible with continuous batching, but in general it is considered that batch weighing is more accurate and reliable.

Batching of cement

Cement is batched either by the bag (normally 50 kg in Great Britain; 94 lb in USA) as weighed at the cement works or, for bulk cement, batched from a silo installed on the site.

Whichever method of batching is used, precautions should be taken to see that cement is not lost as the loading skip empties it into the mixer. Tests on exposed sites with strong winds blowing have shown considerable losses of cement during this operation.

Volume batching of cement should not be permitted; it is subject to serious errors because Portland cement may weigh from 1200 to 1520 kg/m^3 (75 to 95 lb/ft^3) depending upon its fineness and the method of filling the measure. Other errors may result from failure to fill the measure accurately and from warping or denting of the measure.

Batching by bags of cement

Provided whole bags of cement are used, the only error lies in a variation in the weight in each bag as supplied by the cement works. This is not likely to be serious, although variations from 45 kg to 54 kg (100 lb to 120 lb) in single bags have been recorded. Under the ASTM specifications, individual sacks may be as much as 5 per cent below the standard weight of 94 lb, provided that the average weight of 50 sacks taken at random is not less than 94 lb. For important work an occasional check is justified.

The size of the batch of concrete should be adjusted so that full bags of cement may be used. Splitting of bags is liable to cause substantial error and for accurate gauging the part bags should be weighed.

Weighing cement on site

There are many advantages to be gained from using bulk cement stored in a silo and weighing it on site:

(1) Bulk cement is cheaper than bagged cement.

(2) A gang of men is not needed to off-load the cement, as is the case with bags of cement.

(3) The trouble of ensuring adequate protection of the cement during storage is avoided.

(4) The mixer can always be used to capacity without the risk of the errors which may result from splitting bags.

(5) Changes in mix proportions are made more easily.

It is not uncommon for mixers to be used well below their capacity, or to be overloaded, because the amount of cement for a batch to the mixer's full capacity does not coincide with a whole number of bags (p. 143).

However, the weighing of cement usually provides more difficulty than that of aggregate, mainly because it is difficult to obtain a uniform flow, and compressed-air assistance is often needed to overcome arching in the valve opening. When it is appreciated that the normal rate of flow in a large weighing machine is about 25 kg (55 lb) per second, it will be realized that any sudden rush, when the quantity in the weigh bucket is near the correct weight, may result in some excess which will be reflected either in increased workability or in increased strength for the particular batch.

Another difficulty which arises in weighing cement is the tendency to build-up in the weighing hopper, particularly when the atmosphere is damp. If cement and aggregate are weighed together, the dampness from the aggregate causes a serious build-up of cement mortar in the weighing hopper, which measurements have shown to amount to 50 kg (110 lb) or more in a matter of ten batches, even when some shaking is applied to the hopper to hasten discharge.

Weighing machines operating in conjunction with bulk supplies of cement are often used on both large and small sites. They may be of the spring-balance type or the beam and jockey weight type; or they may be of the hydraulic capsule type (referred to on p. 128).

Cement in bulk is delivered in pressurized lorries which blow it into silos which are made in a range of capacities from 12 tonnes upwards (illustrated in Fig. 8.4). These silos generally incorporate a mechanism for weighing the correct quantity for a batch and dispensing it into the mixer hopper or directly into the mixer.

Fig. 8.7 shows a dispenser of the beam and jockey weight type in which the dispensing hopper on its supporting trolley runs along the supporting beam. When the hopper is pushed to the back of the beam directly under the discharge chute of the silo, cement is automatically dispensed and cut off at the correct weight by a trip-mechanism. The hopper is then pulled along to the front of the beam where it can then be discharged directly into the adjacent mixer hopper.

Cement weighing machines should receive regular attention to the following points:

Fig. 8.7 Cement silo dispensing mechanism

(1) Building up of cement in the weigh hopper and discharge chute, which may result in the use of too little cement.

(2) The cement should flow freely and continuously from the storage hopper, compressed air being blown in at the valve, if necessary. If the cement does not flow freely it is difficult for the operator or the weighing mechanism to cut off at the correct quantity.

Vibration of the silo is more likely to result in compaction of the cement and blockage than to cause a free flow, hence the use of compressed air.

(3) The knife-edges and all other moving parts should be kept completely free from coatings of cement in order to avoid friction and stiffness. The leather or plastic apron which prevents the cement from escaping from the dispenser during filling should be kept clean, since any accumulation will be included in the weighing.

(4) If a dashpot damping device is fitted, this should be filled with the correct oil. The connections should be inspected occasionally to make sure it does not become loose or detached.

(5) Levers and rods should be inspected occasionally to ensure that they have not been so bent as to bind on the framework.

(6) Care should be taken to see that spectators do not lean on the weigh bucket, especially when calibrating.

(7) The balance of the beam should be checked frequently, in case the counter-balances become moved, or cement adds to the weight of the beam itself.

Cement weighing mechanisms should always be checked by adding weights to the dispenser to the full amount required in the batch. The necessity for this is demonstrated by the curves given in Fig. 8.8, which are plotted from calibrations of a cement weigh batcher over a period of several months. If weights

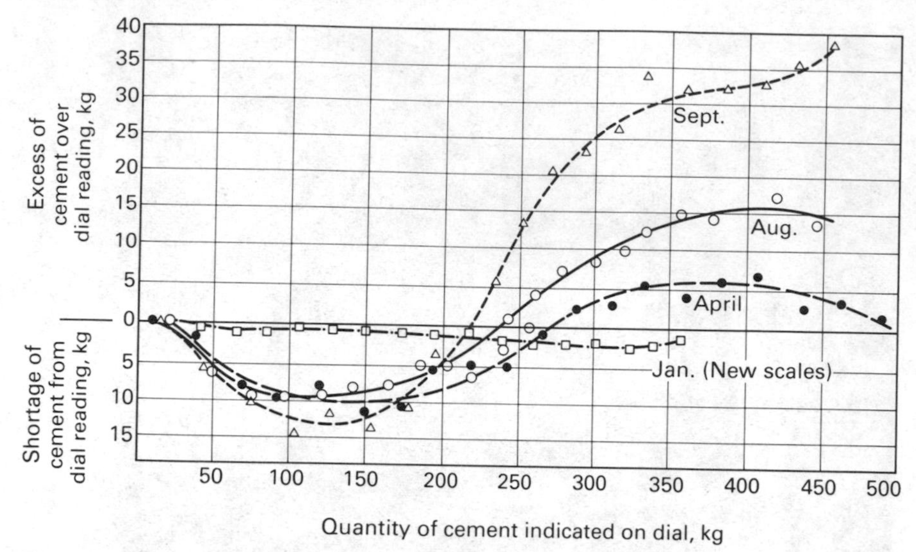

Fig. 8.8 Deterioration in accuracy of cement scales

to a maximum of only 150 kg (330 lb) had been used, it would have been assumed that the scales were weighing light, whereas above about 250 kg (550 lb) the quantity of cement delivered was in excess of that indicated on the dial. The necessity for regular checking is apparent from the extent of the errors.

Cement weigh buckets generally discharge directly into the mixer hopper, but in large plants the cement storage bin is often situated to one side of the plant to save height, and means must be provided to convey the cement horizontally to the weigh bucket. A worm feed is often used for this purpose; but a more efficient system is provided by the air slide, which is a small duct with a mesh or canvas floor through which low pressure air is introduced, causing the cement to flow freely, even with a very small slope.

On sites where more than one type of cement is used, it is necessary to provide a silo for each cement type. In some cases, a silo may be split into two compartments to accommodate two types of cement. Whichever arrangement is used, it is most important that each silo or compartment is clearly labelled to show the type of cement it holds. Otherwise the wrong type of cement may be fed into a particular silo.

Moisture meters for fine aggregate

Electrical apparatus for measuring automatically the moisture content of sand is sometimes fitted to large batching plants. The accuracy of such apparatus is somewhat limited and it is desirable to check it at least daily by measuring the moisture content of the sand by more conventional means. Nevertheless, the apparatus provides a very useful indication of changes in moisture content and gives the plant operator, indirectly, a constant indication of adjustments needed to the quantity of added water required. It can contribute to good control.

Fig. 8.9 Central batching and mixing set-up for large dock contract. A cement silo is provided over each mixer

Water supply

Purity of water

Water to be used for making concrete must be reasonably free from deleterious materials such as silt, clay, organic matter and organic acids, alkalis, and other salts. BS 3148 : 1959 [1] gives methods of testing water and in the appendix are notes on the suitability of water for making concrete.

Silt and clay can be removed by allowing them to settle out in suitable basins or tanks, but the other materials, being soluble, are difficult to eliminate. No specific limit to the quantity of dissolved salts can be given, but unless clear water has a noticeably brackish or saline taste it may be safely used. In any case of doubt it is advisable to send a sample of the water to a laboratory for analysis and comparison with water known to be satisfactory. Care should be taken, when sampling water, that a clean bottle is used, and that the bottle is washed out with the water to be submitted, before finally taking the sample.

The experience and recommendations of a large number of investigators and authorities on the use of sea water are conflicting, some reporting adverse results while others report no harmful effects [2]. Whenever possible the use of sea water for making concrete should be avoided, since it is liable to cause unsightly efflorescence as moisture movements carry soluble salts to the surface. The chlorides in sea water are liable to cause serious corrosion of reinforcements, and unless measures are taken to ensure that the concrete will be dense and impermeable, the use of sea water in reinforced concrete is better avoided.

Drinking water from the mains supply is safe and convenient.

Water supply on the site

Water is required both for mixing the concrete and for curing. When it is considered that each batch of concrete from a 280 litre mixer may require 36 litres of water, and 80 batches a day may be made, giving a total of 2880 litres a day per mixer, it will be realized that the quantities required for mixing alone are considerable. More important, the 36 litres of water in each batch must be replaced in the time taken for mixing, say one minute, in order that the measuring cylinder is full, ready for the next batch. The size of hose-piping connecting the mixer with the source of supply must, therefore, be of sufficient size, having regard to its length and the water pressure, to cope with this intermittent high rate of demand.

The question of an adequate water supply is commonly neglected on concrete constructional works, and results in loss of efficiency in operation of mixing machines. Not only is time lost waiting for the water to refill, but the measurement of the quantity of water discharged into the mixer becomes inaccurate. Most systems of measurement rely on complete filling of the tank, and the mixer driver, when in a hurry, is apt to omit to do so. An adjustment is then made to the quantity of water indicated on the scale, and when a longer period occurs between batches, the mix becomes too wet.

An idea of the water supply requirements on a large contract may be judged from the fact that the quantities used may be equivalent to the normal supply in a small town.

In the hot dry regions of the world, the mixing water has often to be brought from a considerable distance, and if the pipeline or water tanker is exposed for long in the sun, it can become too hot to touch. The use of such hot water, together with hot aggregates and cement increases the rapid evaporation of water from the concrete during transport, and reduces the setting time, so that placing may become difficult. In the hot season, it is therefore necessary to bury the water pipeline, and to cover water tankers.

In cold winter conditions water pipes should be lagged.

Water control at the mixer

With given proportions of aggregate and cement, the water/cement ratio is determined by the workability required for satisfactory placing. With the workability controlled by close observation at the mixer, so that any variation is small, it may be taken that any changes in the water/cement ratio will also be small and the strength of the concrete will be reasonably uniform. Further control may be obtained by making tests on the moisture content of the aggregate, particularly the sand, at frequent intervals, and adjusting the quantity of water added at the mixer accordingly. Except where large stockpiles of aggregate are maintained, however, the variations in moisture content of the sand from one load to another, and from one part to another of a small stockpile, make the value of tests for moisture content doubtful. The cost of making such tests continually is, in any case, excessive when small quantities of concrete only are to be produced, and under these conditions it is suggested that the following measures be adopted in order to secure concrete of the required quality.

The first few batches of concrete should be produced under the supervision of an experienced engineer or inspector, the aggregate and sand being carefully weighed and checked for surface and absorbed moisture content. When the aggregate is saturated the absorbed moisture may be taken as constant and should not be included in the water/cement ratio. The surface moisture will be variable and should be deducted from the theoretical quantity of water required when determining the water to be added at the mixer. If the aggregate is dry, it will be necessary to determine the quantity of water absorbed in bringing the aggregate to a saturated surface dry condition, and to add this additional water to the theoretical quantity required. In the tropics it is advisable to keep the aggregate wet, if an ample supply of clean water is available.

The workability should be checked by means of a slump test, or compacting factor apparatus, and cubes made for crushing strength tests. With most concrete mixers an appreciable quantity of fine material from the first few batches remains in the drum. Therefore only after producing two or three correctly proportioned batches will the concrete have settled down to a uniform consistency. The appearance of the mix should be particularly noted and should be observed by the mixer driver and the placing ganger. This concrete should then be placed and observations made on its suitability for placing in the formwork, on ease of consolidation, and on any tendency to segregate. If the concrete is found to be satisfactory in all respects, the mixer driver and the placing ganger should be instructed that all further concrete produced should have a similar appearance.

Reference is also made to the control of consistency in the field on p. 103.

Allowance for surface moisture

Several different methods of determining surface moisture are available, such as the simple drying method using a frying pan and gas ring, and the rather more difficult procedure using a pycnometer. Both methods are fully described in Chapter 27, p. 405.

Measurement of water at the mixer

Water gauges on concrete mixers are generally one of two main types. Their principal features are given below.

(1) *Vertical tanks* were developed to overcome inaccuracies with horizontal measuring tanks due to changes in inclination. Examples are shown in Fig. 8.10.

The tank shown in Fig. 8.10 (*a*) is set by winding a small handle on top and counting the turns, a procedure which is not entirely satisfactory, since it is not possible to check the quantity delivered without turning back to zero and counting the number of turns. A large number of turns are required and errors

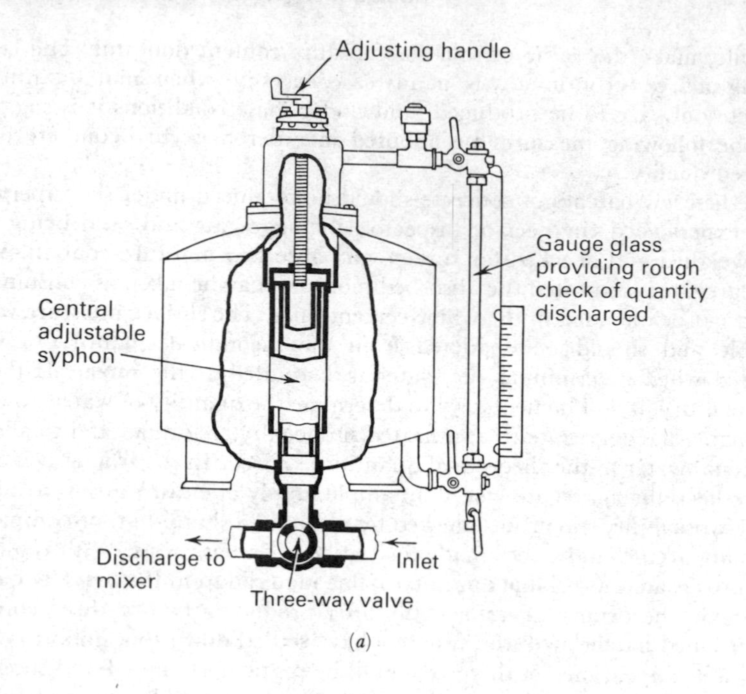

(*a*)

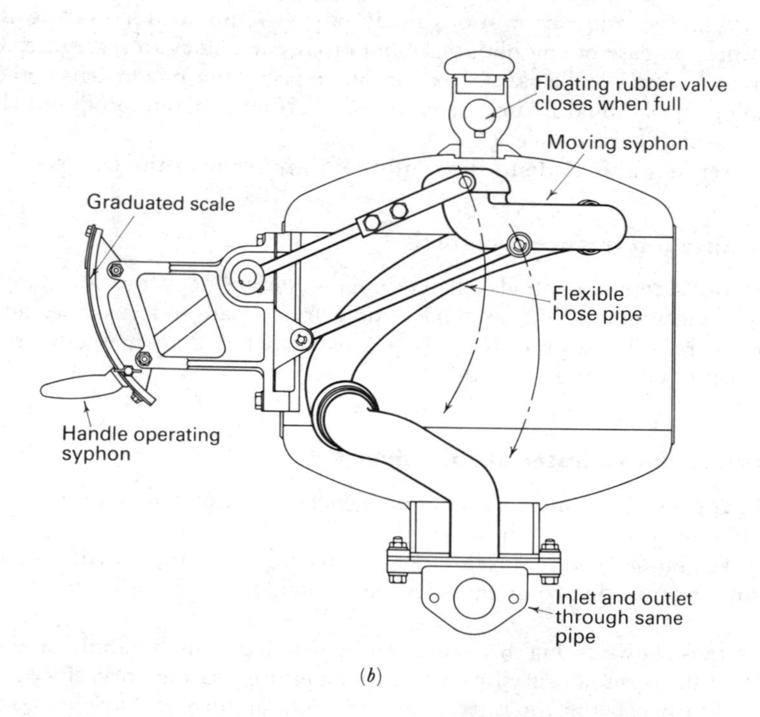

(*b*)

Fig. 8.10 Different arrangement of two vertical water measuring tanks

in counting often occur. The other tank, Fig. 8.10 (*b*), overcomes this difficulty by the provision of a scale at the end of the lever which actuates the central moving syphon. A flexible hose allows vertical movement of the syphon. This is more satisfactory than a bellows for this purpose, being less liable to develop leaks.

(2) *Water meters* fitted in the main water supply line to the mixer. For use on batch mixers, the meters are provided with a reset hand to indicate batch readings, and a register of the total water passed. The latter is particularly useful in determining the average water content of a group of concrete mixes.

Water meters are affected by dirt, scale, and sediment more than measuring tanks, and are therefore less suitable for portable mixers. They are most satisfactory for stationary mixers and central mixing plants fed direct from the mains water supply.

Before setting the water gauge on a mixer for the first time, it is essential to check the calibration. Errors of as much as 9 litres (2 gallons) in 45 litres (10 gallons) have been found with some types of water gauge. Regular attention should be given to the cleanliness of the tank, and to the fitting of any valves or syphoning arrangement.

American machinery is likely to be calibrated in US gallons, equivalent to 3·79 litres of water, compared with 4·55 litres for the Imperial gallon. This point should also be remembered when reading American literature.

REFERENCES

1 BS 3148 : 1959, *Tests for Water for Making Concrete.* British Standards Institution.
2 FULTON, F. S., *Concrete Technology*, 1964. The Portland Cement Institute, Johannesburg.

9

Mixing and Transporting Concrete

THE mixing of concrete is almost invariably carried out by machine, but a very brief reference to hand mixing is necessary to cover those occasions, which occur on many jobs, when hand mixing must be done.

Hand mixing should be done thoroughly. A clean surface should be selected for mixing. If this is not available, a wooden platform with close joints to prevent loss of mortar should be obtained. It is usually specified that the concrete shall be mixed three times dry and three times wet, turning over from one spot to another. The colour and uniformity of the concrete is, perhaps, the best gauge of the efficiency of the mixing.

In order to compensate for the lower strength which usually is obtained with hand mixing, it is advisable to allow an extra 10 per cent of cement above that normally required.

Machine mixing

Many types of machine are available for mixing concrete, some having a revolving or stationary pan, with paddles to mix the materials, others having tilting or non-tilting revolving drums. The continuous mixer, although in a class on its own, relies for the actual mixing process on a revolving drum in one type and on a revolving shaft with 'tines', in a stationary trough, in another type.

Pan mixers with a revolving star of blades are the most efficient mixers from the point of view of the uniformity of the concrete produced. They are especially suitable for stiff and lean mixes, which present difficulties with most other types of mixer, mainly due to sticking of the mortar in the drum. They are, however, somewhat cumbersome and difficult to maintain in good condition.

Uniformity of mixing is difficult to measure, but there is evidence that considerable variations can occur with most types of mixers in the proportions of the constituent materials within the same batch; for instance, variations of as much as 20 per cent in the proportion of coarse aggregate have been measured in a fair proportion of batches mixed (1). In conventional rotating-drum mixers this variation is due principally to the lack of interchange of materials from one end of the drum to the other. Some lack of uniformity of concrete mixed in pan mixers appears to be due to the throwing of stones from one part of the batch to another by the high-speed action of the mixing star.

Mixing time and output

Some research has been made both in America and in this country (1) on the time required to mix concrete intimately and uniformly. It appears that the number of revolutions of the mixing drum or pan is more important than the mixing time, and in general no more than 20 revolutions are necessary for adequate mixing. For most types of mixer up to the 1 m³ size operating at their correct speed, this represents a mixing time of a little over 1 minute. Normally a mixing time of between 1 minute and 1½ minutes may be regarded as adequate. With some high speed pan mixers, 35 seconds is adequate mixing time for most types of concrete.

The minimum mixing time generally specified for Bureau of Reclamation work in America [2], the time to start after all ingredients, except the last of the water, are in the mixer, is as follows:

Capacity of mixer m³	Time of mixing minutes
2 or less	1½
2·5	2
3·0	2½
5·0	3

In considering the potential output of a particular mixer, it is the time of the total mixing cycle, not just the time of mixing, which, with the batch size, determines the output. For instance, with a mixer which requires 1½ minutes mixing time, the design or condition of the loading hopper may be such that the materials flow only slowly into the mixer. The loading time may be 30 seconds or more; and the discharge of the mixed concrete may be slow, taking up to 1 minute, particularly if the mix is dry. Adding the time to raise the loading hopper and allowing for minor delays, the total cycle may well be 4 minutes or more, so that only some 15 batches an hour, or less, can be produced. In comparison, 35–40 batches per hour are possible with some of the more efficient mixers.

The output per hour from any mixer is dependent, to a large extent, on the efficiency of organization, both in charging the mixer and in transporting away the mixed concrete.

Mixer size

Concrete mixers complying with the requirements of BS 1305 : 1974 'Batch Type Concrete Mixers' are designated for size by a number representing the nominal batch quantity in litres up to 1000 litres and cubic metres above this size, together with letter(s) indicating the type of mixer. The nominal batch quantity is the volume of compacted concrete which can be mixed satisfactorily in one batch. The letters used are:

Tilting drum type	T
Non-tilting drum type	NT
Reversing drum type	R
Forced action type	P

The capacity of the mixer may not be fully utilized if cement in bags is used. Full bags should always be used and the batch quantity calculated accordingly. For some mixer sizes and batch quantities it will be found that the mixer will only work at little more than half its rated capacity. Consideration may need to be given to the use of a slightly richer mix or a change of mixer size.

As indicated in Chapter 8, one of the advantages of using bulk cement is that the quantity of cement batched can be chosen, according to the mix proportions, so that the mixer is used to capacity.

General principles in the use of concrete mixers

There are five main types of concrete mixer in use in the United Kingdom, and while their method of operation varies considerably, the following general principles are usually applicable:

(1) It is an advantage to feed the cement, sand and coarse aggregate into the mixer simultaneously and in such a manner that the flow of each extends over the same period. The concrete produced is more uniform than that obtained when the ingredients are introduced one after the other, or in unmatched quantities.

With most mixers it will be found that filling the loading hopper in more or less horizontal layers of gravel, cement and sand, possibly with another layer of a different size of gravel on top, provides the best conditions[1]. Placing of the gravel at the bottom of the loading hopper is more likely to result in a self-cleaning process, so avoiding the formation of a hardened layer of mortar at the bottom of the hopper. The piling-up of any one size of material in the throat of the loading skip should be avoided. With large mixers (3–4 m^3), the sequence of materials into the mixer is of particular importance.

(2) The water should enter the mixer at the same time and over the same period as the other materials. With many mixers this is not possible, since the rate of flow is limited, and the time taken for the full quantity to enter the mixer may be up to $\frac{1}{2}$ minute. In such cases it is advisable to start the flow of water a little in advance of the other ingredients.

If all the water is added before or after the other ingredients, the batch of concrete is liable to vary from a wet portion to a dry portion. The direction of inflow of water also influences the uniformity and may need adjustment.

(3) Mixing should continue until the concrete is of uniform consistency and colour. With drum mixers it is an advantage if the mixer driver can see into the drum at both ends, so that he can watch the consistency, and make any necessary minor correction in the water added.

(4) The mixer should not be loaded beyond its rated capacity. Overloading results in spillage of materials, and less satisfactory mixing, in addition to imposing undue strain on the mechanical parts.

(5) The mixer should be set up accurately so that the axis of rotation of the mixer drum, except in the case of the tilting-drum type, is horizontal. Inaccurate setting may result in poor mixing, or in the case of continuous mixers, too fast or too slow a passage through the mixer drum. The horizontal type of water measuring tank is also affected by errors in levelling.

(6) For satisfactory performance the mixer should be capable of producing uniform concrete throughout the batch, so that the quantity of the largest size of aggregate at the end of discharge does not differ by more than 20 per cent from that at the beginning of discharge. In order to minimize the risk of stone pockets, due to excess of stone in any part of the batch, it is advisable to

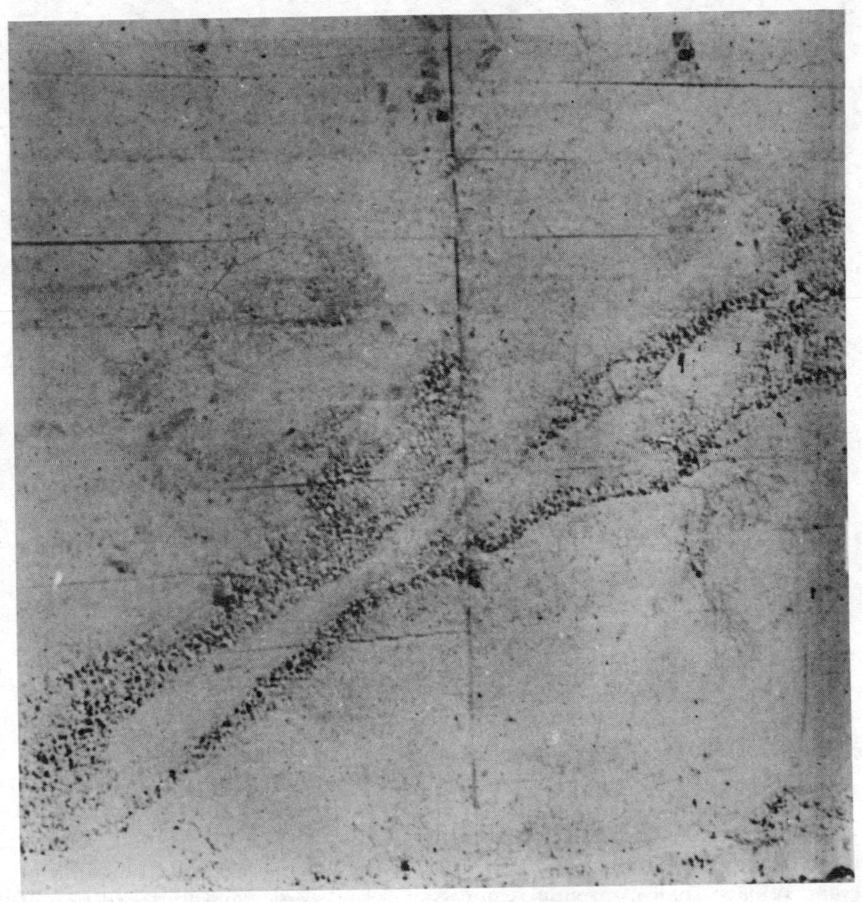

Fig. 9.1 Honeycombing caused by uneven mixing coupled with unsatisfactory placing indicated by the sloping pour planes

discharge the whole batch into a suitable container, rather than to discharge in small separate quantities into wheelbarrows. If the container is suitably designed, the stone will become spread through the batch.

Honeycombing, resulting from uneven distribution of stones and sand in batches of concrete, is seen in Fig. 9.1.

(7) The mixer should be run at the correct speed as stated by the manufacturer. The speed should be checked regularly.

The effect of the speed of the drum on the efficiency of mixing is illustrated in Fig. 9.2[(1)]. The lower the 'efficiency number', the greater the efficiency of mixing.

(8) Some cement mortar from the first batch of concrete mixed is left on the blades and round the drum. In order to avoid difficulties in placing due to shortage of fines an extra 10 per cent each of cement and sand should be added for the first batch.

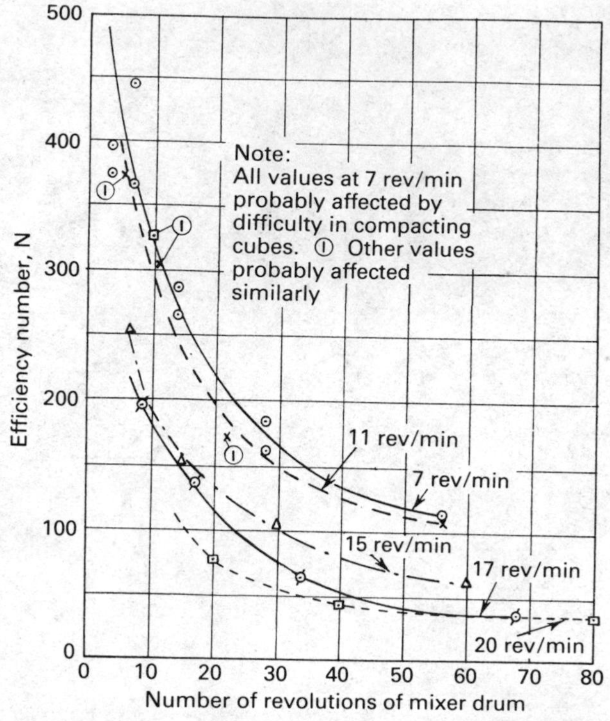

Fig. 9.2 Relation between efficiency number, drum revolutions and mixing speed

(9) Considerable amounts of concrete adhering to the blades or the inner surface of the drum reduce the efficiency of the mixing. Regular cleaning at the end of each spell of mixing is necessary to prevent concrete building up, especially if stiff mixes are in use.

(10) Badly worn and bent blades reduce efficiency, and should be replaced. Wear of the inlet and discharge chutes eventually results in loss of materials, and should be remedied by suitable repairs.

(11) Adherence of cement is reduced by rubbing grease or oil over the mixer after cleaning. Layers of cement are particularly apt to build up in the nose of the loading hopper and should be chipped off at frequent intervals.

Reference should be made to Chapter 10 for comment on the mixing of ready-mixed concrete.

Non-tilting drum mixers

Non-tilting drum mixers normally have a single drum rotating about a horizontal axis. The drum is mounted on rollers and is turned either by a rack-and-pinion drive or by a chain passing round the drum and a cog-wheel driven from the engine. The blades inside the drum are arranged somewhat differently by the various makers of mixing machines, but the general tendency is for the blades to work the concrete towards the discharge end of the

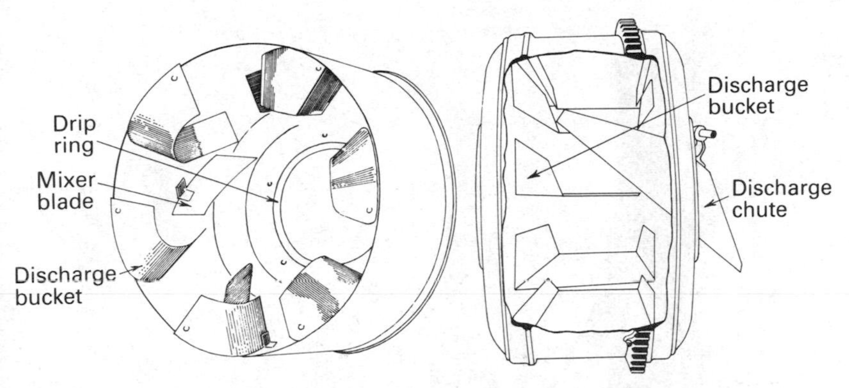

Fig. 9.3 Interiors of two non-tilting drum mixers showing bucket arrangements

Fig. 9.4 Non-tilting drum type concrete mixer
300 litres capacity, and provided with a vertical water measuring tank

mixer, in order to provide a rapid rate of discharge. Examples of blade arrangements are given in Fig. 9.3.

With many rotating drum mixers it is important for uniform mixing that the water delivery pipe be so positioned that the water is directed to the rear part of the drum, i.e. the side into which the dry materials are fed.

An example of a mixer designed to produce 300 litres (about 10 ft^3) of mixed concrete per batch is given in Fig. 9.4. The drum has ten similar blades arranged around the periphery. Discharge is effected by a manually operated chute. Another type of non-tilting drum mixer is shown in Fig. 8.1.

Fig. 9.5 A compact central mixing plant
The cement storage silo discharges into a separate weigh hopper directly over the pan mixer. The aggregates are discharged into a weigh hopper which travels up the inclined rails and discharges into the mixer. The small drag line operates from the end of the boom above the aggregate stock piles. On being discharged from the mixer the concrete falls into a wet hopper.

Non-tilting drum mixers are available in sizes 200, 250, 350, 500 and 750 litres nominal batch capacities.

Split drum mixers

Split-drum mixers have been used for some years in Belgium and a number are now in use in the United Kingdom. Normally of 2 m^3 capacity, their distinctive feature is that the drum separates into two halves along a vertical plane, allowing the mixed concrete to be discharged rapidly.

Tilting-drum mixers

Small tilting-drum mixers of the type shown in Fig. 9.6 are commonly used for all types of building work and in Great Britain are generally available in sizes of 100, 150, 175 and 200 litres batch capacity. Those of capacity up to 150 litres of mixed concrete are often loaded by shovelling straight into the drum, but this method is unsatisfactory for high-quality concrete. Some form of gauging should be adopted.

Medium size tilting mixers are provided with a loading skip similar to that used for non-tilting drum mixers.

The large sizes (Fig. 9.7) may be 1 m^3 to 7·5 m^3. They are common overseas in large scale concrete production plants. Their suitability for large plants is to some extent due to the way in which three or four of them can be arranged

Fig. 9.6 Concrete being discharged from a tilting-drum mixer

radially, such that each in turn can be fed from the same chute bringing materials from overhead hoppers, while each discharges in turn into the same wet hopper. Also tilting drum mixers are the most suitable type for concrete with large sized aggregate, e.g., 150 mm; and because of their rapid discharge they are suitable for low workability concrete.

Reversing drum mixers

Mixers in this category rotate in one direction for mixing and in the reverse direction for discharge. One of the smaller sizes of this type of mixer is shown in Fig. 9.8. The drum has two sets of blades, one for mixing, the other for discharging. When the drum is reversed after mixing is completed, the concrete is discharged quickly and clearly.

Reversing drum mixers give efficient mixing with very little building up and 'balling' of the concrete inside the drum. They are particularly suitable for relatively dry concrete mixes.

A rubber or steel cone extension is generally fitted to the discharge side of the drum to enable the concrete to be discharged freely without spillage into the middle of a skip or dumper.

Forced action mixers

The several types of forced action, i.e. pan, mixer which are made all have the advantage, particularly over conventional non-tilting drum mixers, of

Fig. 9.7 Large size tilting-drum mixer

Fig. 9.8 Reversing drum mixer of 350 litres (12 ft³) capacity with built in batch weigher and powered scraper shovel. The arrangement of the blades in the drum is such that on reversing the direction of rotation the mix is discharged. Note the rubber drum cone extension for discharging cleanly into a rollover crane skip

Fig. 9.9 Action of the mixing star in a 'Cumflow' forced action or pan mixer

Fig. 9.10 600 litre high-speed forced action or pan mixer. The electrical drive for the paddles is housed in the centre of the stationary pan. Steel cover plates removed to show the interior of the pan

being able to mix stiff and lean concrete efficiently. For this reason they are widely used for precast concrete manufacture and are the best type for lean concrete. One common type of pan mixer, with a rotating pan, is fitted with a mixing star of cast-steel paddles (Fig. 9.9), mounted eccentrically to the pan; the star revolves either in the same direction as the pan or in the counter direction, but at greater speed. Another type is shown in Fig. 9.10. This, in common with most modern large pan mixers, has a stationary pan.

The pan mixer is not an easily transportable mixer, and wear is fairly heavy, so that its use is mainly confined to precast concrete factories and large civil engineering works.

A small version of this type of mixer having a rated capacity of 40 litres of mixed concrete has proved very useful for trial mixes in the laboratory, and for the preparation of small quantities of coloured facing for architectural cast stone. Even with this mixer, however, it is found that some segregation of the larger sizes of aggregate occurs, and for trial mixes it is necessary finally to turn the concrete over by hand to ensure uniformity throughout the batch.

Forced action mixers are available in sizes from 200 litres up to 2 m^3 nominal batch capacity.

Continuous mixers

Continuous mixers have been in use for many years for construction work such as dams, foundations, retaining walls and mass concrete filling. They have also been found to be particularly useful for the production of dry lean concrete for road construction.

The earlier type of continuous mixer was fed with aggregates and cement that were carried forward by spiral conveyors. The mix proportions were varied by altering the speed of these spiral conveyors. On reaching the mixer drum, the materials were mixed and transferred to the discharge end by blades on the inside of the drum. The concrete emerged as a continuous stream. These earlier continuous mixers were, therefore, supplied with aggregates and cement batched on a volume basis and this often left something to be desired in terms of the accuracy of the resulting concrete mix proportions.

The disadvantage of volume batching was overcome with later models of continuous mixers which were designed with a continuous weigh batching system (see p. 133). These mixers have proved capable of producing good quality concrete. They have an advantage over batch type mixers in that the operator can readily make minor adjustments to the workability of the concrete as each batch is being produced. Continuous mixers utilizing continuous weigh batching systems have been successfully used at certain ready-mixed concrete plants in recent years.

Transport

The transport of concrete from the mixing plant to the point at which it is to be placed must comply with three main requirements, namely:

(1) Transport must be rapid so that the concrete does not dry out or lose its workability or plasticity during the time which elapses between mixing and placing.

(2) Segregation must be reduced to a minimum in order to avoid non-uniform concrete. For the same reason any loss of fine material or cement and water should be prevented.

(3) The transport should be organized so that during the placing of any particular lift or section, delays will not result in the formation of pour planes, cold joints, or construction joints.

Dry concrete may prove difficult to transport, since it may compact so hard that it has to be hacked from the containers. It is also likely to stick in concrete buckets, skips and transfer hoppers. Providing the container is well designed there is only one remedy and that is to make the mix more workable.

Some of the mortar from the first one or two batches of concrete transported, nearly always sticks to the container. It has been suggested previously in dealing with concrete mixers that an extra 10 per cent of cement and sand should be added to the first batch produced; this should be sufficient to compensate for any loss of mortar in transit as well as in the mixer.

Various methods are available for transporting concrete, of which the following are the most important:

(1) Wheelbarrows, handcarts, light trucks and skips on rails
(2) Dumpers
(3) Lorries
(4) Steel skips and buckets
(5) Chutes
(6) Belt conveyors
(7) Mono-rail transporters
(8) Concrete pumps and pneumatic placers, with pipe lines

A number of general points concerning the handling and transport of mixed concrete are illustrated in Fig. 9.11.

Wheelbarrows, handcarts and light trucks. Wheelbarrows and handcarts may be used over short distances. It is advisable, however, to discharge the whole of the contents of the mixer into an intermediate hopper, from which the concrete can be drawn by gravity into the barrows, in order to overcome variations in the consistency of the concrete as discharged from the mixer, to avoid segregation in long chutes from the mixer, and to minimize delays.

Light trucks or skips on rails may be used for longer deliveries, depositing into wheelbarrows or handcarts which are hoisted to the level of placing. Wheelbarrows are well suited for wheeling along the scaffolding on building work.

With any of the above methods jolting and vibration should be avoided as far as possible, for instance barrow runs should be made as smooth as possible. For longer deliveries, particularly in hot, dry atmospheres, it is advisable to cover the concrete with wet sacks or a tarpaulin. The high rate of evaporation from a heap of concrete is not always appreciated.

Dumpers. Dumpers are useful for transporting many constructional materials and are particularly useful for transporting concrete for small scale roadworks and for foundations and ground floor slabs on housing sites. Segregation or compaction frequently occurs when dumpers are used and care in transportation is particularly necessary.

Lorries. Lorries are commonly used in conjunction with central mixing plants for airfield and other work. They are loaded with 2 or 3 m^3 of concrete from a storage hopper of the type shown in Fig. 9.5. Reference is made to the use of side-tipping lorries for the transport of concrete in road construction in Chapter 21.

High-lift lorries have been developed for circumstances where it is possible and convenient to tip from a height of about 1·5 m.

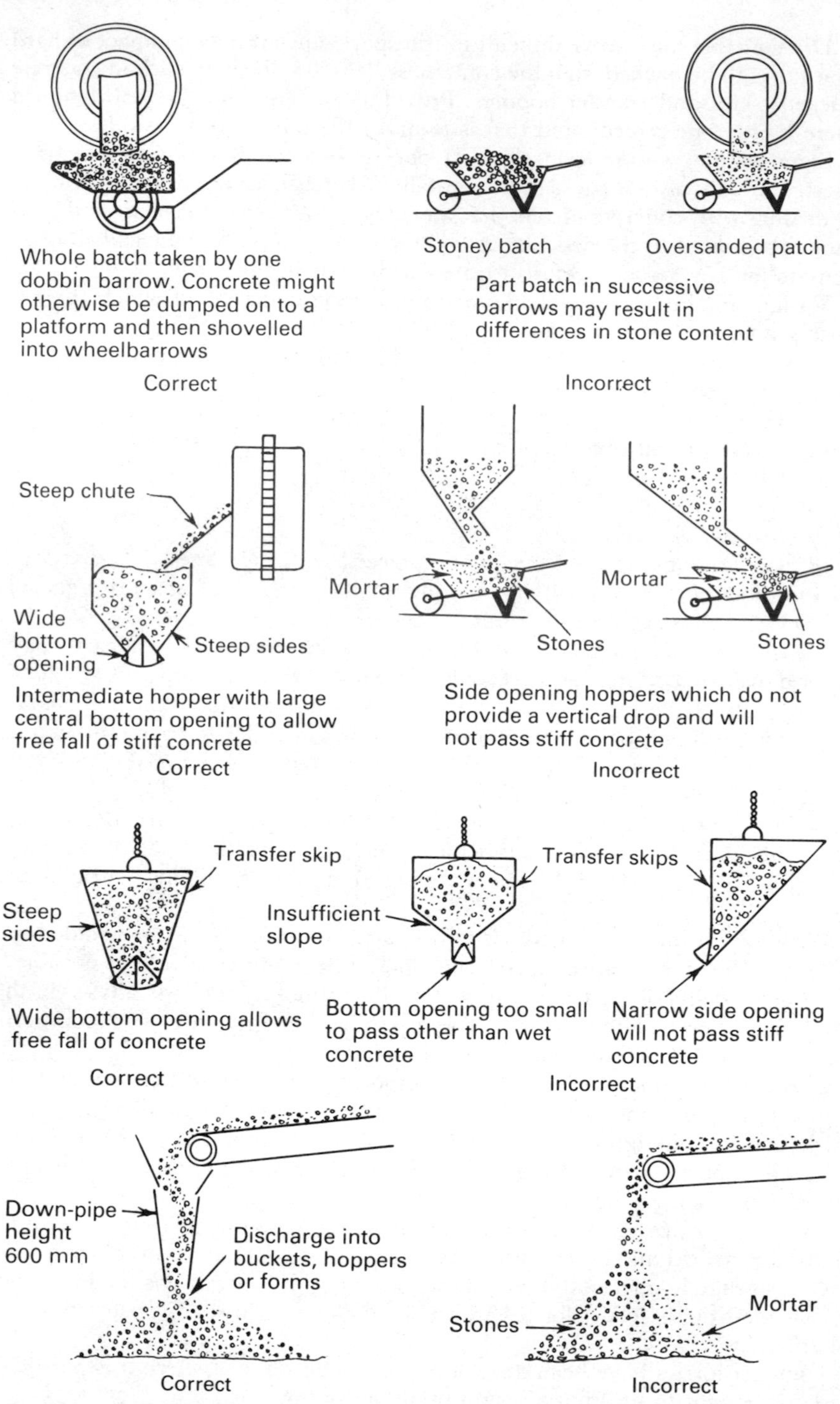

Fig. 9.11 Methods of handling mixed concrete

Fig. 9.12 3 m³ dump bucket, with compressed air operated discharge mechanism

With all types of lorry there is some risk of segregation of wet mixes and of compaction of dry mixes. In general, the slump needs to be between 15 mm and 75 mm if difficulties are to be avoided.

Evaporation of water from the concrete in hot, dry regions during transport can be quite serious, and the only alternative to increasing the water/cement ratio at the mixer, is to provide some cover to the lorry or other transporting medium, unless the distance is very short. Truck or transit mixers, which can deposit the concrete directly into the forms, have definite advantages in this type of climate, and the high cost of this form of mixer may be more than mitigated by the easing of other items in the placing procedure.

Steel skips and buckets. Well-designed steel skips and buckets, properly operated, afford a most satisfactory means of handling and placing concrete.

Transfer skips for use with the smaller mixers are usually made to hold about 0·5 m³. Two types are in use, a wedge-shaped skip with bottom opening, and a lay back skip which is allowed to fall over on its back for filling from a low level mixer discharge. The latter type has the advantage that the impact of concrete discharged into it does not result in compacted concrete immediately over the bottom opening, with resultant difficulty in discharging. With either of these types of skip it is essential that the bottom opening is large enough to allow the easy discharge of reasonably stiff concrete mixes. Any changes of direction or restrictions should be avoided, since these cause arching and difficulty in discharge, and the inevitable result on the site is that more water is added at the mixer. A suitable design of wedge-shaped skip is illustrated in Fig. 9.11.

Fig. 9.13 Mono-rail transporter being used for the placing of foundation bases and other work on a building site

Steel dump buckets for the transport of concrete are usually square or circular in plan, and hold about 3 m^3. Other sizes may of course be adopted to suit the scale of the work being done and buckets up to 6 m^3 in capacity were used in the construction of the Shasta and Boulder Dams in the USA. Discharge may be through manually or mechanically, usually compressed air, operated gates at the bottom of the bucket. Care is necessary to prevent excessive impact and possible damage to the formwork by releasing the concrete from too great a height. Some detail of a 3 m^3 bucket is shown in Fig. 9.12. Buckets are ordinarily transported by truck, derrick and cableway. Some types of hoist have a bucket which is elevated and discharged into a hopper at the level of placing. From the latter it is distributed by wheelbarrow or possibly by chutes.

Chutes. Long open chutes tend to cause segregation and drying of the concrete and are one of the most objectionable methods of transporting concrete. If chutes are used they should be designed so that the concrete is to some extent remixed at the lower end by passing down through a funnel-shaped pipe or drop chute.

Alternatively, they should discharge into a storage hopper, from which the concrete should be transported to the point of placing by wheelbarrows or other means.

The concrete mix should be workable and cohesive, if it is to flow satisfactorily down a chute.

Belt conveyors. Belt conveyors have very limited application in concrete construction. Occasionally a fixed conveyor between two points is useful, which in some conditions needs to be enclosed in a protective cover throughout its

Fig. 9.14 Concrete pump. Mobile versions of the same type are sometimes used

length. One type of conveyor consists of a series of short belts with swivel connections such that the discharge end can be moved about readily if a flat running surface is available for its wheeled support.

Belt conveyors tend to cause segregation on steep inclines; at transfer points; as the belt passes over the rollers; and as the concrete is discharged from the belt. A method of providing a certain amount of remixing may sometimes be necessary at the discharge end.

Mono-rail transporter. The mono-rail transporter shown in Fig. 9.13, is valuable on sites where relatively small quantities of concrete have to be transported over a considerable area congested with other work in process, and crossed by trenches and other obstacles. With suitable support the rail can be inclined so that concrete can, for instance, be conveyed to the upper floor of a two storey structure. The mono-rail skip is started on its journey by the mixer driver and is automatically stopped at the required location by a trip device.

Concrete pumps and pipelines. Pumping through steel pipelines is one of the most satisfactory methods of transporting concrete, particularly in congested areas. Concrete may be pumped through a straight horizontal pipe to a distance of 500 m or more, but curves, lifts, and harsh concrete materially reduce

Fig. 9.15 Concrete being discharged from pipeline

the maximum pumping distance. An allowance of 12 m should be deducted from the maximum horizontal pumping distance for every right-angle bend, and 2·5 m for every 0·3 m of head. Good workable concrete can be raised to a height of 40 m and more within a short distance of the pump.

An illustration of the most common type of static concrete pump is given in Fig. 9.14. The mixed concrete is fed into the conical hopper above the pump, from which it passes through an inlet valve into the pump cylinder during the retreating or suction part of the piston stroke. As the piston moves forward, the inlet valve closes and an outlet valve opens to allow the concrete to pass into the pipeline. In starting the pumping operations a quantity of mortar equivalent to the concrete with the coarse aggregate omitted, is fed into the hopper. This mortar is necessary to lubricate the pipeline, and about 1 m^3 is sufficient for 300 m of pipe.

The most usual size of pipeline is 150 mm, although larger and smaller pipelines are sometimes used. Outputs vary from 6 m^3 per hour for small machines up to 35 m^3 per hour for larger mechanical types of pumps.

A type of hydraulic concrete pump developed on the Continent provides an output of up to 70 m^3 of concrete per hour. The pumping unit is operated by high pressure water which is pumped into the cylinder behind the piston ram on the working stroke and sucked from it on the return stroke. Less maintenance costs are involved than with the older mechanical type of pump. However, there are risks of water leakage past the piston when wear occurs.

Mobile concrete pumps are used with relatively small bore pipes, mainly in conjunction with ready-mixed concrete. These pumps are mounted on a lorry or a trailer chassis (see Fig. 9.16). These pumps are particularly suited to the supply of relatively small amounts of concrete to sites where access is difficult. Flexible rubber piping, which can be handled quickly and easily is being used increasingly to replace partially or wholly the steel piping previously used. A disadvantage is a greater rate of wear and deterioration.

The pumps seen in Fig. 9.16 are fitted with a boom carrying the pipeline,

Fig. 9.16 Mobile concrete pump fitted with booms which facilitate delivery to various locations

giving a great deal of flexibility in delivering concrete to various locations.

Concrete for pumping must be well graded without any tendency toward segregation. The sand content may need to be as high as 40–43%, with 15–20% passing a BS 300 μm sieve and about 3% passing a BS 150 μm sieve.

Mixes richer than about 1 : 6 cement : total aggregate can usually be pumped without much difficulty. Leaner mixes require greater care both as regards the grading of the aggregate and the accurate proportioning of the mix. Weigh batched concrete at 1 : 9 cement : total aggregate, has been pumped successfully over about half the distances mentioned above.

A slump of 40 mm to 75 mm but not more than 75 mm appears most suitable. Segregated or harsh lean concrete is usually the cause of stoppages in the pipeline. Too dry or too wet a consistency, and interruptions which allow the concrete to stiffen are other causes of trouble. Concrete as delivered by pump is generally of uniform quality and practically free from segregation by reason of these requirements. In Fig. 9.15 the concrete is shown being discharged and spread to form a concrete floor.

If the concrete discharged from the pipeline should appear wetter than the concrete fed into the pump hopper, it is possible that water used in the pump to keep the pistons free from grit and so reduce wear, may be leaking past the glands. Good maintenance arrangements are essential when concrete pumps are in use.

Cleaning of the pump and pipeline at the end of concreting operations is very important, and for this purpose a plug or 'go-devil' is inserted at the pump end, and forced through the pipeline with compressed air. A bundle of

Fig. 9.17 A mobile pneumatic concrete placer of 0·15 m³ capacity having an output of 3 to 4·5 m³ per hour. For each batch of concrete air pressure is built up until the concrete is propelled along the pipe line

sacks or hessian forms a satisfactory plug. As the plug may be ejected from the end of the pipeline with considerable force, the pump operator should make sure that all persons are behind the mouth of the pipe during the cleaning out operation.

Pneumatic placers. Another means of conveying concrete through steel pipelines is by means of compressed air, using a pneumatic placer such as that shown in Fig. 9.17. The important distinction between pumping and pneumatic placing is that, while in the former case a continuous plug of concrete moves slowly along the pipeline, in the latter case separate batches of concrete are 'shot' at considerable velocity along the pipeline. In tunnel lining work, the discharge end of the pipeline is best kept just buried in the previously placed fresh concrete between the shutter and tunnel walls (see also 'Tunnels', p. 335). In almost all other circumstances a trap is fitted on the end of the pipeline.

Pneumatic placers vary in size from 0·1 m³ up to about 1·25 m³, the larger sizes seldom being used except in tunnel work. Placers involve much less capital outlay than pumps, but require considerable quantities of compressed air, a costly commodity, particularly when working at the limit of their range, which is of the same order as that of pumps. Although in some circumstances either a pneumatic placer or concrete pump may equally well be used, their different characteristics render each most suited to somewhat different circumstances. For instance, the pump is likely to be more suitable where considerable quantities of concrete need to be delivered at a steady rate into one area of the site. Pneumatic placers are satisfactory for intermittent conveyance of modest quantities, in which conditions pumps are likely to prove impracticable.

With pneumatic placers highly workable concrete of 100–150 mm slump or even more is necessary, and with all mixes there is a tendency for some segregation to take place, though with suitable mixes this need not occur to a detrimental extent.

Small pneumatic placers, fitted with a suitable nozzle, can be used for the pneumatic spraying of mortar or concrete with small aggregate, a process often used in the repair of old concrete, and sometimes in new construction. The results may be somewhat variable.

Cableways. It is primarily in the construction of concrete dams that cableways are used for transporting concrete in skips or buckets, the latter being of up to 6 m³ capacity or more on large projects.

Transit-mixers and transit-cars on rail tracks. In tunnel lining construction the concrete is generally transported along a rail track from the batching plant outside the tunnel by means of transit cars of 1·5 to 6 m³ capacity. If used as transit mixers, the batched materials are mixed with water during the journey to the point of placing; though the capacity of the horizontal axis drum is then less than if it is used for concrete already mixed.

REFERENCES

1 Murdock, L. J., The efficiency of concrete mixing plant, *J. Inst. Civ. Engrs*, **31**, Nov. 1948.
2 *Concrete Manual*, 1st ed. 1938, 8th ed. 1975. US Bureau of Reclamation.

ADDITIONAL REFERENCES

Illingworth, J. R., *Movement and Distribution of Concrete*. McGraw Hill, London. 1972
Crowe, F. W., Successful small-line pumping. *Concrete*, Sept. 1968, pp. 380–386.
Kempster, E., *Pumpable Concrete*. Publication CP 29/69, Building Research Station.

10
Ready-mixed concrete

READY-mixed concrete was introduced into Great Britain in 1930 when the first plant was installed at Bedfont, Middlesex. The initial growth was slow and it was not until after 1950 that a significant expansion occurred. The ready-mixed concrete industry is now a major sector of the construction industry and it is notable that over half of all site placed concrete is now supplied from ready-mixed concrete depots. In 1976 the number of depots supplying concrete in Great Britain had reached about 1200, and the majority of these had outputs of 20 000 to 30 000 m^3 per year.

Factors affecting choice of ready-mixed concrete

On small sites, it is generally more convenient and cheaper to obtain concrete from a ready-mixed concrete supplier than to mix on site. The situation is different on big sites such as motorways, airfields and power stations, where large quantities of concrete are required and where there is adequate room for the contractor to install his own batching and mixing plant. In the majority of these cases, the contractor should be able to produce concrete more economically himself and he also has the operation completely under his own control.

There are, however, many types of sites where careful consideration has to be given to the choice of whether to use ready-mixed concrete or site-mixed concrete. Relative costs which are meaningful can only be obtained after careful planning and these must be considered along with such items as supplies of aggregates, access to site, available space on site, required output of concrete, mixing methods and methods of handling concrete on site. A true cost comparison between site-mixed concrete and ready-mixed concrete can only be achieved by considering the whole chain of events up to the point of placing, taking the quantity placed in a given time as the basis of comparison(1). The final choice will usually depend on the relative costs, but this is not always so.

Many sites, particularly building contracts in large towns and cities, are very short of space and on occasion, there is simply no room for a site-mixing plant and stockpiles of aggregates. In these circumstances, there can be no hesitation in choosing ready-mixed concrete. Adequate access for the truck mixers is essential on any site and there must also be room to manoeuvre in the area of the concrete placing operations. Provided suitable access is available, a point in favour of truck mixers is that they are often able to discharge the concrete directly into place (usually foundations), thereby obviating the necessity for site handling plant.

There is often justification for using both ready-mixed concrete and site-mixed concrete on a site. For example, in the early stages of a contract, concrete may be required before the site plant has been installed. Ready-mixed concrete will therefore be used for the early work before site-mixed concrete can be produced. There is also the situation when the quantities of concrete required in a given time are larger than the site-mixing plant can cope with. Clearly the use of ready-mixed concrete is desirable to supplement the site production in such cases.

There are some situations where ready-mixed concrete is an obvious choice. The placing of large pours of concrete in one continuous operation requires large scale production facilities for a short period and may well require the full resources of at least one and probably two large ready-mixed concrete plants for a period of up to about 24 hours. Another situation favouring the use of ready-mixed concrete is the concreting of large bored piles which requires large but spasmodic amounts of concrete.

The use of ready-mixed concrete brings its own particular problems which should be recognized and catered for. In particular, the responsibility for the quality of the concrete is usually divided between the ready-mixed concrete supplier and the contractor who places it. This can lead to confusion and the client's engineer responsible for the contract may find it less easy to ensure that adequate control of the concrete construction is maintained.

Another important problem is the organization of work on site. Particular attention must be given to the day to day planning of concrete requirements and care is needed to ensure that any shortage, surplus or delay is avoided as far as possible. Breakdown of plant and unforeseen difficulties can, however, upset the most careful planning and such situations are sensitive to the conditions of obtaining a ready-mixed concrete supply. With site-mixing, the supply of concrete can be stopped and started as the immediate situation demands.

It is not always possible to obtain the required type of concrete from a ready-mixed supplier and in these cases it will be necessary to produce the concrete on site. Mixes in this category include those which require a special cement or aggregates.

Operation of ready-mixed concrete plants

There are two main types of ready-mixed concrete depot:

(1) A central batching and mixing plant where all the materials are batched and fed directly into a mixer (which may be a drum mixer, forced action pan mixer, or continuous paddle mixer). The concrete is thoroughly mixed before discharge into a truck mixer (or other vehicle) for transport to site.

(2) A dry batch plant which feeds the batched materials to truck mixers for mixing and transport to site (see Fig. 10.1). It is preferable for the water to be added to the truck mixer at the plant, but some suppliers add water when the truck has arrived on site.

The dry batch plant is usually the cheaper of the two to operate and about two-thirds of all the depots are of this type. The central mixing plant, however, enables better control to be maintained on the quality of the concrete, particularly its workability. One of the reasons for this is that the batcherman

Fig. 10.1 Dry batch plant at a ready-mixed concrete depot, with a truck mixer in position for being loaded

can inspect the concrete before it is discharged into the truck mixer. Central mixing also enables a much wider range of concrete mixes to be mixed quickly and efficiently particularly when forced action pan mixers are employed. The production of dry lean concrete is a typical example. Moreover, the method of feeding the mixer is less critical with the central mixing plant since on dry batch plants, the materials should be fed simultaneously into the truck mixer by ribbon or sandwich loading whilst the drum is turning. A further merit of the central mixing plant is that it enables a customer to collect concrete himself in a lorry.

Truck mixers are basically free fall mixers mounted on a truck chassis. When used to mix dry batched materials, they generally revolve at 10–15 revolutions per minute, but when used as agitators, the drum is required to turn only very slowly, one revolution per minute being adequate. The drum is driven by power take-off from the truck engine or by a separate donkey engine. The size of truck mixers varies, the amount of concrete carried generally varying between 3·5 m^3 and 7·3 m^3.

The limit to the maximum load of concrete which can be carried on public roads in Great Britain is determined by the official weight regulations. The result of these regulations and subsequent refinements in truck design has been that three axle trucks carrying 6 m^3 concrete have become the most common. From a technical point of view, the operating range of a truck mixer is limited by the requirement, based on BS 1926 : 1962 [2], that the discharge shall be completed within two hours after contact between cement and water, either as mixing or water contained in the damp aggregates.

Quality control

The principles of good quality control are the same for ready-mixed concrete as they are for any other concrete. Some aspects however acquire greater importance with ready-mixed concrete mainly because the production of the concrete is an off-the-site operation and the engineer and contractor cannot be sure that the specified requirements of the concrete are being met when it is delivered on site.

Since the industry's profit is obtained solely from the sale of concrete, it is vital that maximum economy is exercised in the use of cement which is the most expensive ingredient. Experience has shown a tendency for ready-mixed concrete suppliers to reduce cement contents below that advisable for the production of dense durable concrete. The concrete may have an adequate cube strength to satisfy the specification, but it may be difficult to compact on site, with a resulting poor finish or even porous concrete. A minimum cement content should be specified in addition to the cube strength requirement.

In theory, concrete should be more uniform and of better quality from a large concrete plant which produces concrete all day. However, this is not always so due to errors by the plant operators when making frequent changes in batch weights to suit different clients. Also deliveries may be made to the wrong discharge point, possibly the wrong site, but more likely the wrong point on a large site requiring different qualities of concrete at a number of points at the same time. Such errors may be serious. On one site for instance, a batch of concrete with a minimum specified strength of 17·5 N/mm² (2500 lb/in²) was deposited in a structural member for which the minimum required strengths was 31·5 N/mm² 4500 lb/in²). The supervision on site and quality control measures should be designed to overcome these problems.

In 1968 the British Ready Mixed Concrete Association (BRMCA) introduced an authorization scheme for companies in the association. The scheme defines minimum requirement for (a) personnel, (b) materials, (c) plant and equipment, (d) documentation and (e) quality control procedures. Any depot which complies with sections (a) to (d) receives a certificate showing that it is a 'BRMCA Approved Depot'. If in addition it complies with section (e) it receives a certificate which indicates it is a 'BRMCA Approved Depot, with Quality Control Procedures'.

The authorization scheme is aimed primarily at quality control and does not cover the standard of service. There is little doubt that the introduction of the scheme has resulted in an overall improvement in the quality of ready-mixed concrete, although many criticisms have continued to be levelled at certain suppliers and the quality of their concrete at times. For example, the documentation section of the scheme states that the supplier shall provide the purchaser with a delivery ticket for each batch of concrete on which, amongst other items, should be given the minimum weight of cement per m³ of concrete where this has been specified. However, some suppliers appear very reluctant to supply this information. It can also be difficult to trace the weights of materials batched for any specific load in the records of a plant in spite of the fact that this is a requirement of the scheme.

Disputes over the quality of ready-mixed concrete frequently involve the questioning by the supplier of the standards of sampling and testing. It is, therefore, important that the specified method of sampling and testing is followed in all details. This should prevent undue emphasis being given to this questioning of the validity of test results, as a convenient way of obscuring the fact that the concrete was of poor quality.

Sampling and testing should normally be done strictly in accordance with BS 1881 : 1970. This specifies that sampling of concrete from a mixer should be carried out by taking not less than four increments at equally spaced intervals, i.e. when four increments are required, about the time one-fifth, two-fifths, three-fifths and four-fifths of the concrete have been discharged. The increments are then mixed together to form a representative sample for the whole batch. In practice this means that all the concrete in the truck mixer must be discharged before a representative sample is obtained. If a workability test on this sample then shows it to be unsatisfactory, it is clearly too late to be able to send it back.

Since at present there is no standard method for measuring the workability of the concrete before discharge of the complete load, it is recommended that the following procedure is adopted: after making sure that the concrete has been uniformly mixed, take a sample, made up of four portions, from the first 0·5 m³ of concrete discharged, and carry out a slump (or compacting factor) test on the sample. If the result complies with the specified requirements then the load should be accepted. If the result falls outside the limits, a further sample should be taken from the second 0·5 m³ of the discharge and if this is satisfactory the load should be accepted. If not, the concrete may be regarded as being outside the specification limits.

This procedure is one that has been accepted by many clients, suppliers and contractors.

The workability of ready-mixed concrete is generally dependent on the judgement of the batcherman at the depot or the truck driver when water is added on site. This judgement is based on visual inspection, and it appears that this is still the most accurate way of controlling workability. Unfortunately, it still gives rise to considerable deviation from the specified workability in a significant proportion of deliveries. An important factor is that the batcherman should have a full view of each batch of concrete before it leaves the plant.

Ready-mixed concrete producers need up-to-date information on the strength of the cement they are using. There has, therefore, been considerable emphasis given to accelerated curing of cubes in order to obtain an indication of strength within 24 hours. The results are used to predict 28 day strength

Fig. 10.2 Truck mixer discharging concrete directly into a foundation slab

which are then coupled with a method of 'cusum' analysis[3] which provides information on which changes in cement content are based.

The production of good quality concrete is clearly dependent on adequate mixing time. There is a tendency to ignore this requirement in the interests of production and a fast turn-round of truck mixers. The mixing time in a truck mixer should be at least seven minutes at a mixing speed of about 12 revolutions per minute. If this time is shortened there is a danger of surface finish and segregation problems.

Handling ready-mixed concrete on site

Careful planning is needed in advance of ready-mixed concrete being supplied to a site. When constructing foundations and works at ground level, ready-mixed concrete trucks can frequently travel to a position alongside the work and discharge the concrete directly into place (Fig. 10.2). For deep foundations it may be necessary to use one or more chutes. For direct discharge in this way, there must be clear access for the truck mixer and there must be room for it to manoeuvre. The average truck mixer is 8 m (26 ft) long by 2·5 m (8 ft) wide by 3·5 m (11·5 ft) high and requires a turning circle of at least 15 m (49 ft). The ground must be firm and capable of carrying a fully loaded truck mixer which will generally weigh about 24 tonnes. Roads and access points must be capable of carrying this load, even in wet conditions, and it is clear that they should be stronger than is necessary for the usual site traffic. It will generally be more economic to provide an adequate access road

Fig. 10.3 Ready-mixed concrete being placed by skips during construction of a bridge deck

at the start of a job rather than add to a weak surface as it fails and after trucks have been bogged down. It is recommended that, for most conditions, 200 mm (8 in) of fully compacted hardcore or equivalent be provided.

Particular attention is required when truck mixers run close to the side of an excavation. Proper shoring of the excavation may be necessary to prevent collapse of the sides due to the weight of the vehicle. A truck mixer can discharge concrete at about 0·5 m³ per minute and when it is possible to discharge a full load continuously, the total time required to do so is about ten minutes. This situation however, is more the exception than the rule.

When the concrete is required to be placed above ground level, the truck mixer can discharge at a convenient point into a skip which is lifted by a crane to the required position (Fig. 10.3). Several skip loads will result from each truck mixer delivery and on occasion the truck mixer will have to wait on site a considerable time to discharge all its load. It is clearly desirable, therefore, to use as large a skip as possible or alternatively to use two skips. If a purchaser causes a truck to be seriously delayed, he may be faced with a bill for demurrage.

Alternatively the truck mixer may discharge into a concrete pump which will convey the concrete through a pipeline to the required position (Fig. 10.4). Several varieties of small bore mobile concrete pumps have been developed in recent years and the larger types are capable of pumping up to heights in excess of 50 m (164 ft). The success of this technique is dependent on the ease with which the pipeline can be laid out and moved from one position to another. The pipeline must, therefore, be as light as possible and the

Fig. 10.4 Concrete from a truck mixer being pumped into place using a mobile concrete pump fitted with a boom

diameter is usually restricted to 75 mm (3 in) or 100 mm (4 in).

If the concrete is to be pumped it is essential that the ready-mixed concrete supplier is informed, since he must provide a concrete mix which is pumpable. This may mean that he must add more cement to his concrete mix or adjust his mix design in some other way. In any event, there is every likelihood that the cost of the concrete will be increased and it is essential that agreement on all aspects of the concrete supply and pumping operations is reached before concreting starts.

Close liaison between the supplier and purchaser is an essential requirement for most situations when ready-mixed concrete is used. Suppliers have to plan a daily schedule for their clients and it is necessary for them to know the amount of concrete required on each site and the rate at which it is required. Difficulties can occur in trying to meet the particular requirements on a given site due to traffic congestion on the roads or due to breakdowns at the plant. Also, if anything goes wrong with the trip to the first customer such as a delay in unloading, the delivery schedule gets out of phase and the second customer's programme of work is disrupted. Purchasers should ensure that they are completely ready to deal with the concrete when it arrives, otherwise the suppliers time will be wasted and delays will occur. These factors emphasize the need for both parties to maintain a close liaison and to work together in a spirit which recognizes each other's problems.

Fig. 10.5 Ready-mixed concrete being used for a large foundation slab pour (2344 m^3 of concrete). Four mobile concrete pumps were employed for placing the concrete

Close liaison and careful planning is vital when large pours of concrete are placed using ready-mixed concrete. In the case of a large pour involving the continuous placing of 2344 m^3 of concrete in a large foundation slab, it has been reported [4] that very detailed planning resulted in the successful completion of the pour within 22 hours. Two large plants supplied the concrete and two further plants were fully stocked in case of emergency. Both plants and truck mixers were planned to operate at no more than 75 per cent efficiency and the police provided the services of traffic wardens to ensure a smooth flow of truck mixers. During the run-up to the operation, which started at 6.00 pm on a Friday evening, staff were kept continuously informed of the methods of planning and each man was made aware of his own part in the operation. Four concrete pumps were used to place the concrete (Fig. 10.5) and two further pumps were provided on a standby basis.

REFERENCES

1 ILLINGWORTH, J. R., *Movement and Distribution of Concrete*. McGraw Hill. London, 1972.
2 BS 1926 : 1962, *Specification for Ready Mixed Concrete*. British Standards Institution.
3 *Code for Ready Mixed Concrete*. British Ready Mixed Concrete Association, Shepperton. 1975.
4 SMITH, G. A., A large concrete pour: Newcastle-upon-Tyne, September 1974. *Proc. Instn. Civ. Engrs*, Part 1, **58**, August 1975, pp. 395–399.

11
Preparation and Joints

Preparation for placing

THE preparations necessary before placing commences are dependent on the type of construction and the position in which the concrete is to be deposited. For roads and foundations it is necessary to prepare the soil surface or sub-base, while in structural work it is essential to clean construction joints thoroughly in order to obtain a good bond between the fresh and the hardened concrete.

Foundations

Methods of preparing foundation surfaces depend on the type of sub-soil, and on the degree of bond required if the concrete is to be placed in contact with rock.

Clay, loam, hoggin and similar materials which have been consolidated by rolling or punning, are commonly used as a foundation for concrete. The surface should be clear of debris, mud and pools of water. In dry weather the sub-base should be thoroughly moistened.

Porous materials such as sand, gravel, and ash, should be thoroughly consolidated and covered with waterproof paper, polythene sheeting, or similar material to prevent absorption of water from the concrete. It is good practice to lay a 50–75 mm (2–3 in) thick layer of blinding concrete over the site in the early stages of construction. This provides a good working platform and prevents contamination of the foundation concrete by the underlying ground. This blinding concrete need only be a weak mix, a typical cement content being 150 kg/m^3.

When it is necessary to obtain a good bond on to a rock foundation, as for instance in the construction of dams, the rock surface should be well roughened, and thoroughly cleaned of all loose rock, dirt, soft pockets and seams, and other foreign material. Cleaning is accomplished by air and water jets, sand blasting, etc. and thorough washing with water and stiff brooms. Immediately before placing, a layer of mortar about 20 mm thick and of similar mix and consistency to that contained in the concrete to follow, should be thoroughly worked with stiff-brooms into all irregularities in the surface.

Types of joints

Joints are included in concrete structures to provide either a convenient stopping point in the sequence of construction or to permit relative movement within the element or structure. The former type are generally referred to as construction joints whilst the second type embrace a group collectively referred to as movement joints. They may be defined as follows:

(1) *Construction joints.* These are joints where fresh concrete has been placed on or against concrete which has already hardened. They are sometimes referred to as daywork joints. The main requirement is that such joints should enable adjacent sections to act monolithically in the completed structure, and it follows, therefore, that there is no provision for relative movement between those adjacent sections.

(2) *Movement joints.* The three main types are:

(a) *Contraction joints.* These joints form a deliberate discontinuity between both the concrete and reinforcement, on either side, in order to permit relative contraction. There is no initial gap at the joint line.

(b) *Expansion joints.* These joints also have a deliberate discontinuity between both the concrete and reinforcement, on either side, and in addition, an initial gap (often filled with compressible material) is provided to accommodate subsequent expansion. They also accommodate contraction.

(c) *Sliding joints.* These are movement joints at which special provision is made to facilitate relative movement between adjacent sections in the plane of the joint.

Examples of these main types of movement joints are shown in Fig. 11.11. Information on the design and construction of movement joints for roads and ground floor slabs is given in Chapters 21 and 25.

Construction joints

A construction joint is a concrete surface which has hardened because of a limitation or delay in construction so that fresh concrete cannot be integrally incorporated with it.

Construction joints are generally necessary because there is a limit to the amount of concrete that can be placed in one operation. However, they may also be required at certain locations in a structure to avoid shrinkage cracking which could otherwise occur with a long continuous casting operation.

Under laboratory conditions it has been shown that fresh concrete can be bonded to hardened concrete in such a way that the strength of the joint can be almost as good as the strength of the concrete itself[1]. Laboratory tests, however, do not necessarily bear any relationship to work done on site under completely different conditions, where the nature of joint preparation and the efficiency of the completed joint is likely to be affected by:

(1) The size of the joint.
(2) The presence of reinforcement and water bars.
(3) Whether the joint is horizontal or vertical.
(4) The degree of restraint or freedom of the members on each side of the joint.
(5) Workmanship.
(6) The required surface appearance.
(7) The degree of bond required.

(1) *Size of the joint.* The specified joint preparation is more likely to be achieved on sections up to about 300 mm square, such as in columns and

beams, than with sections of walls, where in the case of both vertical and horizontal joints the section may be 300 mm by 6 m or even greater. Placing of the fresh concrete at horizontal joints in columns and walls where the height of pour may be up to 4–6 m calls for the exercise of great care in both placing and compaction.

(2) *The presence of reinforcement and water bars.* The preparation of hardened concrete is made more difficult by the presence of projecting reinforcement, and particularly when water bars are present, especially if the preparation is a manual one. Furthermore, the placing of the fresh concrete is likely to make it difficult for full and adequate compaction to be achieved because of the possible restriction of the reinforcement on the flow of concrete. The degree of compaction of the fresh concrete around water bars has to be of a higher standard and quality than may be acceptable for joints without water bars, in order to prevent the formation of paths of weakness along which the water or contained liquid could flow.

Where congested reinforcement makes it difficult to fix stop-ends, expanded metal mesh, suitably framed, may be found suitable instead. This can often be left embedded in the concrete, provided a good bond at the joint is not essential. In general, however, removal of the mesh is preferred. The edge of the expanded metal should be kept about 40 mm away from the face particularly in those cases where the joint line and appearance are important, otherwise the corners or arrisses along the face are likely to be broken off when the expanded metal is removed.

(3) *Horizontal and vertical joints.* The placing of fresh concrete at a vertical joint requires a different technique from that at a horizontal joint. For example, some specifications require a thin coat of sand/cement mortar to be applied to the surface of the prepared hardened concrete before the fresh concrete is placed; this can, under some circumstances, be achieved with horizontal joints but is not practical for vertical joints having cross-sectional areas exceeding, say, 300 mm by 600 mm high. In addition, the placing of mortars on hardened concretes on site is made more difficult and sometimes impossible by the fact that the formwork has to be erected initially. Placing mortar along a horizontal joint at the bottom of shutters only 2 m high is almost an impossibility, and for vertical joints must undoubtedly be so.

(4) *The degree of restraint.* The effectiveness of a construction joint is affected by the tendency of the jointed parts of concrete to move relative to each other. Stresses, usually tensile, at a joint are likely to arise due to temperature changes during the setting and hardening of the concrete and to subsequent thermal or shrinkage movements. The magnitude of such stresses will be a function of the restraint, or lack of restraint, of the jointed parts. For example, if the two portions of a beam are jointed with the ends of the beam free to move on sliding supports, then it is unlikely that stresses due to thermal or shrinkage movements of any significant magnitude will be realized across the joint. If, on the other hand, the ends of the beams are fixed in position, tensile stresses will be set up at the joint, as well as in the concrete member.

There may also be a differential form of restraint in the case of a wall panel which is cast on top of hardened concrete. At the horizontal joint at the bottom, shrinkage of the concrete is restricted by the frictional resistance of the hardened base and this may cause one or more near-vertical cracks to occur at localized places of weakness. Towards the top of the wall panel, the restraint reduces to a negligible value and shrinkage can take place without the formation of cracks.

(5) *Workmanship.* In the laboratory the preparation of the joint and the subsequent placing of concrete are likely to have been done by a technician working under close supervision. He is likely to have been interested in the work. Construction joints on site will inevitably be prepared by unskilled labour, usually disinterested, and with relatively far less supervision. The quality of workmanship and the care which is exercised in joint preparation and the subsequent placing of the concrete will affect the efficiency of the joint.

(6) *Required surface appearance.* However carefully the preparation of the joint and the placing of the concrete are done, the joint line will always show. Therefore, for joints which will be exposed to view, particularly where a high quality finish is specified, it is preferable for the joint to be featured. Chapter 17 describes suitable techniques.

(7) *The degree of bond required.* There seems little point in preparing a joint to give a high degree of bond in compressive members, provided that the joint is capable of transmitting the compressive forces.

Construction joint preparation

The following points are those generally considered to be good practice in the preparation and placing of concrete at construction joints:

(1) The best results are achieved by simplicity of construction, and joggled joints are best avoided.

(2) All foreign material such as laitance, dirt, oil, wood chippings, shavings, and wire tie clippings must be removed.

(3) Roughening of the joint face is essential if a good bond is to be achieved. The basic requirement is that the concrete face must be clean, free from laitance, and slightly roughened to show the tips of the coarse aggregate.

An effective way of roughening the surface of a horizontal construction joint is to brush off the laitance about two to four hours after compaction of the concrete. In this way, the coarse aggregate is slightly exposed, but care is needed to avoid dislodging pieces of the aggregate. A gentle spray of water is sometimes used to assist removal of the laitance. If the roughening of the surface is left until the following day or later, the concrete will have hardened and will require a more vigorous treatment. A roughening treatment using a stiff wire brush may be suitable if the concrete is not too hard but otherwise it will be necessary to scabble the surface mechanically by small percussion power tools. Sand blasting may also be used on a hardened surface but is usually only suitable when large areas have to be treated.

If it is considered necessary to roughen the surface of vertical construction joints, it is preferable to do so soon after the removal of the stop-end, whilst the concrete is still green. If the stop-end is removed the morning following concreting, it should be possible to remove the cement skin to a depth of about 2 mm using a wire brush. If treatment is delayed beyond this time, scabbling of the surface or sand blasting, as described for horizontal joints, will be necessary. The use of retarders on the face of stop-ends can ease the effort required to achieve a rough surface, but a thorough clean-up is necessary and this may be difficult among congested reinforcing bars.

A roughened surface may also be obtained by manual hacking using chisels, but this is a slow process and often tends to be done in a rather half-hearted and unsatisfactory way. There is also a danger of damaging immature concrete by weakening the bond of the coarse aggregate within the matrix.

In all methods of surface roughening, it is necessary to remove all loose particles, preferably by a jet of water, provided this does not cause damage or serious inconvenience to other work below. An air jet provides a reasonable alternative.

(4) The face of the hardened concrete should be kept dry. Research under laboratory conditions has shown that the bond at a joint is strongest when the surface is dry and not coated with mortar[2].

(5) Opinions differ as to whether or not a layer of mortar applied to the surface of the hardened concrete is necessary. If used, the mortar layer should be well worked or scrubbed into the old surface and the first layer of the new concrete should be well tamped into the mortar, especially at the corners, to ensure a good contact.

(6) Particular attention should be given to the compaction of the concrete adjacent to the joint. In the case of a horizontal joint (as, for example, at a kicker), the first layer of concrete at the bottom of the lift should be spread uniformly to a thickness of about 0·5 m and then thoroughly compacted by inserting poker vibrators at about 0·5 m centres (see Chapter 13).

(7) Any cement mortar which has leaked under imperfectly fixed vertical stopboards should be chipped off as soon as possible after removing the formwork. If this is not done, it is liable to spall off later.

(8) The leakage of mortar at the bottom of a lift and at vertical joints may be prevented by the use of a sealing device, such as a foamed plastic strip, compressed between the inside face of the formwork and the previously placed concrete.

(9) The concrete engineer should make a final inspection just before the concrete is placed to make sure that all foreign substances have been removed, that the forms are correctly positioned and tight, and that the reinforcement is accurately and firmly fixed.

(10) Wet and badly graded mixes which readily segregate and 'bleed' tend to produce excessive quantities of mortar and laitance at the top of a lift, as illustrated in Fig. 11.1, and should be avoided. Examination of water-retaining structures has shown that leakage is as likely to occur through the porous layer of laitance at the top of a lift as at the joint itself. The amount of laitance is least when the concrete, especially in the uppermost portion of the lift, is the stiffest that can be compacted properly, and when the surface is worked as little as possible after placing.

Construction joints are potentially the weakest part of the structure, and every effort should be made to ensure that a good joint is obtained. Many concrete buildings are disfigured by the careless preparation of construction joints or lift planes for further placing, especially if they are positioned indis-

Fig. 11.1 Segregation resulting in a layer of mortar with very little coarse aggregate at the top of the pour

Fig. 11.2 Concrete breakwater made with blastfurnace slag aggregate. After 60 years' exposure to sea action cracking has occurred along the construction joints

Fig. 11.3 Severe blemishes at a construction joint caused by leakage and movement of the formwork away from the surface of the previous lift. In this case honeycombing at the joint has been caused by the leakage of grout. This could have been prevented by the use of a compressible sealing strip

criminately without regard to other architectural features. Examples of defective and badly placed construction joints are given in Figs. 11.2, 11.3 and 11.4.

In some cases, e.g. on earth-retaining walls, moisture percolation through an unbonded or poorly constructed joint can lead to unsightly efflorescence on the face, caused by the deposition of salts (see Fig 2.9).

Position of construction joints

In walls and columns which will be exposed to view, horizontal construction joints should be arranged at levels and positions where they will blend in with the general architectural appearance and should preferably be featured. Construction joints can be masked by such devices as the lines of tiles on Twickenham Bridge (Fig. 11.5), and the formation of grooves or projections as in Figs. 11.6 and 11.7. Note that in the right-hand example in Fig. 11.6, the line of the joint is preferably near the top of the groove.

Vertical joints should be formed with properly constructed stop boards, preferably at positions where the length of joint is shortest or where a symmetrical pattern can be arranged. It is bad practice to allow the concrete to take a natural slope as has been done in Fig. 9.1.

It is not possible to give general rules for the position of construction joints covering the whole field of reinforced concrete construction. It is generally accepted that joints in beams and slabs should be formed at the points of

Fig. 11.4 Disfigurement caused by construction joints. The sparrow-pecked finish has not masked the joints

Fig. 11.5 Use of lines of red tiles to mask lift planes. The surface is bush-hammered gravel concrete

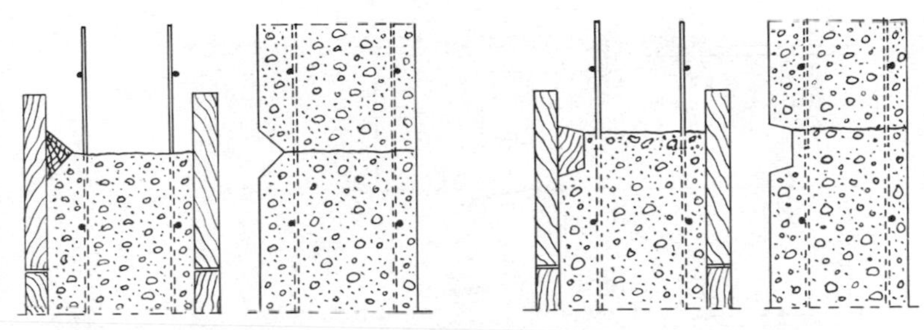

Fig. 11.6 Formation of grooves to mask construction joints. The fillet should be left in until completion of the adjacent pour

Fig. 11.7 Formwork joints deep set to give a panel effect for the in-situ wall of a church at Oslo

minimum shear which occur at the centre of the span or within the middle third. However, recent research work[3] has shown that structural members incorporating construction joints, made with a roughened surface, can behave as monolithically cast members. Tests made on beams with construction joints located in a region of high shear and low bending produced load/deflection characteristics and ultimate moments similar to those of tests on monolithic beams, provided the joint surfaces were roughened. Moreover, the modes of failure were identical. It was concluded that the interlock across the roughened surfaces was sufficient to prevent the beams failing at the joints.

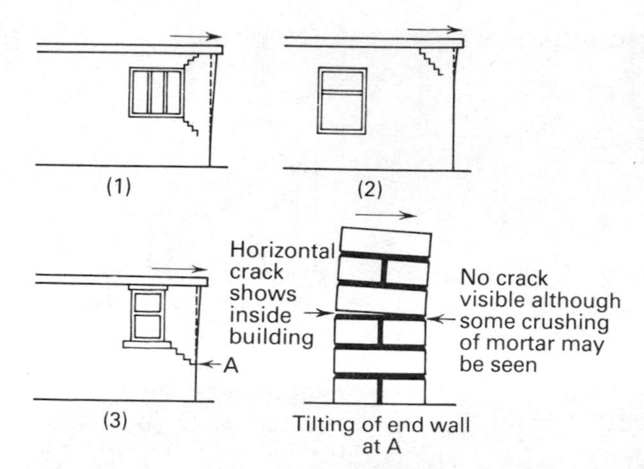

Fig. 11.8 Cracking of end walls of buildings due to the expansion of the reinforced concrete roof

It has been suggested, therefore, that it should be possible for joints in continuous beams to be located over columns and for joints in floors to be positioned on the centre lines of their supporting beams, instead of within the middle third of the span.

Where problems arise regarding the positions of joints in the structure, reference should be made to the designer concerning these details.

Movement joints in buildings

There is no well-established practice for providing movement joints in reinforced concrete structures. A common length between movement joints in a building is 30 m, sufficient reinforcement being used intermediately to control cracking. Wall panels are sometimes constructed to fit into keyways in the vertical columns, leaving only the wall beams subject to movement. With the provision of enough temperature reinforcement to prevent the formation of visible cracks, the distances between movement joints have been in some cases extended to 60 m or 90 m.

When used, movement joints in buildings should completely separate the structure from top to bottom, and should preferably be made by means of a double column or beam. The joint should be waterproofed, the method depending on the water pressure to be resisted, if any.

The thermal expansion of concrete roofs is a common cause of damage to buildings, usually by pushing of parapets and end walls as illustrated in Figs. 11.8, 11.9 and 11.10.

Cracking is also liable to occur, either through the plaster at ceiling level, or of partition walls near ceiling level, particularly near the ends of a long building.

Methods which may be adopted to prevent the defects indicated are:

(1) to allow the roof slab to slide on the supporting walls, with suitable breaking of the plaster finish at the junction;

Fig. 11.9 Cracking of brickwork at the end of a long concrete beam due to thermal movements. An expansion joint should have been provided at the end of the beam, and provision made for sliding of the beam on its brickwork seating

Fig. 11.10 Cracking in brickwork due to movement in reinforced concrete slab above

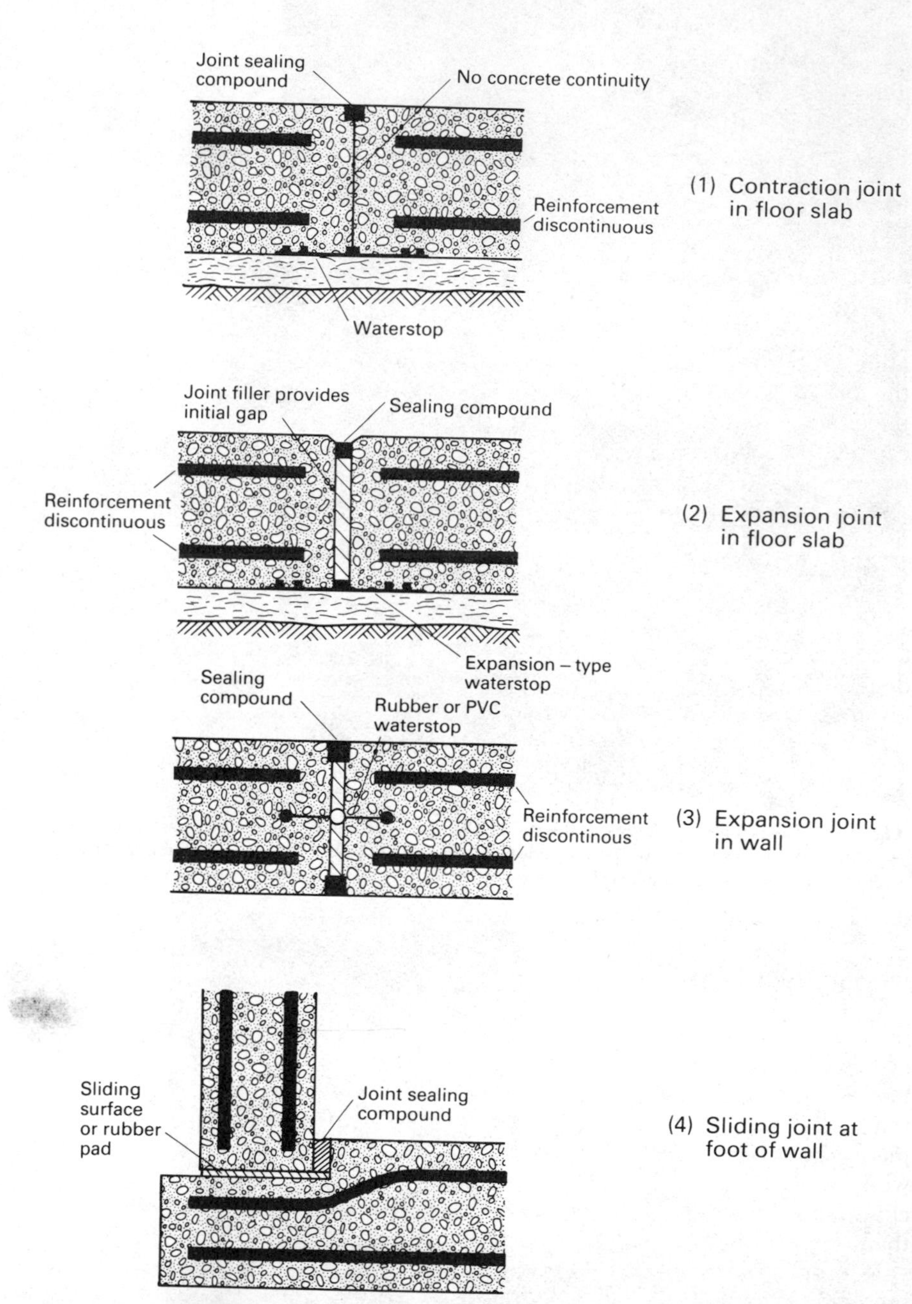

Fig. 11.11 Examples of movement joints, more particularly suited for water-retaining structures

(2) the incorporation of suitably designed expansion joints at convenient intervals;

(3) in conjunction with (1) and (2), the use of a roof covering having a high thermal insulation and the whitening of the roof surface in order to reduce temperature movements.

Movement joints in water-retaining structures

All joints in a water-retaining structure, whether it be a dam, a reservoir, a water tower, or any other structure, are a potential source of leakage, and their design and construction require special attention. They have to accommodate repeated movement of the structure resulting from changes in temperature or moisture conditions, without loss of watertightness. At the same time the joint must provide for the exclusion of grit and debris. This is achieved by the use of a joint sealing compound, which is an impermeable ductile material capable, ideally, of adhering to wet concrete throughout the range of joint movement. This latter requirement is difficult to achieve, so that the use of a sealing compound alone should not normally be relied upon to give watertightness. Useful information on types of sealing compounds and their application is given in reference 4.

In order to achieve watertightness, water bars of the type shown in Fig. 11.11 are nearly always incorporated at movement joints and frequently at construction joints. Other arrangements and a more detailed description of the design and construction of movement joints for water-retaining structures are given in the Code of Practice 'The Structural Use of Concrete for Retaining Aqueous Liquids'[5]. Whichever type of water bar is used, great care is necessary to ensure that it is continuous, with carefully made joints at all junctions. The use of a water bar does not reduce the need for good workmanship if watertightness is to be achieved and it is important that the surrounding concrete should be compacted very thoroughly round the joint. Any failure to achieve good contact with the water-bar may result in leakage, which will be difficult both to locate and to cure.

It is also important for water bars to be held firmly in position since they tend to be displaced during the placing and compaction of the concrete.

The spacing of joints in walls requires careful consideration if random cracking of the concrete due to restrained contraction is to be avoided. BS 5337: 1976 states that the effect of restraints may be reduced by providing movement joints at a maximum spacing of 7·5 m (25 ft). It has, however, been reported[6] that this requirement is not sufficient to eliminate cracks completely in such structures as reservoir walls which are commonly reinforced with 0·3 per cent of relatively large diameter bars. Where it is desirable to eliminate cracks completely, it is advisable to restrict bay lengths to no more than 5 m (16 ft).

If it is necessary to place the concrete in longer lengths in order to improve production, a suitable technique is to place continuous lengths of concrete and to induce subsequent cracking at about 5 m (16 ft) centres within vertical grooves at the location of the water bars. The grooves should be provided on each side of the wall and should be deep enough to reduce the cross-section by about 20 per cent. In the case of thick sections where it would not be feasible to reduce the cross-section solely by grooves, it is necessary to incorporate some form of void(s) on the induced crack path by means of plastic pipes, deflatable duct-tubes or other suitable device. The voids can later be filled in with grout or concrete whilst the vertical grooves should be filled in the normal way with sealing compound.

REFERENCES

1 BATE, E. H., Some experiments with concrete, *Reinf. Concr. Rev.* **4**, No. 7, Sept. 1957, pp. 421–447.
2 BROOK, K. M., Treatment of Concrete Construction Joints. Construction Industry Research & Information Association, Report 16, June 1969, pp. 16.
3 MONKS, W. L. and SADGROVE, B. M., *The Effect of Construction Joints on the Performance of Reinforced Concrete Beams*. Technical Report 42.483, December 1973. Cement and Concrete Associations, pp. 19.
4 *Manual of Good Practice in Sealant Application*. Jointly published by Sealant Manufacturers' Conference and Construction Industry Research and Information Association, January 1976.
5 BS 5337 : 1976, *The Structural Use of Concrete for Retaining Aqueous Liquids* (formerly CP 2007). British Standards Institution.
6 TURTON, C. D., To crack or not to crack? *Concrete*, **8**, No. 11, Nov. 1974.

12
Placing Concrete

Placing concrete

SERIOUS defects in the finished structure may result from the omission of certain precautions in placing concrete in the formwork. Apart from the possible displacement of the reinforcement, prestressing cable ducts, formwork ties, or any other embedded items, careless placing of concrete may cause movement and damage of the formwork.

Separation of the coarse aggregate (i.e. segregation) is a fault associated with poor placing and it is particularly liable to occur when concrete being discharged from a skip or chute is merely allowed to drop continuously and collect in one spot. The result of doing this is that the coarser aggregate particles tend to run down the sides of the cone so formed and collect at the bottom. It is more evident with very wet mixes which have little cohesion, but dry lean concrete mixes are also prone to the same trouble.

An example of segregation resulting from the free fall of concrete from a barrow into a large column base is given in Fig. 12.2. The cone formation can be traced in the honeycombed areas where the stones have collected against the forms.

Segregation can be very pronounced when concrete is discharged off the ends of conveyor belts if suitable funnel-shaped pipes or baffles are not provided.

Suggested details for controlling the fall of concrete are given in Fig. 12.3, which is based partly on the recommended practice published by the American Concrete Institute[(1)], and partly on the authors' own experience.

Among the precautions which should be taken during placing, the following should receive special attention:

(1) The concrete should be deposited as near as practicable to its final position and should not be deposited in a large quantity at any point and allowed to flow along the forms. Failure to do so is apt to cause segregation, honeycombing, sloping pour planes and poor compaction.

Segregation resulting from the flow of concrete along the forms is shown in Fig. 12.4.

(2) Concrete should be deposited in horizontal layers, and each layer should be compacted thoroughly before the succeeding layer is placed. As far as practicable each layer should be placed in one continuous operation, the thickness depending on the size and shape of the section, the consistency of the concrete, spacing of the reinforcement, method of compaction and the necessity that the next layer shall be placed before the previous one has hardened. In reinforced concrete work it is good practice to place the concrete in layers from 0·2 m to 0·4 m thick. In mass concrete thicker layers from 0·4 m to 0·6 m thick are commonly used. Several such layers may be placed in succession to

Fig. 12.1 Careful placing of concrete from a skip

Fig. 12.2 Honeycombing on a large column base. Placing was by tipping from the top, a cone being formed in the centre down which the stones rolled to the face

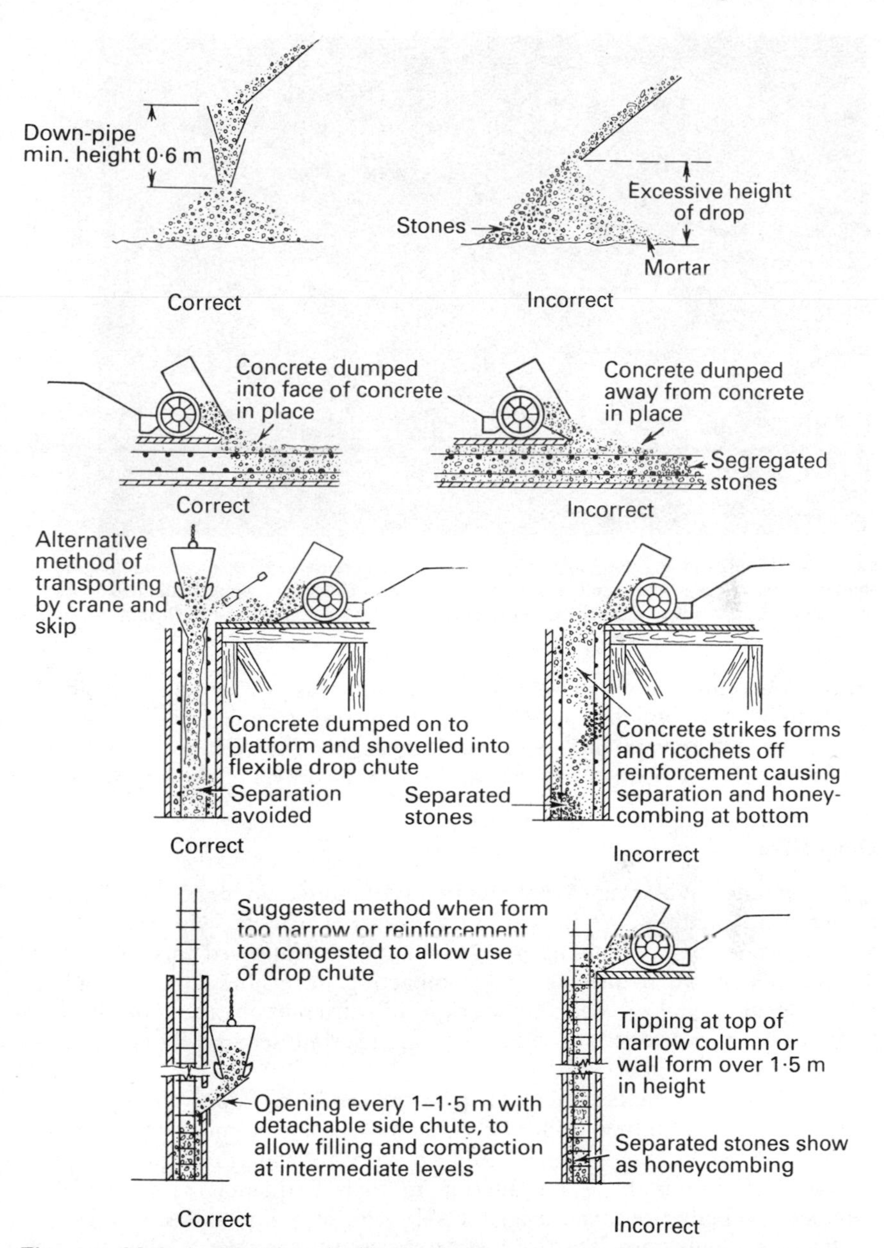

Fig. 12.3 Methods of handling mixed concrete

form one lift, provided they follow one another quickly enough to avoid 'cold joints'. With care and well designed shuttering, concrete can be placed to heights of 10 m or more in one lift, but there must be adequate space within the shuttering and around the reinforcement to allow access for compacting equipment.

Fig. 12.4 Honeycombing resulting from segregation. Concrete was allowed to flow along the shuttering, passing over a stop end on the left. This is part of a large retaining wall in an exposed whinstone finish, the blemishes shown being repeated at several points along the length

(3) Concreting should be carried on continuously in order to avoid the appearance of an unsightly lift line on the finished structure.

(4) The concrete should be worked throughly into position around the reinforcement and bedded fixtures, and into the corners of the formwork.

Deep lifts

At one time the placing of concrete in lifts of more than about 2 m (6 ft) was regarded as unsatisfactory because of segregation troubles with the concrete and the difficulty of obtaining good compaction. Improved mix design and controlled methods of placing and compacting have now largely overcome these objections and good quality work in lifts up to about 10 m and even up to about 15 m can now be produced in appropriate sections by experienced contractors [2].

One of the advantages of deep-lift construction is the improved appearance of the concrete resulting from the reduction in the number of horizontal construction joints. Since these joints are very difficult to disguise and are sometimes very unsightly, a reduction in their frequency, or preferably a complete avoidance of them, is a desirable aim for many classes of work.

However, doubts are sometimes expressed over the effect of allowing concrete to fall more than 2–3 m because the sudden impact at the bottom of the fall will cause the larger aggregate particles to be scattered over a wide area and thereby become separated from the finer material. These doubts may have some justification at the start of a deep lift if the concrete is allowed to fall freely until it hits the hard surface at the bottom (for example, at a kicker or horizontal construction joint). Later, however, when the impact of subsequent batches of concrete is cushioned by the relatively soft bed of freshly placed concrete, the amount of segregation will be insignificant. In these circumstances, any loose stones should become easily embedded in subsequent batches when these are placed and compacted (unless the mix has a low fines content).

Another consideration is that loss of fines from the concrete on to the steel reinforcement may be significant when the first few batches of the concrete are being placed; once the bars have become coated, however, there will be little tendency for fines to adhere to them from subsequent batches of concrete. It is, therefore, only the bottom of the lift which is affected, and one answer to the problem is to decrease the amount of coarse aggregate in the first few mixes. In effect this means that the proportion of fines is increased in order to compensate for the loss of mortar.

If concrete is to be dropped more than 2–3 m, it is clearly preferable that it should fall freely without coming into contact with either the steel reinforcement or the sides of the formwork. This can sometimes be achieved by directing the fall of the concrete through the middle of the member under construction or through other areas sufficiently free of reinforcement. Good results have frequently been achieved simply by dropping the concrete from the top of the formwork, but the success of this technique depends upon careful discharge from a skip and upon negligible obstruction being offered by reinforcement. Careless discharge of the concrete from a badly positioned skip or chute can cause the concrete to strike and bounce off the face of the formwork, resulting in separation of the coarse aggregate from the finer constituents of the mix and possibly in damage to the form face. Particular care is required with members having sloping sides, for example, columns which taper outwards towards the top. In these, there is a greater risk of segregation and damage caused by the concrete hitting the sloping sides of the forms.

A technique which is frequently adopted is the use of vertical trunking or down-pipes to carry the concrete to the bottom. This technique has the advantages of providing better control over the positioning of the concrete at the bottom and of protecting any men working within the forms from being hit by falling concrete. The protection of vibrator operators at the bottom of a large column is a particular case which demands the use of trunking. With the use of vertical trunking or down-pipes, there is no limit to the height through which concrete may be dropped. For example, in the construction of shaft linings for mines and penstock shafts for hydroelectric work, concrete has been dropped several hundred metres through a 150 mm diameter pipe, with no adverse results.

The use of vertical trunking for placing may, however, lead to segregation if the concrete is merely allowed to fall out of the bottom of the trunking in one spot. The formation of a conical heap can be prevented by spreading the concrete over a wider area by moving the bottom of the trunking to different positions. However, this is not always possible where reinforcement is present, since the bars may restrict horizontal movement of the trunking.

Provided there is no obstruction from reinforcement bars, the most satisfactory placing method appears to be by free discharge from a skip moved slowly along the line of the member under construction. With this method, a layer of concrete is spread evenly over a strip, and this concrete should be free from segregation and easy to compact efficiently.

In many cases, openings in the formwork at one side will permit better control over placing and compacting, especially at the bottom part of the lift. These openings which are usually referred to as access doors, portholes or windows, usually range in size from about 0·3 m square to about 1 m long by 0·7 m deep. They are usually provided at about one-third to one-half the height of the lift and at 1·5 m to 2·5 m centres laterally. If concrete is fed through these access doors, any problems which arise by dropping the concrete through the full height of the lift are avoided or, at least, substantially reduced. What is more important, however, is that the access doors enable the concrete to be seen more clearly and the poker vibrators controlled more easily.

Access doors are especially useful for thin walls and columns, where there is generally a fairly heavy congestion of steel reinforcement and where it is virtually impossible to see the bottom of the lift from the top. Moreover, the space available between the reinforcement in thin walls and columns is sometimes insufficient to allow the insertion of any down-pipes. Access doors are indispensable for sloping columns and walls because it is very difficult to pass the concrete down from the top satisfactorily with such members.

Feeding concrete through the access doors is usually achieved by short lengths of chute, which should be so arranged that the concrete is discharged into the member at a fairly slow speed and without striking the form on the opposite side. With suitable design, it is possible to insert panels in the gaps forming the access doors, and to wedge them in place ready for the placing of further concrete in a matter of a few minutes. Good rigid fitting is essential if the line and level of the concrete surface is to be preserved. Badly aligned access doors can spoil the appearance of a concrete structure.

One of the most important properties of concrete required for deep-lift construction is that of cohesion. It must also be ensured that any loss of mortar on the reinforcement or elsewhere, while the concrete is falling, does not cause the mix to be lacking in fines during compaction. The mix should, therefore, have a comparatively high sand content.

The use of a cohesive concrete mix is desirable also to counter any tendencies of water to bleed from the concrete when it is vibrated. The reduction of bleeding is particularly important with deep-lift construction because the effect can be cumulative from layer to layer of placed concrete. Even with a well-designed mix, it is difficult to prevent some excess water or fine material from rising up through the concrete and increasing the water and fines content of each successive layer. One remedy is to adjust the water content of the concrete as the work approaches the top of the lift but, in practice, this requires great care because over-correction can lead to equally undesirable variation in the quality of the concrete. Adjustments of this type are, in fact, rarely made in practice and it seems wise to confine their use to the last batch at the top of the lift and, even then, only when necessary.

Large pours

The problems of placing concrete in large continuous pours have been recognized in the construction of concrete dams for many years. The main problem is the heat developed in the large mass of concrete due to the hydration of the cement. The contraction of the concrete when the temperature subsequently falls can cause serious cracks in the concrete. Precautions to avoid these cracks are described on pp. 332–4.

Recently, there has been an increase in the use of large pours for other classes of work, notably in foundations for high buildings. Many pours have involved the placing of volumes of concrete of the order of 200 to 300 m^3, but considerably larger volumes involving the extension of work well beyond normal working hours have also been placed in a continuous operation.

The choice between a large continuous pour or several small pours for placing a large volume of concrete is influenced by several technical and economic factors, amongst which are the problems arising from the provision of construction joints. Foundations which carry heavy loads can be very heavily reinforced and it is difficult, as well as expensive, to provide stop-ends for construction joints across sections through which large numbers of reinforcement bars pass. There is, therefore, a strong incentive to dispense as far as possible with construction joints and to place the concrete continuously in large pours.

Another factor in favour of large pours is the ability of some ready-mixed concrete companies to supply large quantities of concrete quickly. In addition, the development in recent years of reliable mobile concrete pumps, capable of large outputs, has contributed to the speed with which large pours can be placed.

Where foundations are required to carry heavy loads, the concrete will usually be relatively rich in cement and account must be taken of this when considering precautions to be taken to deal with the large amount of heat evolved.

A primary consideration in placing large continuous pours of concrete is control of cracking. This cracking can be due to both internal temperature gradients set up in the concrete and also the overall thermal contraction when the concrete subsequently cools down. Provision of adequate distribution reinforcement to control the early thermal cracking is, therefore, essential and guidance on suitable design procedures can be obtained from BS 5337 : 1976 Code of practice for 'The Structural Use of Concrete for Retaining Aqueous Liquids' [3] or other references [4],[5].

A recent survey of current practice in placing concrete in large pours [6] showed that factors which appear to be most commonly considered in pour size limitations are:

(1) Availability and reliability of supply of concrete to complete a pour in one single operation

(2) Quality control of the concrete used

(3) Labour availability allowing for temporary importation and possible shift working

(4) Limits on the amount of formwork and reinforcement which can be set up for a single pour

(5) The characteristics of the concrete to be used (i.e. cement content, aggregate type and the possibility of admixtures)

(6) The curing technique to be adopted appropriate to the ambient condi-

tions prevailing at the time of the pour

(7) The sequence to be adopted for placing adjacent pours.

The survey of completed work revealed no significant visual adverse effects caused by the placing of concrete in large pours on the reinforced concrete structures that were examined. The main fault which was found was the presence of plastic cracking due to early shrinkage and settlement of the concrete at the surface. The cracking due to both these causes can be remedied by revibration or tamping of the surface whilst the concrete is still sufficiently plastic to respond to treatment. If this practice is adopted it is important that the time interval between mixing and final vibration or tamping should be limited to no more than two hours, depending very much on the temperature of the concrete, otherwise the disturbance of partially set concrete may cause a substantial reduction in strength.

Different Portland cements have widely differing heats of hydration at various ages when tested under standard conditions (see p. 70) and the choice of cement type for concrete in large pours is often determined by the value of this heat development. Fig. 12.5 shows the average temperature recorded in a part of a large concrete foundation where the concrete contained 326 kg/m^3 (550 lb/yd^3) of ordinary Portland cement. Peak temperatures which are reached in large pours are generally expected to be lower if a cement with a low heat of hydration is used. However, it has been claimed (7) that when rich concrete mixes are used for large pours, the build up of temperature accelerates the hydration, and similar peak temperatures occur for similar cement contents, regardless of cement type. The temperature rise is about 12°C per 100 kg/m^3 of cement content for sections with all dimensions in excess of 2 m.

Experience has indicated that the temperature rise is delayed but not reduced by the use of retarding admixtures in the case of concrete sections which are 2 m thick or more. However, the use of these admixtures can result in a reduced temperature rise for smaller sections which are exposed.

Openings

If a window or door opening occurs in a wall to be concreted, a difficulty arises at the top of the opening, in that if concreting is carried on continuously the subsequent settlement of the wet concrete before hardening causes a crack to form at the corners. The risk of cracking is greatest when wet concrete is used. This trouble can largely be overcome by stopping concreting at the level of the top of the opening, and allowing at least four or five hours for settlement and stiffening to take place before continuing to the top of the formwork. If it is convenient to form a construction joint at the level of the top of the opening, the trouble is eliminated altogether.

Placing under water

Concrete may be placed under water either by the use of bottom dump buckets such as those seen in Fig. 12.6, or by means of tremies. If buckets are used they must not be discharged from the bottom until contact has been made with the foundation or surface of concrete already placed. A certain

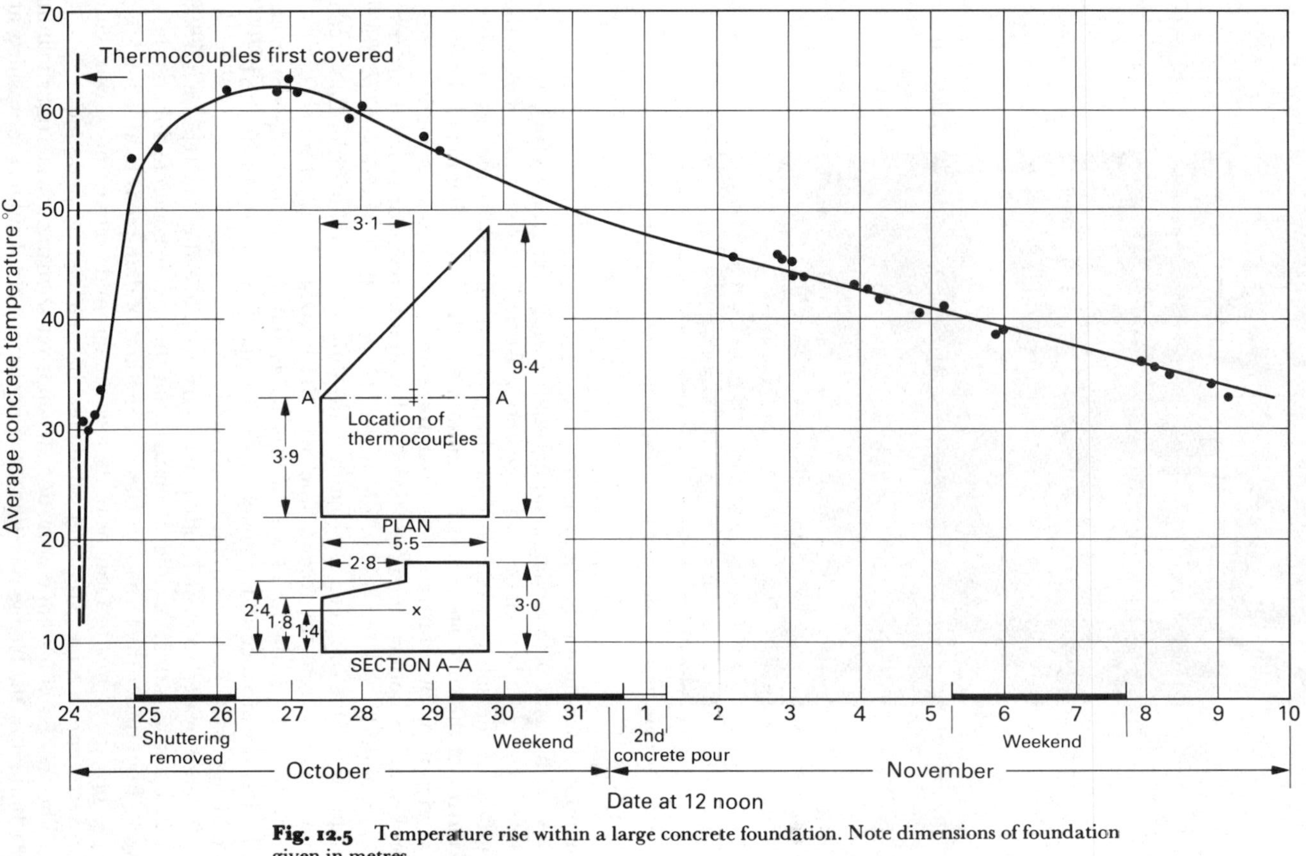

Fig. 12.5 Temperature rise within a large concrete foundation. Note dimensions of foundation given in metres

Fig. 12.6 Underwater concreting by means of bottom dump buckets

amount of washing is liable to occur.

A tremie is a pipe having a funnel-shaped upper end into which the concrete is fed. The bottom or discharge end must be kept continually below the surface of the soft concrete so that there is as little disturbance as possible. Pouring of the concrete proceeds continuously, the pipe being raised slowly as the depth of the concrete deposited increases. If the seal is inadvertently broken at the bottom the succeeding batches should be made richer in order to compensate for cement lost in the water. In general, it is necessary to use a highly workable concrete, of 150 mm (6 in) slump or more, in order that it can flow laterally from the tremie pipe over the horizontal distance necessary to reach the sides of the forms with a reasonably flat and smooth upper surface. The less disturbance to the concrete once it has flowed into place the better. Puddling of the surface should be avoided.

In underwater work it is often necessary to place the concrete on a rock surface and it is then essential to ensure that the tremie concrete does not leak away from under the forms. An effective method of doing so is to attach a long strip of canvas to the form near its lower edge and weight this down with broken stone in bags. Mud and silt should be removed by pumping.

The success of a tremie operation depends on securing a watertight seal at the bottom of the tremie pipe so that the concrete is not subject to washing at the commencement of a pour. One method of sealing tremie pipes is to insert a 'go-devil' consisting of a bundle of a hessian or rubber ball, into the mouth of the tremie after it has been suspended in the form in the position in which it is to be used. This 'go-devil' is forced down the tremie by feeding concrete into the top of the pipe, and in doing so forces the water and air out at the bottom. By the time the go-devil is expelled from the bottom of the pipe the column of concrete following it is descending quickly and rushes out between the bottom of the tremie and the foundation to form a mound round the bottom of the tremie pipe. The flow of concrete is then checked by lowering the pipe to the bottom while more concrete is fed into the top. The pipe is then raised just sufficiently from the bottom to permit a continuous flow of concrete.

The use of tremies presents a problem when frequent removal is necessary as the work proceeds from one point to another, since in situations in which tremies are used there are usually congested conditions due to bracings, reinforcement, etc. Otherwise tremies are preferable to bottom dump buckets.

Rich mixes containing 360 to 400 kg/m^3 of cement are necessary, whether placing underwater by bottom dump buckets or by tremie.

Cement injection into placed aggregate

A method sometimes advocated for concreting consists of placing aggregate of minimum size 20 mm, though often not less than 40 mm, in the formwork, and then injecting cement grout through vertical steel pipes arranged in the aggregate with their lower ends near the bottom of the mass to be concreted. The grout is pumped in and rises steadily through the aggregate. The grout usually contains a plasticizer or air-entraining agent to improve the flow properties, and when mixed in a high-speed mixer is usually described as 'colloidal grout'. When injected into air voids, the grout flows fairly freely, but under water the difference in specific gravity between the grout and water are insufficient to ensure a horizontal flow of more than a few feet. The angle of slope of the grout surface underwater may be expected to be at least 30°.

Such concrete is not always free from honeycombing and the high water/cement ratio necessary to ensure a free flowing grout is not conducive to high strength and impermeability. In underwater concreting the method is also open to objection that mud and silt tend to settle in the interstices of the aggregate before it is grouted.

The method has apparently had some success when used in the dry. A number of failures, including that of a bridge foundation (8) in America, have, however, been reported when grouted concrete has been attempted under water.

REFERENCES

1 ACI 304–73, *Recommended Practice for Measuring, Mixing, Transporting and Placing Concrete*. American Concrete Institute Committee 304, ACI Manual of Concrete Practice, Part 1, 1974.

2 Brook, K. M., *Placing Concrete in Deep Lifts*. London Construction Industry Research and Information Association 1969. Report 15, pp. 23.

3 BS 5337 : 1976, *The Structural Use of Concrete for Retaining Aqueous Liquids* (formerly CP 2007). British Standards Institution.

4 Hughes, B. P., *Design for Early Thermal Stresses in Massive RC Structures. Large Pours for RC Structures*. Proceedings of a One-Day Symposium of the Concrete Society, West Midlands Region, Sept. 25, 1973.

5 HUGHES, B. P., Early thermal movement and cracking of concrete. CPS : 3PC/06/2. *Concrete*, May 7, 1973, pp. 43–44.
6 BIRT, J. C., *Large Concrete Pours—A Survey of Current Practice*. Construction Industry Research and Information Association, Report 49.
7 FITZGIBBON, M. E., Joint Free Construction in Rich Concretes. Appraisal, Decision and Thermal Control. Proceedings of a One-Day Symposium of the Concrete Society, West Midlands Region. Sept. 25, 1973.
8 THORNEY, J. H., Building a foundation through a foundation, *Engineering News Record*, Aug. 28, 1958.

13

Compaction of Concrete

THE object of the compaction of concrete is to eradicate air holes and to achieve maximum density. Compaction also ensures an intimate contact between the concrete and the surfaces of reinforcing steel and embedded parts.

In order to obtain maximum density it is necessary to use a mix which is of adequate workability to enable the operator to place it in position without difficulty with the equipment he has available. On the other hand, it is important that the mix shall not be too wet, as this is liable to result in segregation, excessive laitance at the top of the pour, weakness, and a lower density due to the volume occupied by the excess water. An indication of the loss of strength due to the presence of voids is given in Fig. 13.1. These voids include both the volume of entrapped air, and voids formed by the drying out of excess water, so that it is apparent that a compromise must be reached, depending on the conditions of placing, between the provision of adequate workability and the necessity for reducing the water/cement ratio as much as possible, as discussed in Chapter 6.

When compacting concrete it is important that the reinforcement should not be disturbed and the formwork should not be damaged or displaced. Care must be taken, however, to ensure that the concrete is worked thoroughly in the neighbourhood of the formwork so that the finished surface will be even, dense, and free from honeycombing and pronounced blowholes.

Hand compaction

Ordinary hand methods of compaction consist of rodding, tamping, and spading with suitable tools. For road work, a heavy timber beam prepared to the correct camber and spanning from form to form is sometimes used for compaction. Hand methods of rodding and tamping necessitate the use of a fairly workable mix if the sections are at all narrow and the reinforcement closely spaced. A slump of 100 mm to 175 mm may be necessary if the reinforcement is congested.

Compaction by vibration

Although the use of vibrators has extended into almost every class of concrete work, compaction by hand, properly carried out, gives satisfactory results for many purposes, and because of the simple equipment can sometimes be more economical. Vibration makes it possible to use less workable mixes, resulting in increased strength and lower drying shrinkage for given mix proportions. Even for heavily reinforced sections the slump should not be more than 100 mm. Suggested values for the consistency for various purposes have already been given in Chapter 7. The stiffness of mixes is limited by the

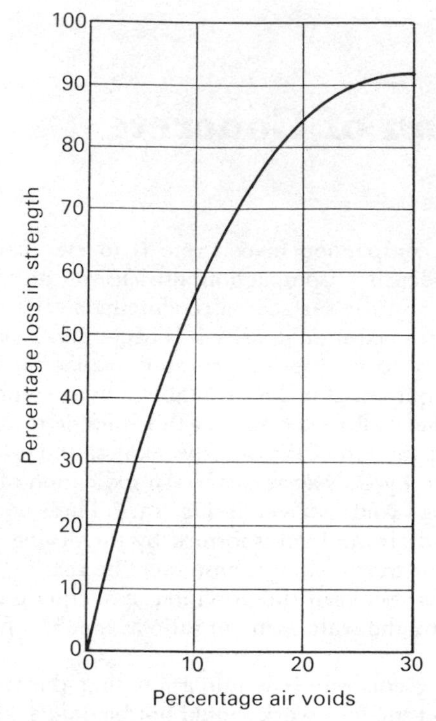

Fig. 13.1 Relationship between loss of strength and air-void space

characteristics of much of the equipment for mixing, handling, placing and finishing the concrete. Difficulties are often encountered in:

(1) Discharging very stiff concrete from mixers.

(2) Discharging stiff concrete from lorries, chutes, skips and other means of transport.

(3) The lack of adequate vibration at points not immediately in contact with the vibrating equipment, resulting in segregation.

(4) The finishing of horizontal surfaces, for instance in road work, to a uniform texture free from ridges and stony patches.

As the stiffness of the mix increases, so also does the cost of the various operations from mixing to finishing; the consistency should, therefore, be carefully adjusted to secure a proper balance between the numerous factors pertaining to the particular job.

Different types of vibrator require different consistencies of concrete for the most efficient compaction, and the characteristics of the machine available can be the governing factor in determining the consistency of the mix to be produced on the site. A joint sub-committee of the Institutions of Civil and Structural Engineers [1], appointed to consider the effect of vibration on concrete, concluded that the process in practice is little affected by the characteristics of the vibration other than its energy. It follows that the vibratory machines in common use, giving 3000 vibrations per minute and an acceleration of $4g$ or more in the concrete, are quite satisfactory.

Fig. 13.2 Pumped concrete being vibrated by poker type vibrators. Since the concrete is quite wet, over-vibration must be avoided

Concrete can be too wet for effective compaction by vibration, as well as too dry particularly when there is an excessive amount of mortar in the mix. Vibration in this case can cause segregation since the larger sizes of aggregate will tend to sink to the bottom and a layer of mortar will form at the surface. The time for which the vibration is applied must, therefore, be limited to a very short period. Fig. 13.2 is an example of a situation where it is essential to avoid excessive vibration.

Vibrating machines are usually operated by petrol engines, compressed air or electricity, and those suitable for site use are of three main types, namely:

(1) Vibrators, used internally.

(2) Vibrators which are attached to the formwork.

(3) Surface vibrators.

Where their use is practicable, internal vibrators are more effective than the other types. Considerable care is, however, necessary with all types to avoid segregation and non-uniform compaction, and the following precautions should be observed:

(1) All forms should be as tight as possible to avoid leakage of mortar and consequent honeycombing. The formwork should also be rigidly braced to prevent displacement, and all wedges and distance pieces nailed. With steel forms, adequate bracing is equally necessary.

(2) In order to avoid air being trapped, the lift should be as shallow as practicable, although not less than 150 mm. The danger of trapping air increases with depth, and a 600 mm layer is about the maximum advisable.

(3) Vibration should not be prolonged in any one position; such a procedure produces honeycombing and lack of homogeneity.

(4) Care should be taken to avoid overworking and thus drawing to the surface an excess of mortar which may later craze, or scale.

Compressed air vibrators have certain advantages and disadvantages compared with electric vibrators. The advantages are that the weight handled by the operator is usually less, and there are no electrical troubles. On the other hand, the air cylinders are apt to freeze in cold weather due to the sudden release of air pressure. This can be prevented by the use of dry air from a larger receiver, by trickling an 'anti-freeze' agent or thin oil into the air line, or by arranging for the air to pass through a coil of pipe which is heated. For comparable service, air compressors are more cumbersome, and expensive than electric generating units, but may be of value for other uses.

Internal vibrators

Internal vibrators, sometimes referred to as poker or immersion vibrators, have a higher efficiency than other types, since all the energy is transmitted directly to the concrete. They are also easier to manipulate and, being portable, can be readily used in difficult positions.

Internal vibrators usually consist of tubes containing small vibrating units. They are generally powered by:

(1) A petrol engine connected by a length of flexible shaft to the vibratory mechanism in the head.

(2) An electric motor connected by a length of flexible shaft to the vibratory mechanism in the head. Alternatively a small electric motor may be situated in the head of the vibrator and directly connected to the vibratory mechanism.

(3) A compressed air motor which, as in the case of the electric motor, may operate through a flexible shaft, or alternatively be housed in the head of the vibrator.

The choice depends on site conditions. Petrol engines are not always easy to start, and failures from damage or inadequate maintenance are not infrequent. Electric motors provide a constant speed and are conveniently portable but, unless a reliable supply of electricity is available, require a generator with its attendant maintenance costs. Compressed air motors are relatively inefficient, requiring considerable quantities of air, causing trouble due to icing in cold weather, and air compressors are expensive to maintain. A poker vibrator driven by an electric motor is shown in use in Fig. 13.3.

Internal vibrators are now available with heads ranging from 25 mm to 100 mm diameter. The 25 mm diameter head has been developed for use in small, heavily reinforced, sections.

Internal vibrators in common use give 3000 vibrations or more per minute with an acceleration of at least $4g$ in the concrete and are quite satisfactory. Machines with frequencies of 5000 to 10 000 vibrations per minute are available, although some types are not entirely satisfactory due to the occurrence of mechanical troubles at the high speeds involved. Higher frequencies are generally not worth while, since they are attained only by loss in mass or

Fig. 13.3 Use of electrically driven internal vibrator in wall shuttering

amplitude and are not more effective in compacting concrete. Mechanical troubles also increase.

With all internal vibrators it is necessary to keep the vibrating head in the concrete while running in order to keep the bearings cool and avoid breakdowns. Flexible drives should not be bent to a small radius, since this causes excessive wear.

The amount of concrete which can be compacted by a vibrator is dependent upon the size and power of the appliance and on the workability of the concrete. With 40 mm slump concrete, a 50 mm diameter immersion vibrator will compact about 6 m^3 of concrete an hour.

Internal vibrators should generally be inserted vertically, or nearly so, at points 450 mm to 750 mm apart. They should then be withdrawn slowly, at about 75 mm per second [(1)]. German specifications are similar and give a speed of movement for the vibrator of 8 cm per second, at which the best results should be obtained. Vibrators should not be used to push concrete laterally in the forms, as this is likely to lead to segregation. Fig. 13.4 illustrates the way in which separation occurs. The vibrator should be at least 1200 mm from the leading edge.

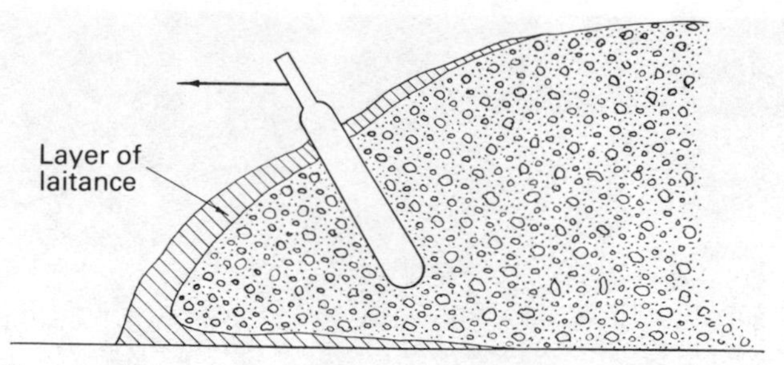

Fig. 13.4 Effect of using internal vibrator to push concrete along form

As a general rule the vibrator should be used not nearer than 100 mm to the form surface in order to obtain a uniform appearance. If the vibrator touches the form surface, a sand streak is very apt to occur; also the surface of the formwork may be damaged. Immersion for periods of 5 to 15 seconds should be sufficient, but the amount of vibration should be judged by the appearance of cement paste next to the forms, and by the change in sound. Fig. 13.5 shows cement paste working up round a vibrator, which has been held over-long in one position. The vibrator should have been withdrawn as soon as the paste began to show, and reinserted a little farther along. A further indication of sufficient vibration is the cessation in the appearance of air bubbles at the surface as the concrete is being vibrated. The vibrators should be immersed to the full depth of the fresh concrete, and preferably 25 mm–50 mm into the previous layer to eliminate any tendency to form a work plane.

Over-vibration should be avoided owing to the danger of segregation and streakiness.

Blowholes on the surface of the concrete are, if anything, more troublesome when vibrators are used than when hand tamping is adopted. No simple procedure is known for eliminating blow-holes formed at the surface, apart from the use of absorptive form linings or vacuum forms. Judicious working of the concrete at the surface of the forms with a thin blade may, however, give some reduction in the amount of pitting. Shutters faced with plywood produce surfaces less pitted than if metal shutters are used. Absorptive form linings have been used to an increasing extent in the USA as a means of preventing blow-holes and otherwise improving surface texture. A few trials have been made in this country with this type of lining and there appears little doubt that it is of considerable benefit. Unfortunately, absorptive form linings can only be used once or twice and are therefore expensive. The types so far tried have been mostly insulation board and similar porous material, sometimes with a covering of hessian or muslin. Work by the Cement and Concrete Association has shown that the number and size of blow-holes can be reduced considerably by selecting the appropriate type of mould oil for the type of formwork in use. This is referred to in greater detail in Chapter 18.

Form vibrators

Form vibrators, or external vibrators, are usually rigidly attached to the forms by means of a clamp or vice and cause a shaking motion of the form

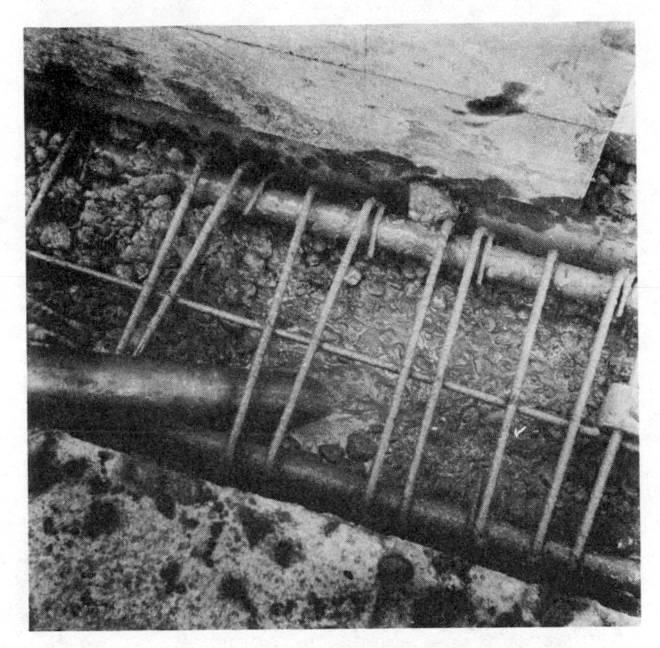

Fig. 13.5 Cement paste beginning to rise round an internal vibrator

through which the pulsations are distributed to the concrete. Form vibration is usually adopted when it is impossible to insert an immersion vibrator, such as into heavily reinforced or small and narrow section members, but since a considerable amount of energy is absorbed by the formwork, more power is required in comparison with internal vibrators. The formwork must be sufficiently strong and rigid to resist the oscillatory action, and especially tight to prevent loss of grout. In their general assembly, timber forms may need screwed connections where nails would normally be used.

Several form vibrators may be seen in Fig. 13.6. They consist of electric motors with unbalanced members, pneumatic hammers or electromagnetic pulsators.

When using external vibrators on the forms, it is necessary to place the concrete in layers no more than about 150 mm deep to provide a uniform build-up. Air holes are more apt to occur with form vibrators than with internal vibrators, particularly near the top of a lift. It is wise to compact the last 600 mm by hand, or possibly by the use of an internal vibrator, if space allows.

Portable types of electric and pneumatic hammers are also used to some extent, being held against the forms at selected points. They are of particular value for the compaction of portions of concrete inaccessible for internal vibration.

Vibrating tables are in effect a type of form vibrator, but with the advantage that the whole mould is vibrated uniformly. They may be operated electrically or pneumatically, and are widely used for the manufacture of precast concrete products. A vibrating table of this type is shown in Fig. 13.7.

Fig. 13.6 Electric vibrators attached to steel formwork. As the height of concrete increases the vibrators will be moved up to the next clips

Fig. 13.7 Vibrating table for precast concrete units

Other types of vibrating table have been developed which compact the concrete by means of a series of jolts rather than by vibration proper. The compaction obtained by this method is quite good.

Surface vibrators

A common type of surface vibrator is the vibrating beam which is used to compact concrete roads and floor slabs. This is described on pp. 317–20 and illustrated in Fig. 21.7.

Another type of surface vibrator consists essentially of a flat horizontal plate or pan on which is mounted one or more electric or pneumatic vibrating units of similar type to those used for attachment to the formwork. These pan vibrators are used for compacting concrete in large structures such as dams, retaining walls, and bridge abutments, where the surface area is extensive. In such instances they usually follow internal vibrators and are used to embed any excess of coarse aggregate and to provide the desired finish.

In general, surface vibrators are not effective beyond a depth of 200–300 mm, depending on the type used.

Workmanship

Skilled workmen are essential to the production of homogeneous and well-compacted concrete. In large measure the correct placing and compaction of the concrete is dependent on their knowledge and experience, and it is particularly important that in teaching workmen, every effort should be made to obtain their close cooperation.

Vibration by unskilled operators is probably more dangerous than hand tamping, since there is a risk of either under-vibrating or over-vibrating the concrete. This may happen in small areas, as, for instance, if the vibrator is left in one position too long through ignorance or diversion of attention on the part of the operator.

In order to obtain the best results using poker vibrators, the concreting programme should be discussed in advance with the operators. They should know how the concrete will be placed and the particular techniques they should adopt with the poker vibrators to achieve full compaction as quickly as possible.

Vibrating equipment generally must be kept clean and free from coatings of set concrete. Frequent servicing and attention to the mechanical parts is also necessary if the machines are to work efficiently.

REFERENCE

1 *The Vibration of Concrete.* Report by the Joint Committee of the Institutions of Civil and Structural Engineers, 1956.

ADDITIONAL REFERENCE

ACI 309–72, *Recommended Practice for Consolidation of Concrete.* ACI Manual of Concrete Practice, Part 1, 1974, American Concrete Institute.

14
Curing Concrete

THE chemical action which results in the setting and hardening of concrete is dependent on the presence of water, and although there is normally an adequate quantity for full hydration at the time of mixing, it is necessary to ensure that the water is either retained or replenished to enable the chemical action to continue. A significant loss of water due to evaporation may cause the hydration process to stop, with a consequent reduced strength development. In addition, evaporation can cause early and rapid drying shrinkage, resulting in tensile stresses which are likely to cause cracking unless the concrete has achieved sufficient strength to withstand these stresses. Methods of curing are therefore designed to maintain the concrete in a continuously moist condition over a period of several days or even weeks either by preventing evaporation by the provision of some suitable protective covering or by repeatedly wetting the surface.

In addition, the concrete should be kept at a favourable temperature for the time specified, and care taken to avoid excessive differences of temperature both within the concrete and in relation to its surroundings. The required extent and nature of the applied curing will be influenced by the presence of either hot or cold weather and reference should be made to Chapters 15 and 16 for further information relating to these conditions.

The proper curing of concrete will improve its qualities in several ways. Apart from being stronger and more durable to chemical attack, it is also more resistant to traffic wear and more watertight. It is also less likely to be harmed by chemical attack.

It is usually required that moist curing shall be carried on for a certain number of days from the time of placing, without any definite reference to the time at which curing shall commence or to the efficiency required. As a result, curing is often commenced too late, and in many cases sporadic spraying with water at infrequent intervals is all that the concrete receives. Since tests are rarely made on the finished structure, little evidence is available of the loss in strength and watertightness which is caused. Tests[1],[2] have, however, been made in the laboratory on the effect of curing on permeability and compressive strength. The results have shown that the permeability of 1 : 2 : 4 concrete, water cured, is high at early ages, but gradually decreases until, after the age of one month, the changes which occur are comparatively small. The days immediately following placing are the most important, for at later periods long curing in water is necessary to improve the impermeability of badly-cured concrete. Wet curing gives the best concrete, and the nearer the curing conditions approach this the more impermeable the product. This is shown in Fig. 14.1[1].

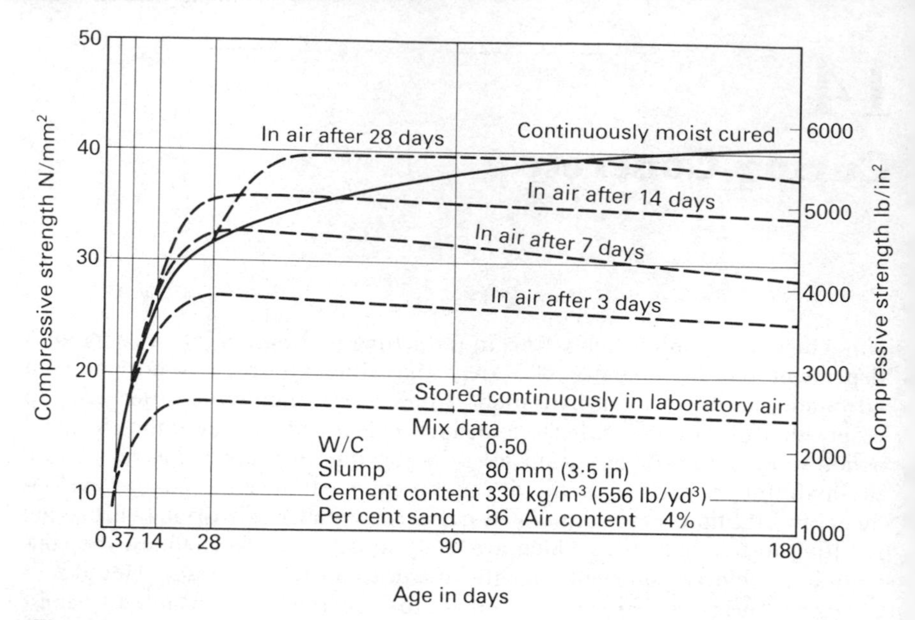

Fig. 14.1 Compressive strength of concrete dried in laboratory air after preliminary moist curing

Fig. 14.2 shows the number of freezing and thawing cycles* required to reduce the weight of specimens of concrete by 25 per cent, compared with the length of moist curing. The curves show that concrete should be prevented from drying out for at least 5 to 7 days in order to obtain the maximum resistance to disintegration by frost. The reduction of permeability with length of moist curing follows a similar curve, and the increased frost resistance with longer curing is probably due in part to the decrease in the water absorbed.

When considering the results of laboratory tests, account must be taken of the fact that the test specimens are generally small and that their behaviour will, therefore, be considerably influenced by changes in the surface layers. Since the surface layers are especially prone to the effects of the curing conditions, it is evident that the deterioration of thick sections would be much less than indicated by small specimens.

A report (5) on curing concluded that whilst the curing of flat slabs and pavement construction is vital, the same necessity for concrete initially protected by formwork is in doubt. Observations made on sites covered in the survey produced no visual evidence of effects in concrete, cast against formwork, through a lack of curing after the forms were stripped. The report concluded that in temperate climates, the subsequent curing of concrete initially protected by formwork is necessary only in exceptional circumstances. It was pointed out, however, that in cases where the appearance of exposed concrete is important, or when forms are removed early, curing for an adequate period should be specified and insisted on.

* A test in which the specimen of concrete is alternately frozen and thawed, the loss in weight being determined as disintegration proceeds.

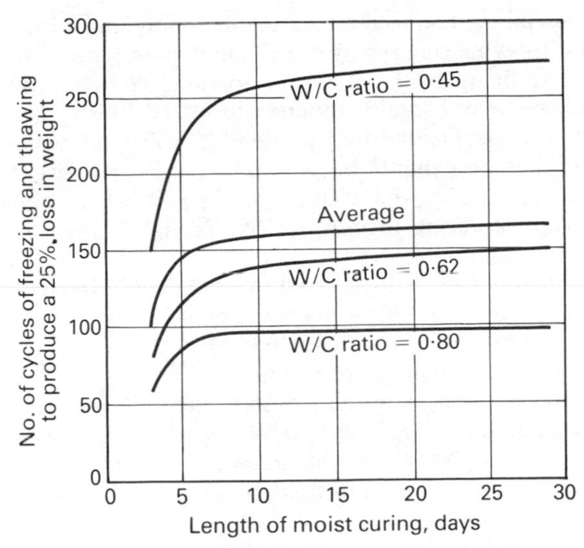

Fig. 14.2 Effect of the length of moist curing on the durability of concrete

CP 110 relates the curing requirements for concrete in buildings and structures to three different environmental conditions; namely (1) damp or indoors, (2) hot weather and drying winds, (3) conditions not covered by the first two categories. In the case of the latter conditions, which represent those most commonly encountered in countries with temperate climates, a minimum curing period of two days is recommended, provided that ordinary Portland cement or equivalent is used and that the temperature of the concrete remains above 10°C. Increases or decreases in the curing period are given for variations in the type of cement used. The curing periods are approximately doubled for each type of cement when the concrete is subject to hot weather or drying winds. No special requirements are called for in the case of damp or indoor conditions.

The actual procedures used for curing concrete vary widely, depending on the type of construction, site conditions, size and shape of the member. However, the treatments commonly used can be broadly divided into two groups, as follows:

(1) Large horizontal surface areas, such as roads and floors.
(2) Formed concrete, such as columns, beams and walls.

Large horizontal surfaces

Careful curing from the time of placement is required for concrete having a large surface area unprotected by formwork, as occurs in the case of roads, airfield paving, floors and dams. Of the several methods of curing available, the best one is undoubtedly that in which the concrete is kept continuously wet by ponding or spraying. Not only does this give good curing, but it also keeps the temperature down as a result of evaporation. Later thermal contraction is therefore reduced.

Ordinary hosepiping connected to a mains supply and with a spray nozzle can be used for spraying concrete surfaces. On certain types of structure it may be economical to fit up banks of sprays operated continuously. Water from spray-pipe systems may contain sufficient dissolved iron to cause staining, a trouble which may be avoided by the use of galvanized or alloy piping. The temperature of the water should be as near as possible to that of the concrete, and should in any case be well above freezing point. It is inadvisable to use water as a means of curing in freezing conditions or when frost is likely to occur.

Spraying with water must be carried out judiciously in structural work in order to avoid interference with other work going on below.

Ponding or continuous spraying is not always a practical method and recourse has often to be made to other means.

Spraying with water as soon as possible after placing the concrete and covering with polythene or building paper is perhaps the next most effective method. However, because of the poor insulating properties of these materials additional shading may be required to reduce the thermal effects of solar radiation.

Alternatively, damp hessian may be laid directly on the surface as soon as the concrete has hardened sufficiently for the hessian not to cause damage or adhere to the surface. Damp sand, in a 50 mm layer has also been used for curing large horizontal surface areas. Both the wet hessian and damp sand method are seldom carried out properly and the spraying or wetting at intervals during both day and night is often forgotten. The damp sand method also suffers from the disadvantage of the extra labour required to place the sand and subsequently remove it.

Although not the most efficient, the most practical and economic form of curing is the use of spray-applied curing compounds especially on large horizontal surfaces. Easily obtained and relatively inexpensive in terms of labour and materials, the more efficient of these compounds undoubtedly provide an effective means of preventing evaporation from flat slabs, assuming that application is thorough and is carried out soon after the finishing operations are completed. Information on these compounds is given under 'Curing compounds' (p. 211).

Floor surfacings dry out very quickly owing to their thinness, and this is often the cause of dusty surfacings of poor wear resistance. Adequate measures should therefore be taken to prevent any initial drying out by giving a light fog spray soon after finishing the trowelling operations, followed by a covering of waterproof paper or polythene sheeting, as soon as this can be done without damage to the surface. Curing membranes, which should be applied whilst the concrete is still moist, are generally accepted as satisfactory means of curing.

The curing of roads and airfields is dealt with in Chapter 21. The influence of damp curing on abrasion resistance of horizontal surfaces is shown in Fig. 25.3.

Formed concrete

Formwork generally provides adequate protection against drying out, but exposed surfaces such as the tops of beams and walls should be covered immediately after casting in dry weather.

Good curing to beam, column, and wall vertical surfaces could be achieved by the continuous application of sprayed water. This is seldom practical, particularly in the case of vertical surfaces, because of the nuisance caused by water dripping on to other work below. Small sections such as columns and beams may be cured most satisfactorily by spraying with water and then wrapping in polythene sheeting as soon as the forms are removed. Alternatively, wet hessian or burlap in contact with the surfaces may be suitable, provided that the necessary arrangements are made on the site to ensure that the hessian is kept continuously wet. It should be realized that wet hessian can quickly dry out during warm days, or even on cool days when a strong breeze is blowing.

For some members, curing membranes may be suitable. However, the application of spray-applied curing compounds on large vertical surfaces can be very uneven and patchy if the spraying equipment is unsatisfactory and if the work is not carefully carried out. Windy conditions also create problems.

It is not usually practicable to wrap concrete walls in polythene sheeting and recourse will probably have to be made to draping wet hessian or polythene from the top of the wall, taking care that all edges are firmly held down to prevent evaporation. In winter in Britain little curing is usually necessary, since in a cold moist atmosphere the rate of evaporation is slow. If polythene sheeting is not available, wrappings of hessian or straw mats may be used to reduce the amount of watering required and to reduce the dripping of water on to other works proceeding below. Tarpaulins and waterproof paper are also of value for reducing the rate of evaporation of water.

Small concrete sections require more protection than larger masses owing to their comparatively large surface area, in comparison with their volume. The corners and edges are most likely to be weakened by early drying out.

Curing compounds

Membrane-forming curing compounds are liquids which are sprayed on to concrete surfaces to retard the evaporation of water from the concrete. A hand-held garden type spray is suitable for most applications. Most of the commercially available curing compounds which have proved satisfactory consist of a resin in a solvent. After application, the solvent evaporates leaving behind on the surface of the concrete a thin continuous film of resin which seals in most of the water in the concrete. This resin film remains intact for about four weeks after which it becomes brittle and begins to peel off under the action of sunlight and weathering. Laboratory tests and field observations have shown that an effective membrane kept intact for four weeks provides the equivalent of 14 days continuous moist curing.

Retardation of evaporation of water from the concrete does, however, reduce the cooling effect and this can be a disadvantage. It may therefore be advisable to provide also for moist curing, or at least shading, in the early stages.

Most proprietary curing compounds are available in different grades, basically a standard grade having what is termed a curing efficiency of 75 per cent and an improved grade having a curing efficiency of 90 per cent. Both are pale straw coloured. Methods of test for determining the efficiency of curing compounds are given in the 'Specification for Road and Bridge Works' [3], issued by the Department of Transport, and in ASTM Designation C 156–74 [4]. The use of a fugitive dye in the resin to provide visible evidence that it has been applied evenly and in sufficient quantity is strongly recommended. The dye quickly disappears after application and will not stain the surface provided that it is not applied to a dry concrete surface. In particular, a white aluminized pigment assists appreciably in reflecting the sun's rays which can otherwise result in a significant amount of heat being absorbed into the concrete. Tests have indicated that the white surface closely approaches the effect of shading, for maintaining low temperatures in concrete exposed to sunlight.

In order to ensure a continuous impervious membrane, it is preferable to apply two coats of a sealing compound. If only one coat is applied, it is important that the compound is sprayed on uniformly so that the complete surface is covered. A uniform coverage can be best achieved by moving the spray nozzle back and forth in one direction followed by moving it likewise at right angles. The nozzle should be held at about 300–500 mm from the surface.

A coverage not thinner than 4 m^2/l per coat is necessary for reasonably smooth concrete surfaces. Rougher surfaces and edges will require a heavier application to ensure continuity of the membrane.

When a curing compound is applied to a vertical surface, it should be applied immediately after removal of the formwork. If the surface is not still 'green' and has dried out, it should be saturated with a spray of water and the curing compound then applied while the concrete surface is still damp. Curing compounds should not be applied to a dry surface because the compound may be absorbed and cause staining.

On unformed (i.e. generally flat) surfaces, curing membranes should be applied soon after the finishing operations on the concrete have been completed when the surface is still moist, but not wet, with free water. The effect of delays in the application of a resin-based compound with aluminium reflective pigment is shown in Fig. 14.3. The results given in this Fig. were obtained from a laboratory investigation [5], in which small concrete slabs were sprayed with curing compound with delays of up to 24 hours after casting. The slabs were weighed at varying ages up to 28 days and the moisture loss was expressed as a percentage of the original weight of the slabs. It is evident if there is a delay of 24 hours the benefit of applying a curing membrane is almost lost.

A disadvantage of curing membranes is that they may prevent satisfactory bonding between the hardened concrete and any fresh concrete or finishes which may subsequently be applied. For concrete surfaces which are to receive finishes such as paints or renderings, it is essential to make certain that the curing membrane will have worn off, or can be removed, before the finishes are applied. Otherwise satisfactory bonding may not be achieved.

Control of temperature variations

Early protective measures to control temperature variations in concrete members are of importance when construction is carried out in hot or cold weather

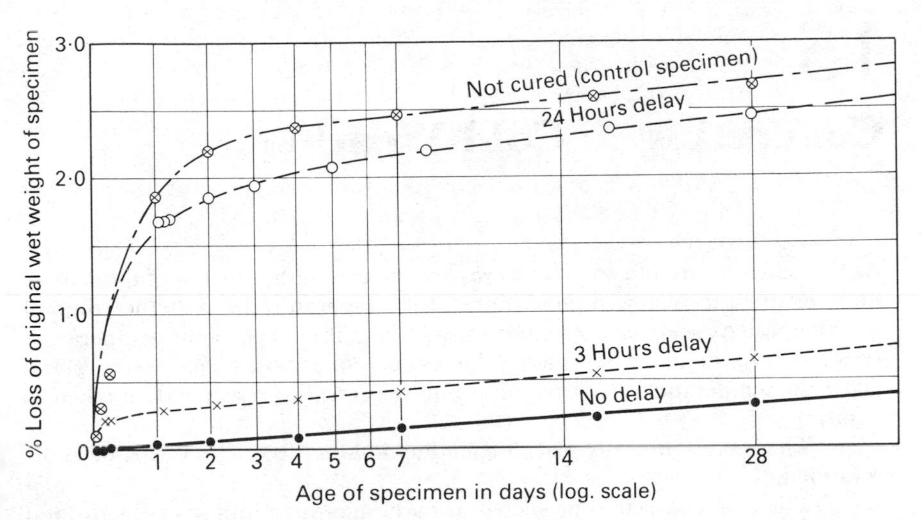

Fig. 14.3 Effect of delay in applying curing compound. The laboratory environment in which these tests were carried out was:
Ambient temperature 20°C; Relative humidity 57–68%.
A resin-based curing compound containing an aluminium pigment was used

(Chapters 15 and 16) and in particular when placing concrete in large pours (p. 190). In the latter case, account must be taken of the high temperatures which are developed in the mass of the concrete by the hydration of the cement.

Subsequent cooling from the maximum temperature results in contraction of the concrete which, if restrained, causes cracking. This effect is often referred to as 'shrinkage', but the use of this term can be misleading since shrinkage is, strictly speaking, the decrease in volume due to drying of the concrete and is a long-term phenomenon.

In the construction of concrete dams, it is standard practice to limit the height of each lift to about 1·5–2 m in order to prevent the temperature rising above an acceptable level. Other measures to control the temperature variations in the construction of concrete dams are discussed on pp. 332–4, and the same principles apply to other forms of construction.

REFERENCES

1 Price, W. H., Factors influencing concrete strength. *J. Amer. Concr. Inst.*, **47**, Feb. 1953, pp. 417–32.
2 Glanville, W. H., *The Permeability of Portland Cement Concrete*, Building Research Technical Paper No. 3, revised edition. HMSO.
3 Department of Transport, *Specification for Road and Bridge Works*. HMSO, 1976.
4 ASTM Designation C 156–74, *Standard Method of Test for Water Retention. Efficiency of Liquid Membrane-forming Compounds and Impermeable Sheet Materials for Curing Concrete*. American Society for Testing Materials.
5 Birt, J. C., *Curing concrete—An Appraisal of Attitudes, Practices and Knowledge*. CIRIA Report 43, February 1973. Construction Industry Research and Information Association, London.

15

Concreting in Cold Weather

As the temperature of concrete decreases, the rate of hardening and development of strength is slowed down until at a temperature below freezing point the chemical process of hardening ceases altogether. When the temperature rises again hardening is resumed. When concreting during cold weather and when the temperature falls below freezing point, measures should be taken to ensure:

(1) That water in newly placed concrete does not freeze and expand as ice is formed.

(2) That the concrete is protected at early ages, i.e. until it will withstand cycles of freezing and thawing without deterioration.

(3) That the strength development is maintained even though at a lower rate than at higher temperatures.

Unless some form of heat is applied and the concrete can be adequately protected, it is advisable to stop concreting when the temperature reaches about 2°C (36°F) on a falling thermometer. With certain precautions, nevertheless, concrete can be continued at freezing temperatures.

During frosty weather, when the temperature rises above freezing point during the day, it is usually sufficient to use a rapid-hardening cement and to protect the concrete by coverings of straw, hessian, or tarpaulins. Timber formwork gives reasonable insulation, except to any exposed surface, against an occasional frost. During continued cold weather further measures are needed and these are described under various headings below.

In planning suitable precautions on a daily basis during winter concreting, advantage should be taken of the weather forecasting services provided by local meteorological stations.

Temperature of materials

The temperature of the concrete when first made depends on the temperature, specific heat, and weight of the constituent materials. The approximate temperature of the concrete can be calculated from the following formula:

$$T_c = \frac{t_c + At_a + 5Wt_w}{1 + A + 5W}$$

where T_c, t_c, t_a and t_w are the temperatures of the concrete, cement, aggregate and water respectively,

A is the aggregate/cement ratio,
W is the water/cement ratio.

As an example, for a mix with a cement/aggregate ratio of 1 to 6 by weight and a water/cement ratio of 0·6 by weight, with the cement and aggregate at a temperature of 2°C and a water temperature of 50°C, the concrete would have a temperature of:

$$\frac{2 + 6 \times 2 + 5 \times 0{\cdot}6 \times 50}{1 + 6 + 5 \times 0{\cdot}6} = 16°\text{C}.$$

When using a cold concrete mixer and cold transporting plant, the heat transfer in the first few batches will lower the temperature of the concrete considerably below the calculated figure.

Water will hold five times the amount of heat held by aggregate or cement, and it provides the most convenient way of introducing heat to a concrete mix. As can be seen from the above example, water heated to 50°C can produce a temperature of 16°C in the resulting concrete even when the cement and aggregates have an initial temperature of only 2°C.

Cement

Provided that the cement, either in bags or silos, is kept free from moisture it will not be affected by frost. Cement in bags should be stored under cover in a dry building. Where this is not possible it should be stacked out of contact with the ground and completely covered. No attempt should be made to heat the cement.

Aggregates

For air temperatures above 0°C, when aggregates are free from ice and frozen lumps, the desired temperature of the concrete can usually be obtained by heating only the mixing water. For air temperatures below 0°C or when the stockpiles containing frozen lumps, ice, or snow, it may be necessary to heat the aggregates.

Sufficient aggregate for each day's pouring should be thawed out as far in advance of batching as possible so as to attain a uniform moisture content and temperature; 24 hours is considered the minimum time for this.

Stockpiles of aggregate are somewhat difficult to heat satisfactorily. Several methods are available, such as steam coils, hot-water coils, steam jets, fire flues, and the erection of covers over the stockpiles inside which hot air blowers are used to keep the air warm. The adoption of any particular method will depend on the relative economy, having regard to the quantity of aggregate to be heated and plant arrangements. Live steam is generally satisfactory if adequate drainage is provided in the stockpile to keep the moisture content uniform. It has the advantage that the steam percolates through the stockpile and heats the aggregate fairly uniformly.

In England, particularly in the South, arrangements for steam heating the aggregates are generally only necessary on large contracts where concreting must be carried out during the winter without interruption. In countries with colder climates, however, there is a greater need to heat the aggregates using steam. The small quick-firing type of boiler is adequate for the majority of sites and conditions. This type of boiler raises steam in less than half an hour and consequently does not require attention during the night, when it can be drained and shut down. The heated aggregate should of course be protected during the night by tarpaulins.

The sand bay should always be kept nearest to the supply end of the steam as this ensures more even distribution of the steam to the remaining bays. A coarse aggregate bay nearest to the boiler means that the bulk of the steam pressure is lost through it before reaching other bays. The exposed surfaces of the aggregates should be covered with tarpaulins or polythene sheets to maintain a uniform distribution of heat and to prevent the formation of ice crusts. It is not possible to generalize as to how much steam will be required, since this will be dependent on factors such as the amount of aggregate, the temperature required, and the efficiency of the tarpaulins. However, as an example, a steam boiler producing 1500 kg of steam (1500 litres of water being used per day) is sufficient to heat 150 m³ of coarse aggregate, 85 m³ of sand and 500 litres of water to produce a concrete at the mixer having a temperature 17°C above the air temperature. Frequent use should be made of thermometers to check the temperature of the aggregates which generally, should be heated up to a temperature within the range 10°C to 20°C.

For smaller contracts where the cost of installing a boiler is not warranted the aggregates should be covered by tents over the stockpiles and the enclosure heated by braziers or low output space heaters until the aggregate is completely unfrozen. For small heaps of aggregate, thawing may be accelerated by turning over the heap at frequent intervals. If this is done fumes from the heaters must be vented outside the enclosure to avoid the danger of carbon monoxide poisoning. This method is unlikely to heat the aggregates evenly to a temperature much above 7°C and should only be regarded as a means of keeping the aggregate in a frost-free condition.

During frosty weather, all aggregate stockpiles should be covered when not in use; this is particularly important overnight. Tarpaulins or polythene sheets are useful for temperatures only a few degrees below freezing, but it is recommended that insulated blankets are used, since these are more efficient.

Mixing water

It is generally found that the cheapest and easiest way of preheating concrete is to heat the mixing water. The required temperature of the water should be between 50°C–60°C, and care should be exercised to ensure that the temperature of the water does not exceed 70°C.

If mixing water is heated above 60°C, overheating may produce quick setting of the concrete or reduce the workability sufficiently to make placing difficult. There is also the danger of a flash set if the mixing water is unduly hot. Because of this danger, in those cases where the water is above 60°C, it is advisable to mix the water and aggregate just before adding the cement.

Where steam is available the mixing water may be heated by means of a pipe, about 40 mm diameter, run from the boiler to an insulated water supply near the mixer. This water tank is additional to the measuring tank on the mixer, and should be positioned above the mixer on scaffolding. It has been found satisfactory to have a 450 litre tank for use with a 350 litre mixer. This steam pipe, which is controlled by a valve placed near the mixer, terminates a little way above the bottom of the tank. Steam can be periodically injected in sufficient quantities to heat the water as required. Alternatively, the water may be heated by a steam coil in the bottom of the supply tank, but this method does not permit close control over the water temperature.

An alternative to steam is the use of calor gas which has been quite widely adopted in recent years. The calor gas is piped to burner units which heat the air which is fed to a heat exchanger in the water supply tank.

The heated water is fed into the measuring tank on the mixer, which should also be lagged externally with foamed polystyrene, fibreglass or similar insulant to avoid heat losses. Constant checks on the temperature of the water in the tanks are necessary, particularly if the mixer is stopped for longer than the normal batch cycle time when the water temperature may rise unless the steam supply is reduced. Whenever the mixing water temperature exceeds 60°C some cold water should be added or some run off to waste until the temperature is below 60°C. Coal or diesel fired hot water boilers may also be used for producing hot water. Electric immersion type heaters have been used but these are expensive to run for long periods.

Batching and mixing

Mention has already been made of placing an additional water tank for heating above the normal mixer measuring tank. The whole of the mixing plant should be kept under cover as much as possible. Sometimes it will be possible for the whole set-up, including aggregates, to be placed in a part of a building; where this is not possible the plant should be placed in as sheltered a place as possible, such as the lee of a wall. Windbreaks should be used to give protection against high winds and driving rain or snow.

The temperature of the concrete leaving the mixer should not be lower than 10°C, as there is an unavoidable loss of heat during transport and placing (see below). On the other hand, the temperature should not be unduly high, since the thermal shrinkage is increased by any increase in temperature drop. It is rarely necessary to raise the temperature to more than about 20°C. As a general rule the temperature of the concrete as mixed should be 3°C–8°C higher than the required temperature after placing; the allowance is dependent on the ambient temperature[1].

The concrete can be warmed by heating the water and the aggregate or by applying heat in the mixer. The simplest method is to heat the water as described above, and this will normally be sufficient for most jobs in the UK. Methods which rely on the application of heat in the concrete mixer, as for instance by a torch or other flame, are not entirely reliable, both from the point of view of heating efficiency and variability in moisture loss.

When it is considered uneconomic to produce heated concrete from a site mixer, it may be possible to buy it from a ready-mixed concrete supplier. Even during very cold weather, some ready-mixed concrete plants will provide concrete on site at a temperature of about 10°C or more.

Placing

The concrete should be placed as soon as possible after mixing since heat is lost quickly, and even when the work proceeds quite efficiently the drop in temperature from mixing to placing may be as much as 3°C to 8°C. Under normal conditions a loss of 3°C can take place during transportation alone, but when excessive hauls are unavoidable the loss may be as much as 5°C.

Open chutes and conveyors should be avoided since a good deal of heat is lost and they also present difficulties owing to the possible formation of ice between runs.

The required temperature of the concrete after placing will be partially dependent on the mass of concrete. For example, in the case of thin sections, the aim should be to achieve a temperature of 10°C to 15°C in the concrete immediately after placing, and for the first three days. For mass concrete, the temperature should preferably be between 5°C and 10°C.

The principal danger during low temperatures is the possibility of the water in the newly placed concrete freezing. Frozen concrete may be mistaken for concrete which has set normally; it is not unknown for the shutters to have been removed from such concrete, and collapse not to have occurred until a subsequent thaw. Such a happening can, of course, have serious consequences. To avoid this, the temperature of the concrete when deposited in the forms should be in no case less than 5°C, and the temperature of the concrete should be kept above this until it has hardened. Research has shown that, provided it is not saturated with water, normal structural concrete is resistant to damage from freezing when it has reached a minimum cube strength of 5 N/mm^2.

Equipment

Skips, wheelbarrows, chutes, and other concreting equipment should be protected from frost when not in use and outside working hours. Any ice that has formed in them should be removed before concreting starts by the use of hot water or, if available, a steam lance from a boiler.

Protection of concrete after placing

Protection of concrete after placing may take the form of insulation of the surface itself or the erection of temporary covers with internal heating. The choice of method in any particular circumstances must depend on relative costs in relation to the rate of progress required and the amount of work to be done. Thin reinforced concrete members normally require more protection over a longer period than larger masses. The corners and edges are particularly vulnerable. Mass concrete, nevertheless, requires protection during the early stages of setting until the quantity of heat evolved is sufficient to offset the heat losses. Under normal circumstances this is not fulfilled until 1–3 days after placing.

Strip and pad foundations below the surface of the ground can usually be given sufficient protection by covering with a straw or fibreglass quilt to conserve the heat which will be generated within the mass of the concrete. It cannot be overstressed that concrete should not be placed on frozen ground.

Ground floor slabs placed on unfrozen ground or hardcore and covered with a protected insulating layer of fibreglass or straw mattress will normally be sufficient to conserve the heat generated in the concrete when the daytime temperature is above freezing point. Severe weather conditions should be countered by covering the slabs with low tents of polythene sheeting or tarpaulins and placing heaters inside the area in order to maintain the temperature of the concrete above 5°C. Care should be taken to avoid early drying out of the concrete and it may be necessary to cover the concrete surface with polythene sheeting. The use of water for curing concrete in cold conditions is not desirable.

Suspended concrete floors may be kept warm after laying by erecting covers on frames over and down the sides of the slab, with a propane gas hot-air blower discharging warm air into the air space thus formed. Coke braziers, paraffin heaters or propane heaters below the slab will keep the underside of the concrete warm. Under less severe conditions, heat applied from the underside of the shuttering with the top surface insulated will generally be sufficient. The top surface insulation should consist of fibreglass or two 25 mm thicknesses of woven straw matting between two layers of polythene. An alternative is a thick layer of loose straw. Electrically heated insulated mats are now available and have been found especially useful for keeping slab concretes at the required curing temperatures.

Tops of columns, beams and walls should be covered with an insulating material immediately after completion of placing in very cold conditions. Tarpaulins stretched across from wallhead to wallhead with paraffin or propane heaters within the cells will help to maintain the temperature at the required level, but precautions must be taken against the possibility of carbon monoxide poisoning. Provided that sufficient heat is applied for a further two days to maintain the concrete temperature above 5°C vertical shuttering may be struck within 24 hours.

Floor toppings require special protection during cold weather owing to their thinness. It is preferable to enclose the space above and below the floor and to maintain a temperature of 5°C or over for the first few days after laying. Heaters should be well insulated from the floor topping by a thick layer of sand, and the area round them prevented from drying out prematurely.

Maturity of concrete

The ability of concrete to resist damage from freezing temperatures is related to the strength of the concrete and this in turn has been shown to depend on the temperature of curing and age. Several investigators have attempted to express the strength of concrete as a function of the maturity which is the summation of the integrals, the product of time and temperature.

Plowman[2] found that the datum temperature for maturity (i.e. the temperature below which the concrete will not increase in strength) is −12°C (11°F).

Therefore, maturity = (temperature + 12) × age in hours when the temperature is in °C

or

= (temperature − 11) × age in hours when the temperature is in °F

Table 15.1 Plowman's Constants for Use in the Maturity Equation

Strength after 28 days at 18°C (64°F) (maturity of 19800°C.h (35600°F.h)) N/mm²	Constants		
	A		B
	for °C.h	for °F.h	
Less than 17	10	−7	68
17–35	21	6	61
35–52	32	18	54
52–69	42	30	46·5

Plowman also developed [2] an easily used and practical 'law' which states that the strength of a particular concrete can be expressed as a percentage of that obtained at a maturity of 19800°C. hours (35600°F. hours). This value is the maturity at 28 days for curing at 18°C (64°F).

The ratio of strengths, as a percentage

$$= A + B \log_{10}\left(\frac{\text{maturity}}{10^3}\right)$$

The datum temperature for maturity calculation is −12°C (11°F). The values of the constants A and B depend on the strength level of the concrete and those suggested by Plowman are shown in Table 15.1. The value of A depends on whether the maturity is based on °C. hours or °F. hours.

The values of the constants for the four strength ranges specified are sufficiently accurate for all normal work when the initial temperature of concrete is between 16°C and 27°C (60°F and 80°F), and given the strength at a maturity of 19800°C. hours (35600°F. hours), the strength of the grade of concrete at any other maturity may be calculated.

Formwork

Before concreting it is essential that all shutters and reinforcement are free from ice. Where the formwork and reinforcement are assembled the day previous to concreting, sheeting over and applying heat from hot-air heaters will be sufficient to keep the formwork and reinforcement free from ice so that concreting may commence first thing the following day.

Where steam is available an additional pipe from the boiler may be run up the building with valve connections at each floor level. A steam lance may then be used for heating formwork and reinforcement.

Provided that precautions have been taken to keep them warm, shutters to vertical surfaces may be struck after 24 hours if heat is to be continued, but whenever possible it is recommended that 48 hours be permitted to elapse before vertical shuttering is struck. It may be desirable to retain external shuttering for a longer period than is strictly necessary, so that it can provide the required heat insulation for internal work, including floor construction.

For multistorey work the external forms can readily be left in position for as long as the pouring cycle permits before they are moved up to the next floor. The external forms, together with the floor shutters erected immediately after striking the internal forms, provide a good means of 'tenting' the newly poured walls.

Electrically heated formwork is now available and where practicable affords a convenient and effective method of increasing the rate of hardening of the concrete. The cost should be considered in relation to any expected savings.

Striking of formwork

The minimum time for striking can best be determined by trial on the job, since it is influenced by factors such as the type and amount of cement, curing temperature, size of member, and the type of formwork. The strength of test specimens cured under similar conditions to the structural members provides a conservative basis for determining the safe time for removal of formwork.

The strength of a typical 1 : 6 by weight concrete mix kept at 2°C for seven days is less than half the strength of similar concrete maintained continuously at 18°C, and remains lower for several weeks. Therefore, even at temperatures above freezing point but below say 10°C, care must be taken not to strike shuttering too soon nor to load the concrete before it has sufficient strength to carry such loads.

Test cubes

Test cubes can be used as a guide to the early strength of the in-situ concrete in order to provide information on which to base the safe removal of formwork. Such cubes are additional to those for checking compliance with the specification. In the former case, they should be placed in contact with the forms and given the same protection as the mass of the concrete, with a little additional insulation to compensate for the fact that they are cast in steel moulds and have a greater relative surface area from which heat losses can occur. Cubes made for compliance or quality control purposes must be cured strictly in accordance with the requirements of BS 1881 : 1970.

REFERENCES

1 ACI 306–66, *Recommended Practice for Cold Weather Concreting*. ACI Manual of Concrete Practice, Part 1, 1974. American Concrete Institute.

2 Plowman, J. M., Maturity and the strength of concrete. *Mag. Concr. Res.*, **8**, No. 22, March 1956.

16

Concreting in Hot Weather

In Great Britain, concreting operations are generally carried out in mild weather and the possibility of difficulties occurring under hot conditions frequently tends to be overlooked until the sudden and unexpected arrival of a spell of hot sunny weather. In countries with hot climates, the problems of concreting demand that suitable precautions are carefully considered in the planning stages of the work and are fully implemented if the work is to be of a good standard.

The precautions apply to concrete both in the fresh and hardened states. One of the main objectives is to control, as far as possible, the evaporation of water from the concrete which can be excessive at high temperatures. It should be noted that the conditions become increasingly critical when the high temperature is accompanied by a low relative humidity and by drying winds. Such conditions are especially conducive to the formation of cracks in the concrete, both before and after hardening.

In order to avoid adverse effects on the quality and durability of the finished concrete structure, many specifications impose a maximum limit on the temperature of the concrete when placed. A maximum concrete temperature of 32°C is recommended as a reasonable upper limit by the American Concrete Institute[(1)]. Most specifications of the US Bureau of Reclamation require that concrete, as deposited, shall have a temperature no higher than 27°C for concrete placed in hot arid climates and 32°C for most other concretes. In certain hot desert areas, some American specifications have prohibited concreting during the summer months.

Mixing and handling concrete in hot weather

The materials for concreting in hot weather should be kept as cool as possible and the following measures are recommended. These measures apply to work carried out in countries with hot climates and they do not, therefore, generally apply to work in Great Britain.

(1) Stockpiles of aggregates should be shaded from the sun.

(2) Periodic sprinkling of the aggregates with water is particularly beneficial since the evaporation of water is a very effective cooling process. The use of cool water, where available, is clearly preferable to achieve the maximum effect. Care is necessary to avoid large variations in moisture content of the aggregates since this will adversely influence the control on the workability of the concrete produced.

(3) The mixing water should be kept as cool as possible by storing in tanks which are shaded or painted white. Supply pipelines should be buried in the ground or insulated.

(4) In regions with particularly hot conditions, it may be necessary to cool the water by adding crushed ice, or by refrigeration.

(5) The use of hot cement should be avoided where possible. The temperature of the cement has, however, less effect than the temperature of the aggregates and water.

The resulting temperature of the concrete can be calculated from the temperatures and weights of the individual constituents by the formula given in Chapter 20 (p. 214). For a normal concrete mix, it can be shown that in round terms a 1°C reduction in the temperature of the concrete may be achieved by either a 2°C reduction in the temperature of the aggregates or a 4°C reduction in the temperature of the water.

One of the problems experienced with hot weather concreting is the rapid stiffening of the concrete after it has been mixed. This stiffening creates difficulties in handling and compacting the concrete within the normal period allowed for the transporting and placing operations. The usual approach to this problem is to attempt to reduce the time of transporting, if this is possible. Other precautions include covering the concrete with damp canvas covers during transit and spraying the transporting containers periodically with water in order to cool them and prevent water being drawn out of the concrete. All mixing and transporting plant should preferably be painted white or a light colour on all exposed surfaces in order to reduce the temperature increase caused by exposure to the sun. White surfaces may be as much as 17°C cooler than black or dark grey surfaces.

It is frequently necessary to increase the water content of the concrete at the mixer to compensate for the loss during transport. This is a reasonably satisfactory procedure although it involves some loss in control over concrete quality. There is a danger that shrinkage cracks may occur after placing if the mixes become too wet. Also the use of a high water content may lead to serious segregation of the concrete whilst it is being handled and transported.

Another approach to the problem of preventing rapid stiffening of the concrete during handling is to use a retarding admixture in the concrete. Admixtures containing retarders are often used in the concrete in countries with hot climates although the reasons for their use are also associated with their water-reducing and plasticizing properties. Retarders are of special benefit in counteracting the accelerating effect which high temperatures have on the hydration of the cement. Since the performance of retarders is sensitive to a number of factors, notably the type of cement, it is essential that mixes are made well in advance of the work in order to determine the effect of the retarder on the characteristics of both the fresh and hardened concrete.

If the transport of the concrete over relatively long distances is unavoidable, the use of truck mixers is preferable in hot weather. Not only is the concrete protected in the drum, but also mixing can be delayed, if desired, until the discharge point is reached.

An important consideration is the time of day when the concreting operations are carried out. In hot climates, it is advisable to avoid working during the severe midday conditions and concentrate on concreting in the early morning or evening. In very hot climates, working at nights is sometimes necessary.

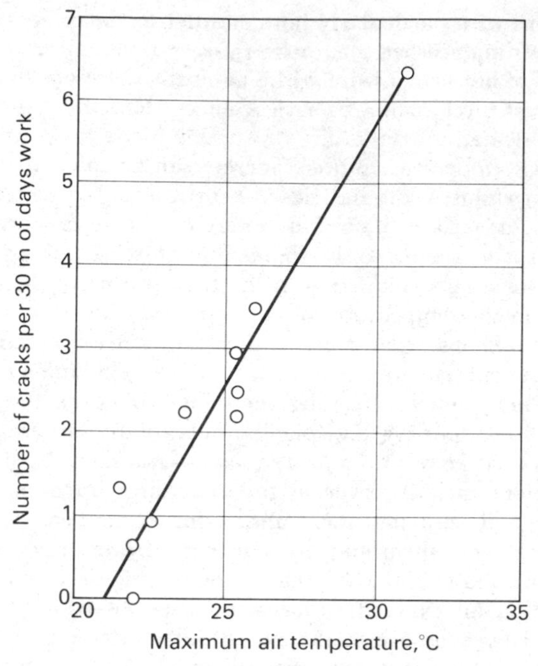

Fig. 16.1 Number of cracks per 30 m of day's work plotted against maximum air temperature during construction of a concrete road

Precautions should also be taken to avoid the heating up of formwork in exposed hot sunny conditions prior to the placing of the concrete. This is particularly important for steel shutters which can become quite hot during exposure to sunshine. Concrete placed directly against these shutters will be of questionable quality since it will tend to dry out quickly. Formwork should be covered or else sprayed with water to cool it before the concrete is placed.

Concrete protection and crack control

Concrete gains heat not only from its own heat of hydration but also from the direct rays of the sun, particularly when used in floors, roads and airfield pavings. On a hot day in Great Britain, it can experience a temperature rise of 10 to 15°C in a paving slab within 4 or 5 hours of it being laid. When the concrete cools overnight, the resulting contraction can give rise to high tensile strains which cause cracking.

An investigation into the crack pattern of a concrete road laid in Leicestershire[2] showed that the main factors responsible for these cracks were the continuous sunshine and the high temperatures which occurred during certain periods of its construction. It was found that cracking had occurred on all the bays in the nearside lane that were laid in the mornings during periods of continuous sunshine and maximum day temperatures exceeding 21°C. The intensity of this cracking was proportional to the maximum day temperature, as shown in Fig. 16.1. Cracking was almost confined to work done in the morning, and this was attributed to the fact that concrete was exposed to the sun and a high air temperature for a longer time and therefore became hotter than the concrete laid during the afternoon. The concrete was cured by spraying the surface with a standard curing compound and it was later considered that, in addition to this curing membrane, a more satisfactory protection would have been achieved if the concrete placed on days when the air temperature exceeded 21°C had been protected by opaque covers supported above the surface of the concrete or the curing compound had contained a reflective pigment to reflect the sun's rays off the surface.

Similar observations have been made on other paved surfaces, as, for example, in France where, as a consequence, the start of concreting operations on motorway work has been deliberately delayed until later in the day.

In addition to cracking of concrete caused by overnight cooling, there is also the long-term behaviour of the concrete to consider. There is an increased risk of future cracking if concrete hardens in a hot environment which becomes cold at some future date.

Apart from cracking associated with temperature changes, there is also the problem of plastic shrinkage cracking associated with rapid evaporation of water from the exposed surface of the concrete. This type of cracking is not confined to work carried out at high temperatures although it is more frequently encountered in hot weather. The cracks are generally almost straight, and their length can range from about 50 mm up to about 2 m. They are often seen as a series of parallel cracks with an alignment of about 45° relative to the direction of construction. The plastic shrinkage cracks usually occur about one to three hours after the concrete has been placed. Although they are normally thought of as surface cracks and do not necessarily impair the performance of the concrete, they have been found on occasion to penetrate the full depth of floor slabs. Further information on plastic shrinkage cracks is given in Chapter 2.

The severity of plastic cracking will depend on the rate of evaporation of water from the surface which in turn will depend on the air temperature, the concrete temperature, the relative humidity and the wind velocity. The effect of these factors may be determined from Fig. 16.2 produced by the Portland Cement Association[1]. It is recommended that when the rate of evaporation is expected to approach 1 kg/m^2 per hour, precautions should be taken to prevent the occurrence of plastic cracks.

Measures to avoid plastic cracking are directed to preventing, as far as possible, evaporation of water from the surface. Early curing measures are essential and the application of a curing membrane, as soon as the concrete surface ceases to glisten with moisture, is generally satisfactory. Where practicable, polythene sheeting laid over the concrete surface is recommended, provided it is in place within half an hour of the concrete being finished. Since it is also desirable to counter the rise in temperature of the concrete and prevent cracking due to thermal gradients and subsequent contraction, the curing membrane should contain a white or aluminized pigment to reflect the sun's rays. It may be necessary to provide covers over the concrete for about four to five hours to prevent excessive gain of heat from the sun, but it is important to allow adequate ventilation under these covers to ensure that the

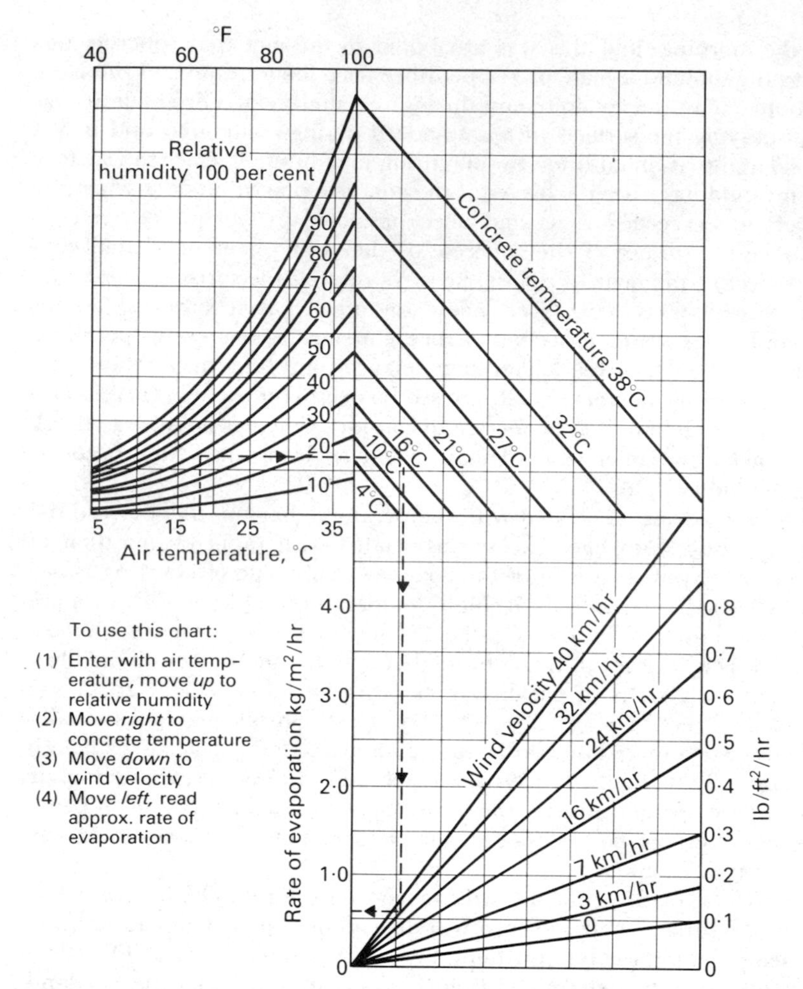

Fig. 16.2 Effect of concrete and air temperatures, relative humidity and wind velocity on the rate of evaporation of surface moisture from concrete. This chart provides a graphic method of estimating the loss of surface moisture for various weather conditions. If the rate of evaporation approaches 1 kg/m² per hour, precautions against plastic shrinkage cracking are necessary

temperature under them does not become excessive.

Another recommended method is to cover the concrete surface with hessian which is kept continuously moist; alternatively a mist spray of water may be frequently applied to the concrete surface. Both these methods have the advantage of keeping the concrete cooler because of evaporation, but they suffer from the disadvantage that a constant supply of water is required. In practice it is found that the diligent attention required to keep the hessian damp is often lacking. Also, in hot dry regions, water is usually at a premium and its use for curing may, therefore, be out of the question. Even if supplies are available, they are likely to contain salts and the continual spraying and evaporation leaves concentrations of such salts on or near the surface of the concrete, causing efflorescence, and a possible weakening of the surface layer. It is advisable in such cases to drape hessian or similar covering a few cms away from the concrete surface if at all practicable, and to spray the hessian, without the water reaching the concrete at all.

Strength of concrete placed in hot weather

It is well established that the strength of concrete a few days old is increased as the curing temperature is increased. Information on the effects of high early temperature on the ultimate strength of the concrete is less clear, but it does appear that in practical terms, high early temperatures will reduce the 28-day strength. A 15 per cent reduction in the 28-day strength was found during an investigation(3) into the manufacture of high-strength concrete in a tropical climate, when the concrete was mixed at a high temperature, but cured at about 23°C. Similar results have been obtained elsewhere in tropical zones.

Results from laboratory work(4) have shown that concrete made at 38°C gave 28-day cube results which were about 15 per cent lower than concrete produced at 18°C. In this work, all the cubes were cured for one day at the temperature at which the concrete was mixed before being placed in water at 14 to 19°C.

REFERENCES

1 ACI 305–72, *Recommended Practice for Hot Weather Concreting*. ACI Manual of Concrete Practice, Part 1, 1974. American Concrete Institute.

2 PULLER-STRECKER, P. J. E. H., *Crack and Condition Surveys of a Section of Trunk Road A46 near Narborough in Leicestershire, Overslabbed in Reinforced Concrete*. Cement & Concrete Association, Technical Report TRA/357, 1961, pp. 19.

3 RIDLEY, T., An investigation into the manufacture of high-strength concrete in a tropical climate. *Proceedings Institution of Civil Engineers*, May 1959, pp. 23–24.

4 KEENE, P. W., *An Investigation of the Effect of Cement Temperature and Ambient Temperature on the Workability and Strength of Concrete*. Cement and Concrete Association, Departmental Note DN/17 (unpublished), April 1962.

17
Surface Finishing of Concrete

MANY concrete structures have a satisfactory appearance, but unfortunately others are deplorable from the start, and yet others, which were quite pleasing when new, have weathered badly. There is, therefore, considerable justification for the view that concrete is not pleasing to the eye. Many of the unsightly features of concrete structures result from contraction cracks, carelessly constructed and badly placed construction joints, patching of damaged or honeycombed areas, poor formwork, and lack of adequate cover to the reinforcement. All are faults which can be prevented.

If an efficient structure, which is also pleasing in appearance, is to be produced, cooperation between the architect, the engineer and the contractor is essential. A close liaison is especially necessary in arranging such details as the positions of construction joints, the construction of the shuttering and formwork, and the method of finishing.

Considerable care needs to be given to the many details involved in the production of concrete if a uniform appearance and a surface free from blemishes and honeycombing is to be achieved. The aggregates should be well graded and of one quality. Fine sands and rich mixes assist in reducing blowholes and colour variations, but can lead to crazing if the water content is excessive. The proportioning, water content, mixing and placing should be constantly watched to ensure that such blemishes as segregation, honeycombing, blowholes, sand streaks and colour changes do not occur. A fast rate of placing in excess of 2 m per hour is advantageous in eliminating air bubbles [1] from vibrated concrete and reducing blow holes on the surface. Immersion vibrators are to be preferred because air bubbles then tend to travel inwards towards the source of vibration. Construction joints should receive particular attention. Bleeding should be prevented as far as practicable, as the concentration of fine particles of cement with a low water content gives a darker colour.

Much depends on adherence to good practice, described in previous Chapters, and on the use of suitable release agents (see p. 269).

The design of the formwork should be such that it can easily be removed and used repeatedly if required; if removal is difficult or is done carelessly, the use of crowbars and other tools may mar the appearance of the concrete, especially at the edges and corners. Sharp arrises and internal corners should be chamfered whenever possible. The accuracy of line and level in concrete structures depends on the accuracy of the formwork and its rigidity. Recommended tolerances are discussed in Chapter 18 on formwork.

A committee appointed by the Institution of Civil Engineers drafted a guide to specifying concrete [2] which is the basis of the Department of the Environment specification for road and bridge works [3]. The following descriptions of finishes for formed surfaces follow the lines of the Institution guide; the Department of the Environment descriptions are slightly different.

Class F.1. This finish is for surfaces against which backfill or further concrete will be placed. Formwork may consist of sawn boards, sheet metal or any other suitable material which will prevent the loss of grout when the concrete is vibrated. In other words, while the concrete may need to be of good quality, the finished appearance is of no importance.

Class F.2. This finish is for surfaces which are permanently exposed to view but where the highest standard of finish is not required. Forms to provide a Class F.2 finish should be faced with wrought and thicknessed boards with square edges arranged in a uniform pattern. Alternatively, plywood or metal panels may be used if they are free from defects likely to detract from the general appearance of the finished surface. Joints between boards and panels should be horizontal and vertical unless otherwise directed.

This finish should be such as to require no general filling of surface pitting, but fins, surface discoloration and other minor defects should be remedied by methods approved by the engineer.

Class F.3. This finish is for surfaces prominently exposed to view where good appearance and alignment are of special importance. To achieve this finish, which is intended to be free of boardmarks, the formwork should be faced with plywood or equivalent material in large sheets. The sheets should be arranged in an approved, uniform pattern. Wherever possible, joints between sheets should be arranged to coincide with architectural features, cills, window heads, or changes in direction of the surface. All joints between panels should be vertical and horizontal unless otherwise directed. Suitable joints should be provided between sheets to maintain accurate alignment in the plane of the sheets. Unfaced wrought boarding or standard steel panels are not permitted for Class F.3 finish. Internal ties and embedded metal parts are also not allowed. The requirements for a class F.3 finish are so onerous that it is rarely achieved as specified.

The Department of the Environment have added a class F.4 with similar requirements to F.3, except that internal ties and embedded metal parts are allowed. Another classification is included in CP 110 : Part 1 : 1972. Five finishes are listed but the classification is not much used.

Many architects and engineers prefer to use their own descriptions of the finish they require. Reference may also usefully be made to 'Recommendations for the Production of High Quality Concrete Surfaces' (4).

Rough finished surface textures have gained favour. They mask minor surface blemishes and imperfections; if properly detailed, they weather better than the smoother finishes. Such patterns in bold relief, as for example that shown in Fig. 17.1, are particularly suited to concrete, and weather better than smooth surfaces on which defects become all too evident. In sooty towns the effect of light and shade is retained, whereas the colour and texture of finer finishes are lost.

Other types of special finish may be summarized under the following headings:

(1) Board-marked finishes, where emphasis is placed on the pattern produced by the boards, and the rough finished surface texture.

(2) Finishes from special linings.

Fig. 17.1 The elephant and rhinoceros pavilion, London Zoo, showing the attractive finish obtained by casting the external walls in-situ with vertical ribs subsequently hacked to expose the aggregate

(3) Tooled finishes.
(4) Scrubbed and exposed aggregate finishes.
(5) Grit-blasted finishes.
(6) Finishes produced by grinding.
(7) Finishes most suited for precast units.

Board-marked finishes

The formwork pattern is of more importance than any other single factor to the production of a monolithic concrete structure having a pleasing appearance. Well-designed formwork can provide an excellent finish without further treatment, as may be seen in the examples given in Figs. 17.2 and 17.3. Even rough boarding of uneven thickness can give an attractive general appearance with careful planning. A finish from rough boarding masks defects and horizontal construction joints, and weathers well in polluted atmospheres. It is necessary either to have tongued and grooved joints, or to plane the edges of the boards, so as to obtain a close fit which will obviate loss of cement mortar with consequent honeycombing.

A more uniform board-marked finish is obtained by using planed boards. As the finish becomes smoother greater care is necessary to preserve a balanced pattern, and to avoid surface defects. The board pattern may be either

Fig. 17.2 Effective use of board-mark finish on the soffit, mushroom headed columns, and slab edge to walkways at the Royal Festival Hall, London

subdued by removing all projecting fins with chisels, carborundum stones, or other approved means, or it may be accentuated by the use of tongued and grooved boarding with chamfered edges.

The following points should receive attention:

(1) The formwork should be tight and in true alignment.

(2) The boards should be evenly matched in width. Narrow boards usually give better results owing to less warping and the finished appearance is more pleasing (Fig. 17.4). Wrought boards are more easily stripped from the concrete and more easily cleaned.

(3) Old and new boards should not be mixed, since old boards are often more absorbent and have a rougher surface. Therefore, variations in texture occur.

Fig. 17.3 Exceptionally high standard of boardmarked concrete finish to mushroom headed columns

Finishes from special linings

Smooth finishes are normally produced by lining the formwork with metal sheets, plywood, or compressed fibre boards treated with oil or shellac to prevent the concrete adhering, see p. 269 dealing with mould release agents. These sheets should be arranged in panels so that the joints form a pleasing pattern on the building. The useful life of plywood is dependent on the type used, the best being that which has a waterproof binder. The care used in handling during stripping and in storage on the site also materially affects the number of times it can be used. Hardboards also require care in handling and can normally be used only three or four times before fraying at the edges

Fig. 17.4 Use of narrow boards provides pleasing finish to church in Finland. The roof is a copper covered prestressed concrete parabolic shell

occurs. Chipboards deteriorate rapidly, due to swelling of the individual chips, and are much weaker in bending than timber, particularly when wet.

Highly absorptive insulation boards about 12 mm thick have been used as lining in order to reduce the number of blowholes on the surface. Tests have indicated that when concrete was placed against absorptive form linings, practically all pitting and voids were eliminated, whether compaction was by hand or by vibration. However, differences in absorption can cause colour variations. Muslin and hessian are sometimes used as a covering to the insulating board. The use of such linings is expensive due to the limited re-use possible.

Generally, nonabsorbent linings give the most uniform colour to the concrete. Barrier paints and lacquers may be used to reduce the permeability of linings, a suitable mould oil then being used to reduce blowholes and ensure good release.

Smooth finishes are best avoided externally, since uneven weathering, small defects in the concrete and all joints (whether in the shuttering or in the concrete) are plainly visible. Internally, treatment of smooth finishes with alkali-resistant paints can give satisfactory results.

Other form linings which have been used to produce special finishes include crêpe rubber, waterproofed serrated cardboard and expanded polystyrene. The rubber sheeting can be used several times, but the cardboard can normally be used only once. A difficulty arises with rubber sheets and cardboard in that they are difficult to fix without creasing and it is also necessary to

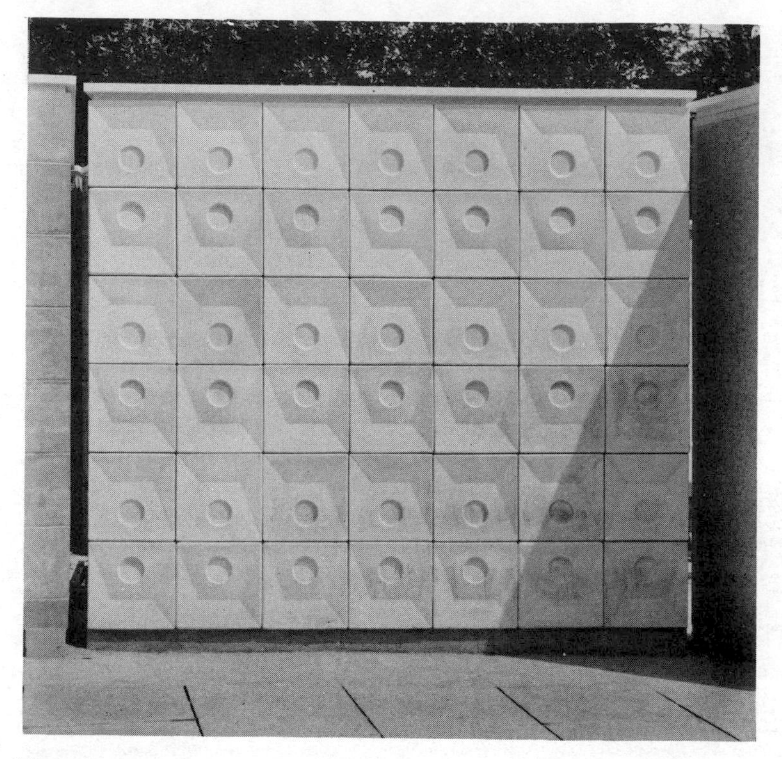

Fig. 17.5 Profiled patterned blocks with an 'egg-shell' finish obtained by casting against thermoplastic formwork linings

arrange them in small panels with suitably placed joints, in order to avoid unsightly joint-lines. Expanded polystyrene is sufficiently rigid to avoid these problems, but jointing needs care in detailing and in fixing and concreting.

Awkwardly shaped mouldings with a smooth surface finish have been satisfactorily produced using moulded resin-bonded fibreglass.

An example of a wall built with concrete blocks cast against thermoplastic formwork linings is shown in Fig. 17.5.

Tooled finishes

A number of methods of producing tooled finishes are available, and with the use of suitable aggregates a very pleasing appearance can be obtained which retains its character even after a number of years of weathering. Bush-hammering is the most popular type of tooled finish, and electrical and compressed air equipment is available for this work. It is advisable when bush-hammering to leave a narrow margin at the edges and corners of the area untreated in order to avoid the danger of spalling. Bush-hammering is done with either a disc type of hammer head or a roller type, as shown in Fig. 17.6.

Fig. 17.7 shows the effect of heavy bush-hammering on ballast concrete. Other forms of tooled finish include sparrow-pecking (Fig. 11.4), which is

Fig. 17.6 Electrically driven roller type bush-hammer tool in use on a column

Fig. 17.7 Point-tooled white concrete finish

done with a pointed hammer, comb chisels used to give various types of pattern, and other effects produced by special types of hammerhead. It should be noted that construction joints, honeycombing, and segregation, cannot be masked by tooling and it is very necessary that every care should be taken in placing the concrete.

Fig. 17.8 Failure of bush-hammering to mask a honeycombed patch of concrete just above a construction joint. Honeycombing was similar to that shown on the left and was patched with mortar before bush-hammering

An example of an attempt to mask a badly made joint with honeycombing at the bottom of the upper lift is shown in Fig. 17.8. The patch in the bush-hammered section is clearly seen. A further example of the failure of bush-hammering to hide defects in construction is seen in Fig. 17.9. Not only do joints show, but there are sundry mortar patches, and the joints between the steel shutter panels are still apparent. Evidence of lipping where the shuttering was not tight against concrete previously placed, is seen at window head and upper sill levels. A contraction crack has occurred below the lower right hand window.

Since construction joints always show in tooled work, it is advisable to arrange a panelled effect, emphasizing the joints by methods such as those shown in Figs. 11.5, 11.6 and 11.7, or by leaving untooled margins or recesses at the joints.

Attempts to adjust the line and level after the bulging or slipping of form-work always result in differences in the tooled texture, a bad example being seen in Fig. 17.10.

Patches always show through tooling sooner or later.

Fig. 17.9 Bush-hammered gravel concrete wall to a block of flats. Note the failure of the bush-hammering to remove the joint lines of the steel shutter panels, numerous mortar patches and construction joints at irregular heights. Shrinkage cracks have occurred below the two windows on the right

Scrubbed and exposed aggregate finishes

Removal of the cement skin may be effected while the concrete is green by means of a stiff fibre brush or a wire brush. The earlier the scrubbing is done the deeper the aggregate is exposed. Plenty of water, preferably as a spray, is needed to wash away the cement paste and clean up the surface of the aggregate as it becomes exposed.

Patching scrubbed work is difficult if the patches are to be indistinguishable from the rest of the surface. The best method is to patch immediately after scrubbing, using the largest aggregate which can be managed. After a number of hours depending on the weather, the patches should be scrubbed to the same texture as the general surface.

Two methods of avoiding the ragged corners which this type of work produces are to use rounded mouldings in the corners of the formwork or to have rebated corners and stop the scrubbing at the edge of the rebates.

Fig. 17.10 Failure of bush-hammering to mask the effect of shutter slipping, even where there is no accompanying loss of fine material

Fig. 17.11 A good example of an exposed aggregate finish on precast concrete units forming the parapet of a bridge

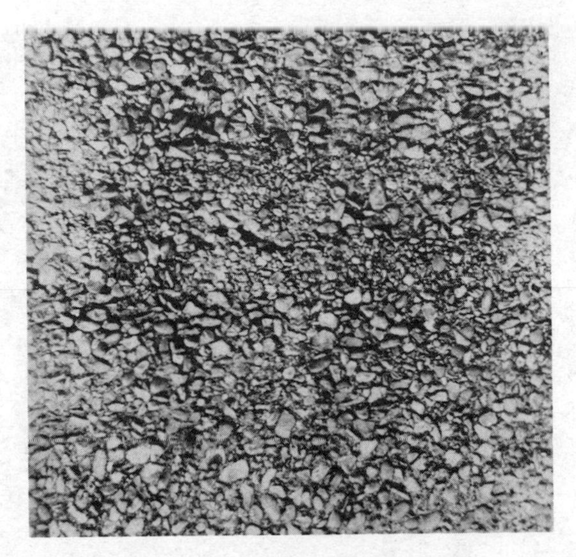

Fig. 17.12 Scrubbed gravel concrete after weathering for some years

For buildings higher than one storey there are practical difficulties which increase the cost of scrubbing and make it almost impossible to produce a clean, uniform surface. This type of finish is, in fact, more practicable for precast concrete units than for monolithic concrete. A good example of precast unit construction with an exposed aggregate finish is shown in Fig. 17.11.

Removal of the cement skin is open to the objection that the surface is not then so impermeable and, therefore, less resistant to the weather. An added disadvantage is that the rough surface texture is apt to become grimy in a polluted atmosphere. Fig. 17.12 shows the appearance of a scrubbed gravel concrete after a number of years' weathering in a polluted atmosphere. The gravel has become dull and lifeless, while a number of holes in the surface show that some disintegration has taken place.

Etching with hydrochloric acid produces a surface similar to that given by scrubbing, except that the aggregate usually appears brighter. Plenty of water must be used after etching in order to prevent a continuation of the erosion. Acid treatment is undesirable on reinforced work with little cover or on low-weight concrete.

Preparations containing substances such as sugar, which retard the setting of cement, may be painted on the inside of the formwork. When the formwork is stripped the surface skin may then be scrubbed away to expose the aggregate. The preparations at present in use are not entirely satisfactory either on account of staining or failing to produce a uniform surface (see Fig. 17.13). The latter trouble seems to be due to the ease with which the paint rubs off when fixing the reinforcement or when filling the forms. Neither are such preparations satisfactory for producing a rough surface as a key for renderings, since there is always a risk that traces of retarder or weakened cement at the surface will seriously reduce the bond. The staining may also penetrate the rendering.

Fig. 17.13 Surface obtained by scrubbing after the use of a chemical retarder on the shuttering. Note the difference in texture between the bottom pour and the top pour. Slight lipping is seen at the joint

Aggregate transfer method

Another method of producing an exposed aggregate finish which has given satisfactory results involves the sticking of the aggregate on to the formwork. The adhesive is one which softens under the action of water from the concrete, so that removal of the formwork, and a wash down, leave the aggregate exposed.

Grit-blasted finishes

Grit-blasting is sometimes used but does not usually give a good finish since it tends to erode the softer portions of the concrete surface and so produce a non-uniform appearance. The process is also relatively expensive.

Fig. 17.14 An example of profiled precast horizontal cladding units

Finishes produced by grinding

The surface skin may be removed by the use of grinding stones or carborundum discs. A grinding stone is necessary for removing projections; discs may then be used over the surface as a whole. Unless carefully operated the grindstone, which is usually about 175 mm diameter when new wearing down to 100 mm or so, is likely to produce a scalloped appearance on the surface. Any making good must be done before grinding, and every effort should be made to see that the patches match the main concrete.

Several concrete structures in London have been finished with carborundum discs, and on all of them the patches are becoming increasingly obvious, even though they were not discernible when the concrete was new.

Carborundum discs were used to grind down the precast concrete slab exterior of a hotel in London which had become discoloured after about three

Fig. 17.15 Coarse pebble exposed aggregate slabs

years' weathering. Although the surface was afterwards given a coat of cellulose, it again lost its brightness and grime collected, particularly at the joints. An example of weathering of a ground surface is given in Fig. 17.16.

Grinding down of fins and other projections is often necessary, but needs to be done with great care if unsightly scarring, by the edge of the grinding wheel, is to be avoided.

Precast units

Precast concrete units can be finished by any of the methods outlined above and have the advantage over cast in-situ concrete in that they are made under more closely controlled conditions such that fewer defects are likely to occur. In addition, with precast units the joints are generally a feature rather than a disfigurement. Examples of precast units are shown in Figs. 17.11, 17.14 and 17.15.

Grouting and cement washes

Cement washed or grouted surfaces cannot be considered as good practice. They are usually of a much richer mix than the concrete underneath, and after exposure to the weather the differential movements due to moisture and temperature give rise to fine cracks, and eventually the skin peels off or is washed away. Cement grouts are satisfactory for filling small blowholes, but they should not be used for filling larger patches or to mask honeycombing.

Fig. 17.16 Repairs to concrete surfaces. Patches which have become increasingly obvious with exposure to the weather. The surface was originally ground with carborundum discs and the patches hardly discernible. Shrinkage cracks are also apparent

Painting concrete surfaces

Paints based on linseed oil cannot satisfactorily be applied to new concrete, since the linseed oil is saponified by alkalis from the cement. However, linseed oil is sometimes advocated for improving the weather resistance of old concrete, and experience in the USA indicates that a linseed oil application can effectively reduce deterioration due to the use of de-icing salts.

Paints having a plastic binder and giving a rough sanded finish have been more successful. They should be applied to a dry surface in warm weather, and may then be expected to last five to fifteen years. Much depends on surface preparation and the possibility of moisture getting behind the paint film.

Repairs

Repairs to concrete surfaces may be avoided to a large extent by arranging the shuttering so that it may be removed easily without the use of crowbars or other tools likely to damage the green concrete. Honeycombed patches should not occur where the supervision is good.

Small blowholes are almost unavoidable, but they may be filled without much fear of them becoming a source of trouble. Larger patches are usually troublesome and since they become more apparent as time goes on, every effort must be made to make a good job of them. Patches which are carelessly done begin to crack from the main concrete after a few years' weathering and finally become detached.

Repairs should be made as soon as possible after the forms have been removed and before the concrete becomes too hard. Stone pockets, segregation patches and damaged areas should be chipped out and the edges undercut slightly to form a key. The area to be patched should be cut to a size which will marry with the general appearance without undue disfiguration, since the patch must be expected to have a slightly different colour and texture from the remainder of the concrete.

All loose material should be washed out before patching. No excess water should be left in the cavity, but the concrete should be damp while retaining some of its natural suction.

A good bond between the patch and the parent concrete may be obtained by sprinkling dry cement on the wet surface just before applying the patching material. Other methods are to throw mortar forcibly on to the wetted concrete, or to brush in a coat of thick cement grout of about 1 : 1 cement : sand. Before this has dried, the remainder of the patch should be filled with mortar or concrete, depending on the size of the repair.

The filling material for very small patches should consist of mortar of the same materials and proportions as those used in the mixing of the concrete. The mortar should be mixed as dry as possible, well compacted into the cavity, and finished in a manner which most nearly conforms to the adjacent surfaces. The colour match may often be improved by adding a proportion of white cement to the mix.

If larger repairs are to be made to hardened concrete it is usually necessary to provide some formwork bearing tightly at the edges of the cavity. For such repairs it is necessary to chip out concrete to a depth of at least 100 mm, and preferably 150 mm in old concrete. While undercutting should generally be adopted, an upward slope from the back of the hole may be necessary at the top to facilitate placing of the concrete. Mortar should be scrubbed into all surfaces with a wire brush before placing the concrete. Any of the original reinforcement that may have been damaged by corrosion or from other causes should be adequately spliced with new steel so as to maintain the original strength. In some cases it may be advisable to include additional reinforcement in the patch.

Patches must be kept moist for several days or all the care taken in the preparation will be defeated. A sheet of polythene stuck or otherwise fixed over the patch is a great help.

Pneumatically applied cement mortars (gunite or shotcrete) are frequently used in the repair of concrete structures. When properly proportioned, mixed, placed, and cured, the coating is dense, hard, and very strong. Problems may, however, occur due to the drying shrinkage of the gunite, resulting in failure of the bond to the old concrete. Normally this work is done by specialist firm.

Epoxy resins are sometimes used for concrete repairs. Generally, preparation should be similar to that for other methods of repair, except that the concrete should be as dry as possible at the time the repair is made. The epoxy-bonded mortar or concrete is compacted into the hole and finished as with cement patches. Temperatures are important since epoxy resins cure and harden more quickly as the temperature increases. Hardening is too slow for the practical use of epoxy resins below about 5°C.

Problems have arisen with epoxy repairs due to:

(1) Inaccurate proportioning. This must be far more accurately done

than is usual for cement mortar.

(2) Dampness from added sand. If sand is added it must be really dry. Slight dampness is sufficient to affect setting.

(3) Not keeping well within the 'pot life' after mixing. This becomes shorter as the temperature increases. Failure to use epoxy mortar soon enough after mixing leads to weak friable patches with poor adhesion to the concrete.

Plastics age and shrink to some degree with time. This shrinkage is liable to produce high stresses at the boundaries of patches, which because of the high strength possible in the epoxy mortar, may cause pulling away in the concrete itself. Generally, epoxy mortar patches should only be made to strong concrete.

Epoxy mortar, in common with many other plastic based materials, loses strength considerably as its temperature rises above about 50°C. As this sort of temperature may arise on a dark surface exposed to the sun, epoxy mortar patches are not suitable where this condition may arise, especially if abrasion or other mechanical wear is to be resisted.

Repairing and Filling Cracks

In such structures as dams, reservoirs and culverts, the formation of cracks due to shrinkage of the concrete or to settlement of the structure is liable to have serious consequences. Sealing of these cracks is difficult, perhaps the best method being to pump liquid cement under pressure into specially prepared boreholes.

Liquid cement will not, however, penetrate cracks less than about 0·5 mm wide. For smaller cracks, chemical solutions may prove satisfactory as a means of reducing seepage of water. The plastics chemist is providing an increasing variety of materials for injection processes. Careful selection is needed if such a material is to provide long-term impermeability.

Other methods have been suggested from time to time including the use of lead-wool caulking, red lead and other materials, but have not proved satisfactory. The cutting of a groove which is patched with cement mortar is more likely to result in two cracks at the edges of the groove than to provide a permanent seal.

It is of little use repairing cracks due to subsidence without treatment of the foundations.

REFERENCES

1 BLAKE, L. S., KINNEAR, R. G. and MURPHY, W. E., Recent research into factors affecting the appearance of in situ concrete, *Symposium on Surface Treatment of In Situ Concrete*, Sept. 1964. Cement and Concrete Association.

2 *A Guide to Specifying Concrete*, Part 1—*Type 1 Specification*, A Report of the Concrete Specification Committee, Institution of Civil Engineers.

3 DEPARTMENT OF THE ENVIRONMENT, *Specification for Road and Bridge Works*, 5th ed. 1976.

4 *Recommendations for the Production of High Quality Concrete Surfaces*, Publication 47.019, reprinted 1974. Cement and Concrete Association.

18
Formwork

THE economical design and construction of formwork is of great importance since its cost may be one-thrid, or even more, of that of the whole concrete structure.

In addition, the appearance of the finished concrete surface and the speed with which the construction is done are largely dependent on the use of properly constructed forms arranged in a manner to give the most efficient usage.

It is not possible here to cover the many problems which commonly confront the designer and builder of formwork. There have been many developments in the design and construction of formwork in recent years, and in particular many proprietary systems have been introduced on to the market. A significant trend has been the growing use of large panels which can be erected and stripped simply and quickly. The scope and specialist nature of the work may be appreciated by reference to the 'Joint Report on Formwork' issued jointly by the Concrete Society and the Institution of Structural Engineers[1].

Design and planning

The hydrostatic pressure of wet concrete depends mainly on the rate of filling, the rate of setting and hardening of the cement, temperature, the richness of the mix, and the water/cement ratio. In other words, the pressure is dependent on the comparative rates of filling and initial stiffening and setting of the concrete. The rate of pouring is therefore important, especially in walls and columns, since the faster the forms are filled, the higher the hydrostatic pressure becomes at the bottom.

Other factors, including the method of compaction, and the size and shape of the formwork, also influence the pressures. Closely spaced reinforcement may be expected to have a similar effect.

A very simple approach to the design of formwork is to regard the concrete as a fluid with a density of 2400 kg/m^3, which exerts a hydrostatic pressure corresponding to this density. Thus, the pressure exerted by the concrete on formwork whether vertical or horizontal at a depth of H metres will be 2400 H kg/m^2. However, as already stated, the pressure on the formwork will be modified by a number of factors and these should be taken into account in the economic design of formwork.

The results of detailed research into these factors were published in CERA (now CIRIA) Research Report No. 1[2], which included a pressure design chart built up from the test results. This chart, which is reproduced in metric form in Fig. 18.1, is described in the report as follows.

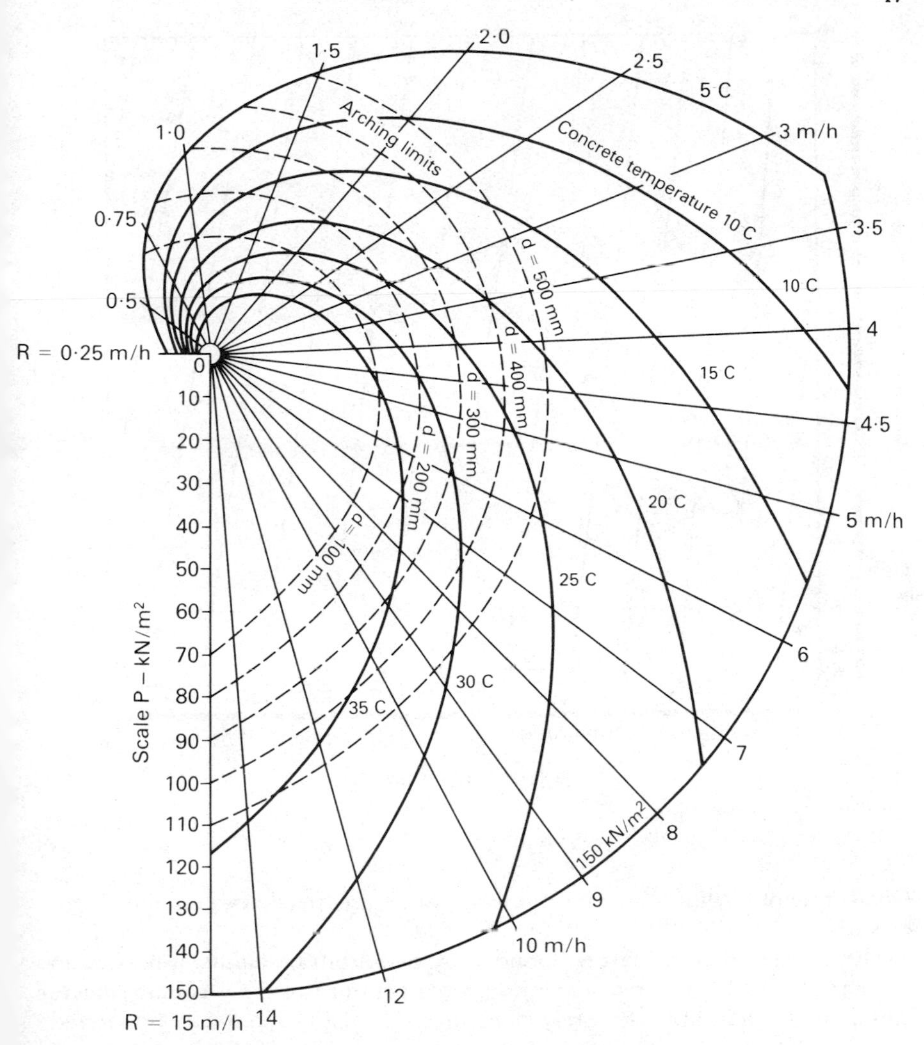

Fig. 18.1 Pressure design chart

The radial lines represent rates of placing *R* from 0·25 m/hour to 15 m/hour while the full spiral lines represent different temperatures of concrete from 5°C to 35°C. The broken spiral lines *d* are arching limits for minimum section dimensions from 100 mm to 500 mm, whereafter arching effects are no longer applicable to most concretes.

The diagram is used by selecting the appropriate rate of placing and temperature of concrete and then measuring the pressure radially, from point O in accordance with scale *P*, to the point at which the lines representing the above criteria intersect. The basic pressure thus derived is then modified by multiplying by a factor *F* (described in the next section) which takes account of concretes of different workability and different degrees of continuity of vibration. The arching limit for the appropriate section is then checked on the same

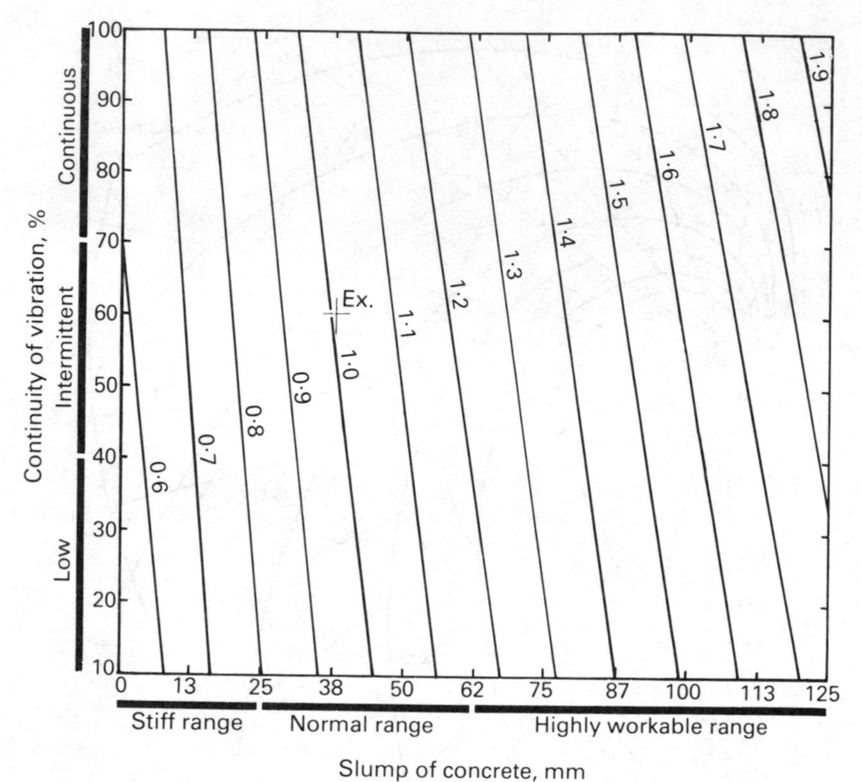

Fig. 18.2 Values of factor *F*

radial 'rate of placing' line and the lesser value used to determine the design pressure.

The pressure design chart is bounded by two arbitrary limits, a maximum pressure of 150 kN/m² and a minimum temperature of 5°C. Circumstances outside these limits are very rarely encountered in practice but may be catered for by use of the pressure formulae in a previous section of the report.

Correction Factor, *F*

The basic design Pressure P_{max} is finally corrected by a factor *F* (see Fig. 18.2) to compensate for concretes of different workabilities and for different continuities of vibration. The design chart in Fig. 18.1 is based on a slump of 40 mm and a continuity of vibration of 55%. An addition of 10 kN/m² is made after the application of factor *F*, to take account of pressure surcharges due to impact.

The shape of the design curve is determined only by P_{max} and the total height of the structure. The upper portion of the curve is triangular in shape and P_{max} was found to occur for all practical purposes at a depth of P_{max}/Δ ft below the top of the lift (Δ being the density of the concrete). The shape of the curve below P_{max} depends upon the degree of movement of the forms relative to the partially stiffened concrete. Where the formwork is highly rigid, little reduction in pressure below P_{max} occurs; where the formwork flexes under load, a redistribution of pressure occurs and the pressure influence line may begin to tail off below P_{max}. The mechanism of the pressure reduction is highly complex, and in the absence of detailed information regarding the degree of rigidity of the forms it is recommended that a rectangular shape be adopted for the lower part of the pressure curve. It should be noted, of course, that P_{max} cannot be greater than the total head of concrete H multiplied by its density, i.e.

$$P_{max} \ngtr \Delta H$$

A worked example follows:

Calculate the design pressure on the formwork for a concrete column 0·75 m × 1 m × 5 m high, to be cast at a vertical rate of placing of 4 m/hour in the summer at an estimated concrete temperature of 15°C. A single 75 mm poker vibrator will be used with about 60% continuity to compact concrete having a slump of 40 mm. The concrete will be discharged from a skip without using trunking. The concrete will be site-mixed and there will be no appreciable delay between mixing and placing.

Δ = 2400 kg/m³
H = 5 m (therefore maximum possible hydrostatic pressure
= 2400 × 5 = 12,000 kg/m²
= 118 kN/m²)
R = 4 m/hour (therefore placing time is 5/4h = 75 min)
d = 750 mm (minimum width of column)
concrete temperature = 15°C
slump = 40 mm
continuity of vibration = 60%

From Fig. 18.1, intersection of R of 4 m/hour and concrete temperature of 15°C gives P = 102 kN/m².

From Fig. 18.2, correction factor F = 1·01

$$\begin{aligned}\text{Design pressure} &= (P \times F) + \text{impact allowance} \\ &= (102 \times 1{\cdot}01) + 10 = 103 + 10 \\ &= 113 \text{ kN/m}^2\end{aligned}$$

The shape of the design curve is, therefore, as shown in Fig. 18.3.

There are several requirements in the design of the formwork which must be met in the interests of efficiency and economy. The formwork must be designed to carry all the necessary loads with a reasonable margin of safety. This will cover all the details from the forms to the joists, studs, shores, yokes, walings, props, etc.

Every effort should be made to avoid the continual remaking and changing of forms, in order to prevent undue waste of material. To this end, the formwork should, as far as possible, be made simple and to a symmetrical design. Structural design engineers should be aware that frequent changes in sizes of columns and beams should be avoided whenever practicable, and in many cases it will be found that a standardization of size, which may result in a

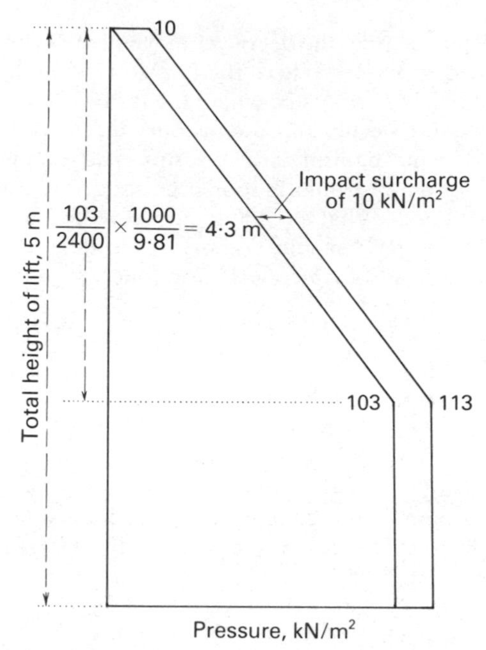

Fig. 18.3 Design curve for worked example

slight excess of concrete, will save many times the cost of labour and formwork materials.

For further information on the details of formwork design, reference should be made to the several publications on the subject[3],[4],[5],[6].

General requirements

General requirements for forms are:

(1) The sheeting should be durable and rigid, and should be braced sufficiently to prevent bulging or twisting. The sheeting should be thick enough to withstand the pressures of wet concrete, and the edges should be true. Warping can be minimized by careful storage when not in use.

(2) The forms and falsework should have sufficient strength to support the loads which they are called upon to carry, without undue deflection. The load will, in certain cases, include not only that of the concrete, but also that of men, equipment and materials used in the construction. It is recommended that an allowance of 1·5 to 3·5 kN/m² is made for the live load on soffit forms during concreting operations by traditional methods[1]. This extra load is, of course, additional to the load from the wet concrete. Sufficient allowance should be made for any settlement or deflection, so that the hardened concrete conforms to the specified line and level. The engineer may also require a camber to allow for the elastic deflection of structural members, and that due to creep of the concrete. It is recommended that beams and slabs in excess of 6 m shall have the formwork pre-cambered by 2 mm for every 1 m of span.

Table 18.1 Examples of Permissible Tolerances for Various Grades of Precision

Type of structure	Nature of irregularity	Tolerances (mm) Grade[1] 2	1	Special
Reinforced concrete foundations (including rafts, ground beams, column bases, pile caps and strip foundations)	Position on plan PD[2] in plan of any point measured from the nearest building grid line	± 50	± 35	± 25
	Dimensions on plan			
	PD per 300 mm	± 15	± 10	± 5
	Maximum permissible deviation	± 50	± 35	± 25
Concrete elements above foundations (beams, slabs, columns and walls)	Position on plan PD in plan of any point measured from the nearest building grid line	± 15	± 10	+ 5
	Verticality Plumbness in height of			
	Over 1·5 m up to 3 m	± 20	± 15	± 10
	Over 3 m up to 30 m	± 30	± 20	± 15
	Cross section and linear dimensions PD from dimensions			
	Up to and including 300 mm	± 10	± 5	± 3
	Over 300 mm up to 600 mm	± 15	± 10	± 5
	Over 600 mm up to 1·5 m	± 20	± 15	± 10
	Over 1·5 m up to 3 m	± 25	± 20	± 15
	Over 3 m up to 15 m	± 40	± 30	± 20
	Abrupt changes of a continuous in-situ surface (e.g. at joints in formwork)	6	4	2

Notes

(1) Grade 2 accuracy represents tolerances regarded as normal good practice.

Grade 1 accuracy is achieved at some expense by using either special materials for the formwork or special care in the use of normal materials.

Special grade tolerances are even more refined and can only be achieved at great expense by extra care in the use of high quality formwork materials and elaborate inspection procedures.

(2) PD indicates permissible deviation (i.e. tolerance).

(3) Forms should be correct to line and adequately braced. Reasonable tolerances are necessary in order to avoid the considerable expense involved in working to close limits, and it is advisable to specify them whenever possible. The tolerances given in Table 18.1 for the finished concrete are taken from PD 6440[7], to which reference should be made for more detailed information. These tolerances are intended only as a guide to general practice, and it is important to remember that the cost of producing close tolerances is considerably greater than is generally appreciated. The definition of the grades is based on the information given in the report on 'Formwork' produced by a joint committee of the Concrete Society and the Institution of Structural Engineers[1]. This report also contains useful tabular information on examples of facilities required to produce different grades of accuracy.

(4) The junction between the formwork and any previously cast concrete

Fig. 18.4 Concrete wall at a power station. Slight lipping has occurred at several of the joints where the shutter panels were not tight against the concrete previously placed

should be tight in order to avoid any leakage of cement paste or mortar and subsequent ridging at this point. Lipping due to slight movement in steel forms is shown in Fig. 18.4. The provision of foamed rubber strips on the forms will reduce the loss of cement paste.

(5) The joints between the various members making up a form should be tight enough to prevent any material leakage of the cement mortar, as such leakage causes defects in the appearance of the finished work. It must, however, be remembered that high concrete pressures will expose any weakness in the formwork, such as joints which may appear tight on inspection prior to casting but which may open as the result of deformation of the forms under pressure.

(6) Forms and formwork must be of reasonable size and weight for easy handling with the equipment available. The forms should be capable of being easily and speedily erected and should be designed so that striking can be done easily and in the required sequence.

(7) The pattern of the formwork, arrangement of vertical stop ends and the general quality of the form surface should conform with the requirements of the finished appearance of the structure. Straight and true angles, arrises or edges are particularly important.

Timber for formwork

At one time the use of timber for formwork sheeting and props was very widespread but for both technical and economic reasons, it has been used to a much less extent in the United Kingdom in recent years. It is, however, still used extensively in the USA and Western Europe. In the case of formwork sheeting the use of timber boards has been very largely replaced by plywood whilst steel hardware has become the accepted replacement for many structural accessories, e.g. props.

Redwood (fir) is generally regarded as the best all-round timber for formwork, having a higher resin content than whitewood (spruce), which tends to make it more durable, and gives more uses. Hemlock is also used to some extent. Hardwoods are difficult and costly to work and are, therefore, little used for formwork. They are, however, very valuable for caps and wedges and for details where a large number of repetitions are to be made.

Timber which is only partially seasoned, is the most satisfactory for use in formwork sheeting. Kiln-dried timber has a tendency to swell when soaked with water from the concrete, and if the boards for formwork sheeting are tight-jointed this swelling causes bulging and distortion. On the other hand, green timber dries out and shrinks, causing fins and ridges on the concrete, although this tendency can be checked to some extent by keeping the boards wet until the time of placing.

When using seasoned timber the joints should not be made too close, since provision should always be made for some slight swelling without any undesirable results. Old timber which has previously been used in concrete forms is usually more absorbent and has a rougher surface than new timber, and, therefore, gives a different surface texture.

Timber may be rough sawn or wrought (planed). Wrought timber gives more accurate assembly and a smoother surface finish if that is desirable. As indicated in the chapter on surface finishes, the carefully planned use of rough sawn boards gives an interesting finish.

Every effort should be made to arrange the work in such a way that cutting of timber is kept to a minimum since the cost of timber is considerable and a surprising amount of waste can occur.

Timber joists are commonly 100 mm × 50 mm but may vary in size up to 225 mm × 100 mm, according to the class of work. Studs and walings may vary from 100 mm × 50 mm to about 150 mm × 150 mm, whilst bearers usually vary from 150 mm × 50 mm to 225 mm × 50 mm. Timber sections about 75 mm × 50 mm are useful for general bracing and light forms.

Plywood for formwork

Plywood sheeting is now extensively used for many categories of concrete work. It is generally stronger than softwood timber and it gives a surface to the concrete which is smooth in comparison to that given by timber boards. It is readily amenable to being bent to form curved surfaces (see Fig. 18.16).

The thickness of plywood generally used varies from 4·5 mm to 32 mm, but for most work a thickness of 18–19 mm or 15–16 mm is used. Sheets of plywood are made up in the factory by superimposing a number of plies with a coating of adhesive between successive plies. The plies are laid so that the grain runs in directions at right angles to each other in adjacent sheets, but the grain of the two outer sheets is always in the same direction. Most plywood is,

Fig. 18.5 Steel-framed Acrow U-forms being used for wall construction

therefore, significantly stronger when spanning over supports laid at right-angles to the direction of the grain of the facing veneers.

Sizes of sheets vary from 2240 mm to 3360 mm in length and from 840 mm to 1400 mm in width, but the standard stock size sheet is 2240 mm × 1120 mm.

Commonly available types of formwork plywood in the United Kingdom are made from Douglas fir from North America, birch from Finland, and certain hardwoods from Africa and Malaysia. The properties of these plywoods may be found in BS Code of Practice CP 112 : 1971 (8).

Plywood has considerable strength in itself and close boarding is not required behind it. Some indication of the spacing of the supports may be obtained from the following figures when using 4·5 mm plywood. For floor slabs 75 mm thick, the supports should be spaced 125 mm apart, while if the slab is 250 mm thick, the spacing should be reduced to 75 mm. A suitable spacing in wall forms where the lift is not more than 900 mm high is 50 mm. With 15 mm plywood these spacings can be multiplied by about ten quite safely.

Plywood may be used with untreated surfaces (apart from the application of a release agent), but where a large number of uses is required from each panel, a surface coating should be applied. Surface coatings should preferably be applied by the plywood manufacturer. Surface treatments with barrier paints or varnish can be expected to give up to about 20 uses with reasonable care whilst over 30 uses may be obtained with the more expensive plywoods which have been overlaid with plastic facing. Surface coatings can give a very smooth finish to the concrete, but the glossy finish produced by some of the

Fig. 18.6 Steel-framed Acrow U-forms being used for construction of floor slab

plastic faced plywoods is unfortunately often associated with unsightly dark areas on the surface of the finished concrete.

Since the joint lines between adjacent panels will show on the concrete surface, it is desirable that the plywood should be fixed in panels conforming with the general pattern of the structure. The joints between the panels are usually simple butt joints but they should be sealed with foamed plastic strip or other suitable sealing material, where prevention of grout loss is important. The edges of plywood panels are vulnerable to the access of moisture and consequently to subsequent splitting and fraying. It is, therefore, desirable to seal all edges and this may be done by the application of a synthetic resin or other waterproofing agent. Attention should be given to the maintenance of this edge sealing if maximum usage is to be obtained from the panels.

It is advisable to test plywoods of unknown performance before large scale use since unsatisfactory surface finishes have occasionally occurred with some

Fig. 18.7 Travelling formwork for a large concrete culvert

uncoated plywoods because of the presence of sugars in the wood. These sugars retard the setting of the cement in the surface skin and cause dusting of the concrete surface. This problem is associated with the plywood being exposed to strong sunlight over a period of several days. It may be minimized by giving the surface of the plywood an initial wash of neat cement grout which is brushed off after it has been allowed to set. The grout must be applied and removed before the application of the release agent.

Metal forms

Various types of metal form are in common use in constructions which allow a large number of repetitions. Some successful proprietary formwork employs standard individual plywood panels supported in metal frames. The application of such a system is shown in Figs. 18.5 and 18.6. Steel forms have been commonly used for tunnels, sewers, culverts and circular columns, and often for floor and column construction. Steel moulds are used extensively for the manufacture of many types of precast concrete products.

An example of the use of steel forms in the construction of a large culvert is shown in Fig. 18.7. The internal and external forms in this instance are carried on steel railway tracks along which they are moved as concreting proceeds.

It is advisable when using steel forms to protect the back from hot sun.

Formwork hardware

Common round wire nails are used in sizes ranging from 50 mm to 100 mm, a common rule being to use a nail at least twice as long as the thickness of the timber nailed on. The holding power of nails is indicated in BS Code of Practice, CP 112 : 1971 [8].

Tie bolts are generally used in sizes from 12 mm to 20 mm for walls and columns. Rods which are threaded at both ends and fitted with nuts are sometimes used as an alternative to bolts. Plate washers not less than 50 mm square should be used in conjunction with bolts or rods since otherwise the bolt heads and nuts are apt to bite into the timber and cause gradual movement which will affect the finished dimensions of the wall.

A wide range of wall ties is available for connecting opposite formwork faces. A commonly used tie, which also serves as a spreader to keep the forms apart, has a metal centre section with screw sockets which remains embedded in the concrete whilst the bolts and cones on each side are removed and re-used. The holes left in the surface of the concrete by the removal of the bolts and cones have to be made good. It is advisable for the ends of the metal centre section to be at least 40 mm from the surface of the concrete in order to avoid possible corrosion trouble. There are several types of wall tie which are obtainable, of which the less elaborate types can be installed more quickly.

Many devices have been developed in order to facilitate the erection and removal of forms. These include such devices as the many and various sizes and shapes of column clamps and yokes, anchor ties, hanger ties, adjustable steel props, centerings for floor slabs, and jacking arrangements. Most of this hardware is satisfactory and economical provided it is limited to work for which it is intended and that care is taken to see that the various components are not lost or damaged since their cost is usually high.

Stop ends

Vertical stop ends cause problems and considerable costs when they have to be constructed in and around congested reinforcement. Often pieces of timber cut to shape are wired to the reinforcement or wedged in place. Another method is to use expanded metal wired between the reinforcement and to leave it in place. In order to avoid corrosion it should not project beyond the reinforcement, and timber should be used near the surface.

Effects of vibration

Vibratory methods of placing concrete produce some vibration in the form-work, especially when external vibrators are attached to the forms themselves. This vibration increases the tendency for joints to open and close, the nails to draw out, wedges to loosen, and bolts and nuts to slacken. Extra care must, therefore, be given to these points.

Wedges should be nailed to prevent them from becoming loose. If used in conjunction with steel forms, they should be continuously watched. When external vibrators are used the form bracing should be increased by about 25 per cent, preferably by closer spacing.

Beam and slab forms

The formwork to beams is usually constructed so that the sides may be struck before the soffit, since the latter must be left in place until the beam can

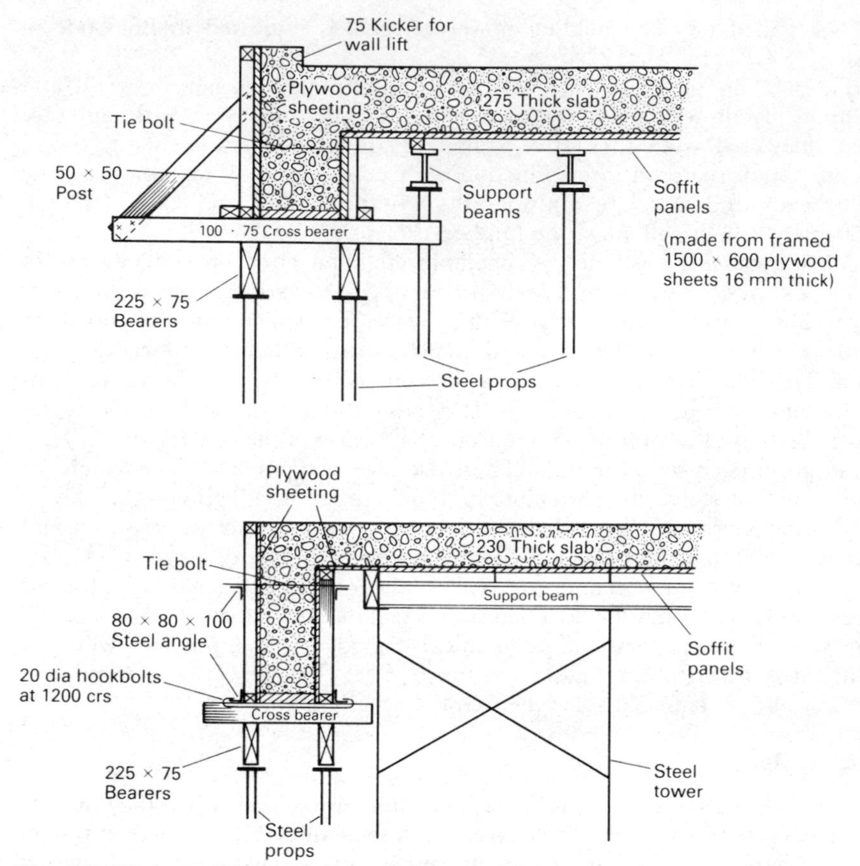

Fig. 18.8 Examples of formwork design for beams and slabs

safely support its own weight. The slab formwork is usually supported by that to the beam.

Examples of forms for beams and slabs are given in Figs. 18.8 and 18.9. Figs. 18.10 and 18.11 show the construction of forms on site.

Column forms

Typical formwork details for a small column and a high column are shown in Figs. 18.12 and 18.13 respectively. The column side panels are made from plywood sheeting and it can be seen that two sides have a width equal to the width of the column face whilst the other two sides have a width which provides an overlap to suit the formwork arrangement. The panels are held together in this instance by adjustable steel clamps which are fitted around the column formwork after erection. These clamps consist of four arms of flat steel which are provided with slotted holes for wedge fixing. This design of clamp permits an infinitely variable adjustment up to a specified maximum size. Other methods can be equally suitable.

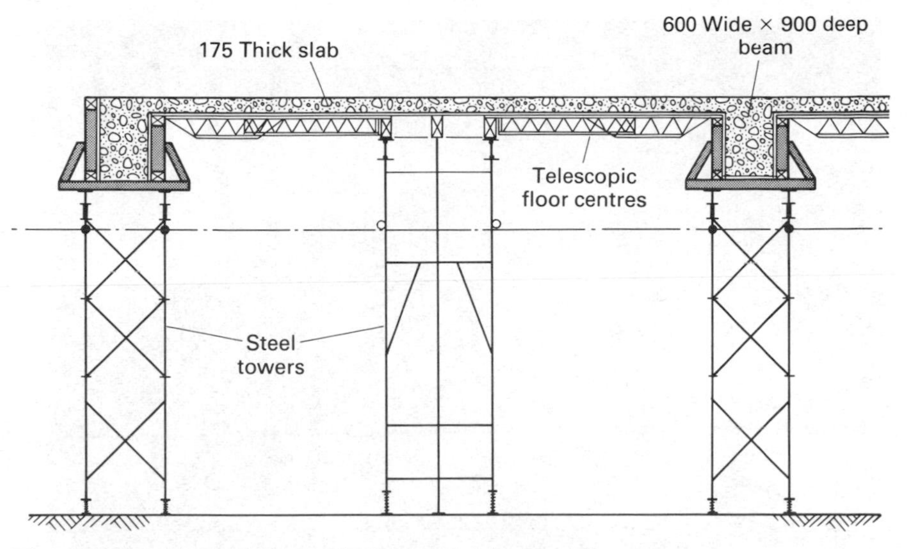

Fig. 18.9 Formwork and falsework for beam and slab construction

Fig. 18.10 The construction of wooden forms for reinforced concrete beams and slabs

In the case of high columns, as in Fig. 18.13, the spacing of the steel clamps must be carefully calculated to take account of the pressure exerted by the concrete. A closer spacing will obviously be required at the bottom of the column to cater for the higher pressures which will arise when the concrete is placed to the full height of the column in one operation.

For the construction of circular shaped columns, proprietary steel forms are often employed although plywood sheeting or hardboard sheeting are used in certain cases. Purpose-made glass reinforced plastic column moulds have also been used.

Fig. 18.11 Combination of wooden and steel shuttering for floor and beam construction

Wall forms

Fig. 18.14 shows a simple wall form held together with wall ties which pass through the timber studs on each side. The position of the wall ties at the top of a lift is common to the position of the wall ties at the bottom of the next lift. Simple bolts with nuts and washers may be used instead of wall ties but distance pieces to keep the forms the required distance apart will then be necessary. Wire ties should not be used since they are apt to stretch. Also the ends of the wires rust producing unsightly stains.

A large wall form is seen in practice in Fig. 18.15. In this case the panel is built up from plywood sheeting fixed to timber studs and supported by pairs of steel channels, back to back, for walings. Wall ties pass between each pair of channels at intervals.

With most wall forms, it is necessary to provide a certain amount of bracing by means of steel props or timber struts. This bracing is required to resist pressures resulting from uneven placing of the concrete and to keep the formwork in alignment.

Special form linings

Apart from the usual formwork facing materials, i.e. plywood, timber and steel, there are a number of materials which may be used as form linings to produce either smooth surfaces or specially patterned surfaces. Paper is the cheapest type of form lining, but it is difficult to avoid cockling and creases. Also concrete tends to find its way under the paper at edges and laps. Similar problems occur with polythene sheeting and showerproofed cardboard.

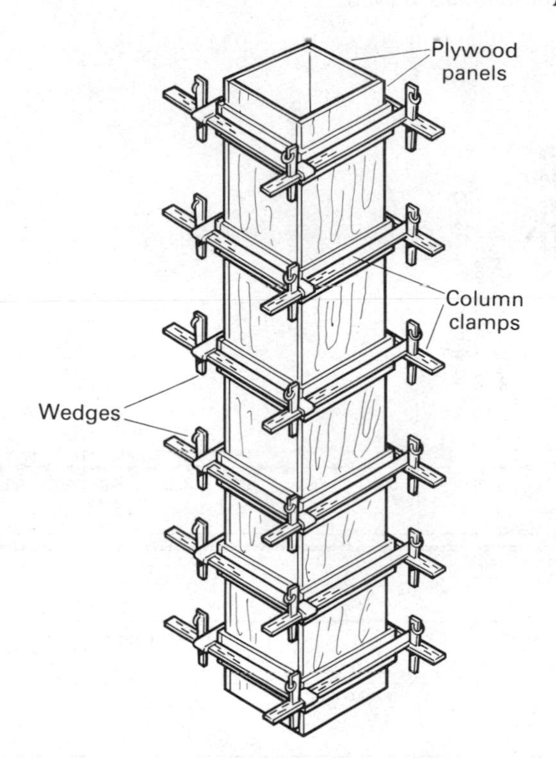

Fig. 18.12 Formwork for small column.

NOTE: (1) Use of standard column clamps to hold vertical panels together.

(2) Props are required on all four sides. These should be raked back to the floor or propped off other members.

(3) A kicker should be provided at the bottom of the column against which the vertical panels can bear

Various types of hardboards, made from wood fibres and resinous glues highly compressed, have been used as liners for flat and curved formwork. Generally, however, the results have been disappointing, mainly because the hardboard swells when it becomes wet and as a consequence tends to ruckle.

Absorptive form linings consisting of softboard about 12 mm thick have been used to eliminate blowholes on the surface of the concrete. These linings can only be used once and are too expensive for general use.

Linings of expanded polystyrene and foamed polyurethane sheets may be used to produce profiled or sculptured surfaces. These materials are easily carved to the desired shape, but care is necessary to avoid damage to them when the concrete is placed and compacted. They can only be used once.

Proprietary rubber linings with various kinds of profiled patterns are available, or can be made specially for particular patterns. They are expensive and are only economical when a high number of uses are anticipated. The profiled pattern should be one that does not present problems in stripping from the concrete. They have to be carefully fixed to a backing of timber or plywood in such a way that the timber can be removed before peeling off the rubber lining.

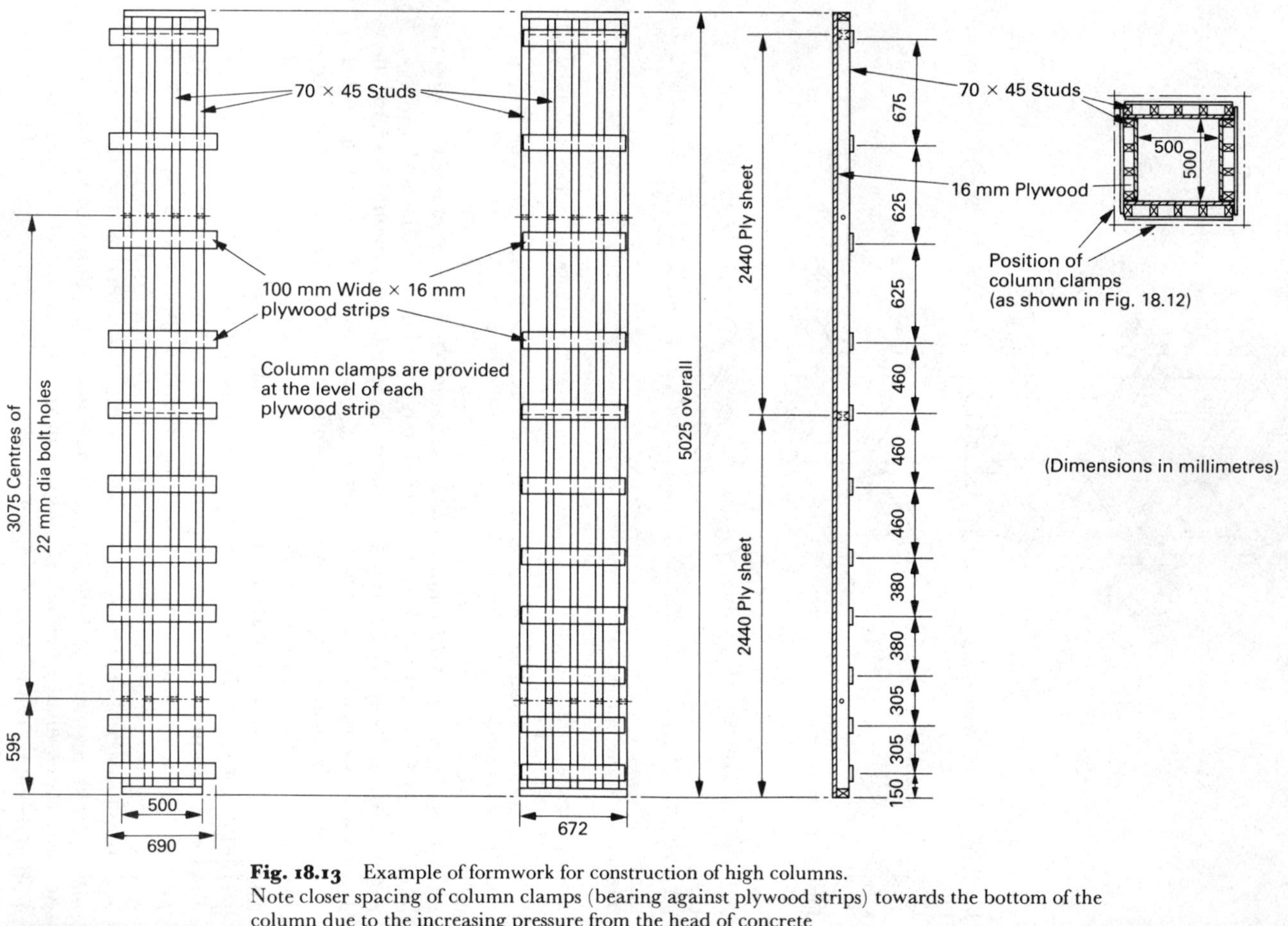

Fig. 18.13 Example of formwork for construction of high columns.
Note closer spacing of column clamps (bearing against plywood strips) towards the bottom of the column due to the increasing pressure from the head of concrete

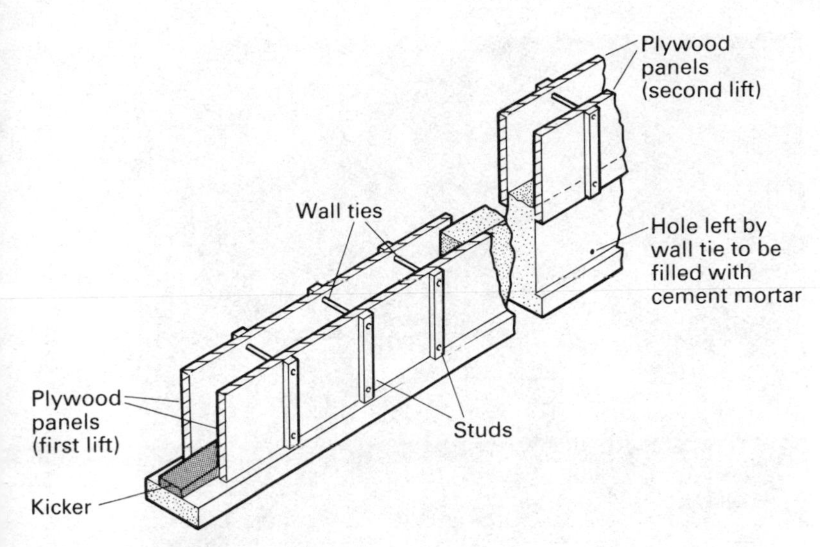

Fig. 18.14 Simple formwork for walls

Purpose-made glass reinforced plastic liners can be used for heavily sculptured finishes, but the initial cost is high and a large number of uses are essential for economic use. Careful attention to the detailing at joints between panels is necessary.

Slipforms

Slipforming is used for continuous vertical construction on such structures as silos, tanks, chimneys, lift shafts and, in certain cases, walls. It can become economic for structures more than about 12 m high. Slipforms are made to slide up the concrete continuously without stopping, and, therefore, avoid loss of time during stripping and resetting. Moreover, continuous pouring day and night is an advantage structurally when watertightness is important as in the case of water tanks.

With slipforming, the forms are lifted by hydraulic jacks operating on steel rods or tubes which are left in the concrete. The forms should be well-oiled and accurately centred before commencing to place the concrete. The jack rods or tubes must bear on the foundation slab and must be kept vertical. Very close control must be maintained on the quality of the concrete. A good cohesive concrete which is not prone to segregation is required and the mix should, therefore, be designed to have a slight excess of well-graded sand. The coarse aggregate, which should not be greater than 20 mm, should be rounded and as free from crushed materials as possible. The workability of the concrete should be of the order of 100–125 mm slump, although concrete with slumps ranging from 0–175 mm has been slipped successfully. A high early strength is desirable.

The forms should be filled about three-quarters full before starting to raise them, and care must be taken to see that the concrete is not dragged up with

Fig. 18.15 Typical large plywood-sheeted formwork panel for wall construction

the forms and is sufficiently stiff at the bottom not to flow out. The raising of the forms should be kept about 0·3 m to 0·4 m ahead of the concreting, the level of the concrete being uniform all round the form.

The rate of climb is dependent on the depth of the forms, the characteristics of the concrete and the ambient temperature. Forms are generally between 1 m and 1·2 m in depth, and rate of climb usually varies between 150 mm and 600 mm per hour.

Some trowelling of the concrete surface soon after it has become exposed may be necessary to remove unsightly slide marks.

An example of the slipform construction using hydraulic jacks and steel yokes is given in Fig. 18.18. Further information can be found in reports dealing specifically with slipforms[9],[10].

Fig. 18.16 Timber and plywood formwork for curved wing walls at Winthorpe Bridge over River Trent

Moulds for trough and waffle floor slabs

Trough and waffle (or coffered) floors are generally adopted for spans which would be uneconomic as solid slabs because of the large dead weight. Standard sizes of inverted U-shaped moulds are available to provide the required shape, the square shapes being required for forming voids in a waffle floor, whereas the rectangular shapes are used for the trough floors. The moulds are usually made of glass fibre reinforced resin, plastic, or steel. The use of glass fibre reinforced polyester moulds, which are particularly light to handle, is shown in the construction of a waffle floor in Fig. 25.1.

Fig. 18.17 Various types of formwork in the construction of a pier. Part of the work was covered at each tide

Falsework

A report on 'Falsework'(11) produced by a joint committee of the Concrete Society and the Institution of Structural Engineers has defined falsework as 'any temporary structure used to support a permanent structure during its erection and until it becomes self-supporting'. This definition covers a wide range of falsework structures and includes such major works as temporary support for a major bridge, and such minor works as temporary props under a window lintel. The report includes a great deal of useful information on falsework and serves as a very useful source of reference. Particular note should be paid to the recommendations on loading, since falsework is often subjected to conditions liable to cause overloading and accidental damage. Even when the heaviest loads are applied in one direction only other effects, such as wind, mobile loads, and rapid placing of concrete, may cause lateral forces. These forces must be resisted by bracing, shores, guys, etc.

Guidance is given on the use of tubular steel scaffolding which, because of its simplicity and adaptability, is widely used for the support of formwork. Fig. 18.19 shows the use of tubular members as falsework for the construction of a bridge. It does, however, suffer from the limitation that it is relatively weak when used as a beam. Special attention must be given to ensuring that all the couplings are fully tightened, otherwise there is a danger of failures which could be serious.

An investigation into the performance of adjustable steel props, which are extensively used for supporting formwork, has been reported by CIRIA(12).

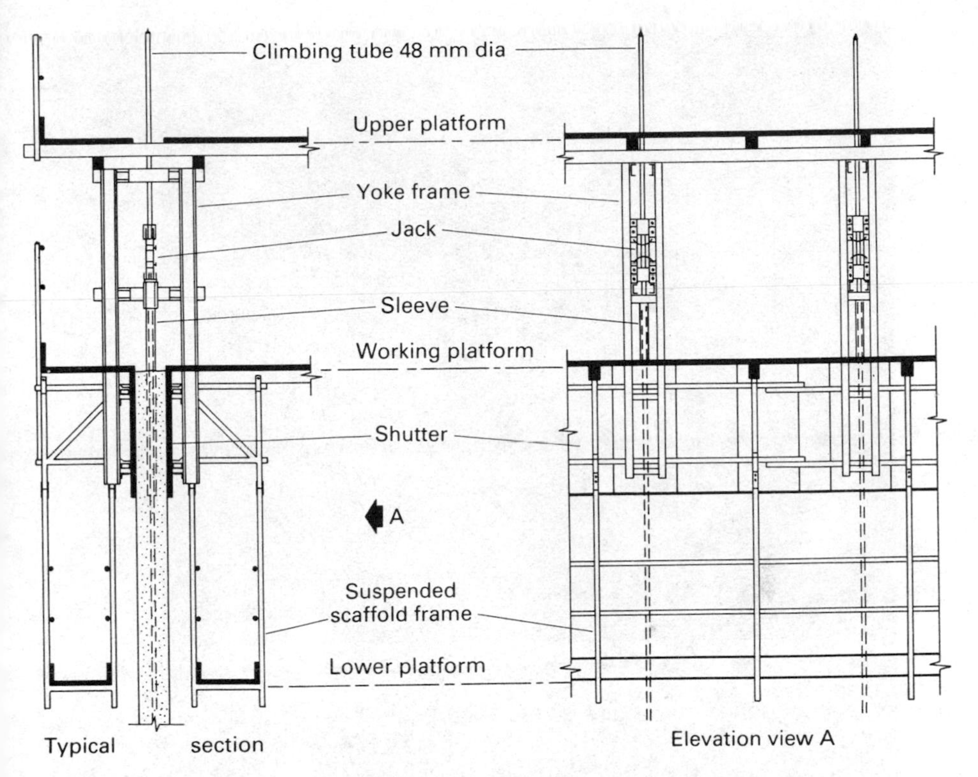

Fig. 18.18 'Siemcrete' system of slipforming showing main components

The tests made in this investigation indicated that the load carrying capacity of props is considerably reduced when they are erected more than 1½° out-of-plumb, or when loaded eccentrically by more than 38 mm. The report recommends that all props should be checked on site for plumb, and that fork-heads or similar attachments should be used in the top of props when they are required to support timber runners.

Removal of forms

Formwork should be designed so that striking may be carried out safely and without damage to the concrete. The length of time that forms should remain in place depends upon the strength of the concrete and the loads to which it will be subjected on removal of the formwork, and will vary with different parts of the structure. In general the formwork can be removed earlier with richer mixes and with rapid-hardening cement.

Forms can be removed more rapidly in warm weather than in cold weather and unless special precautions are taken, it is necessary when the temperature remains round about freezing-point to leave the forms in place for some time. Formwork should not be removed without specific instructions from the engineer in charge, who should make sure that the concrete is strong enough

Fig. 18.19 Tubular members used as falsework for the construction of a bridge

to carry its own dead load and any other loads upon it. In CP 110 : 1972 it is stated that formwork supporting cast in-situ concrete in flexure may be struck when the cube strength is 10 N/mm² or twice the stress to which it will be subjected, whichever is the greater. Some authorities recommend three times the stress.

Table 18.2, which is taken from CP 110 : 1972, gives guidance on striking times for formwork when using ordinary Portland cement concrete. Where it is not possible to determine the surface temperature of the concrete, air temperatures may be used instead, but they are less reliable. Adjustments to the striking times should be made for other types of cement, depending on their rate of strength gain (see Figs. 4.6 and 4.7).

In the case of arches and similar structures, the striking time with ordinary Portland cement concrete should be at least 7 days for spans or diameters up to 3 m and, thereafter, one day per 0·3 m increase up to a maximum of 30 days.

Table 18.2 Minimum Period before Striking Formwork

Type of formwork	Minimum period before striking	
	Surface temperature of concrete	
	16°C	7°C
Vertical formwork to columns, walls and large beams	9 hours	12 hours
Soffit formwork to slabs	4 days	7 days
Props to slabs	11 days	14 days
Soffit formwork to beams	8 days	14 days
Props to beams	15 days	21 days

In cold weather, the recommended striking times should be increased according to the reduced maturity of the concrete. In cases of doubt, it is advisable to err on leaving the forms in place too long, especially as regards the soffits of beams and slabs.

ACI recommendations(13) are not very different from the times given in Table 18.2.

Soffit formwork may be designed so that the soffit form itself may be removed without disturbing the supporting props. Advantage is often taken of such design in proprietary formwork systems, since it is more economic to be able to remove the special form panels for rapid use elsewhere.

In addition to considerations of concrete strength, it may be necessary to give attention to the colour and appearance of the concrete when deciding on striking times. When uniformity of appearance is important, formwork should not be removed until four days after casting the concrete.

Release agents

Numerous materials have been painted on forms to prevent sticking of the concrete and to facilitate removal, such as whitewash, soft soap, grease and oils. Oils are generally regarded as the most satisfactory, and various types have been used extensively on wooden, metal, concrete, and other form materials. The requirements of good release agents are that they shall prevent sticking, and they should not harm the concrete either by staining or by softening of the surface.

The Cement and Concrete Association have undertaken comprehensive investigations into factors affecting the appearance of in-situ concrete and these were reported at a symposium on surface treatment of in-situ concrete, held in London in September 1964. The main conclusions on release agents are given below.

Straight oils may be of vegetable, mineral or fish base and should contain no additives. While they are particularly useful for impregnating timber surfaces and so give to the concrete uniform colour with little staining, they do tend to leave such blemishes as blowholes and pocking.

Oil-phased emulsions, which consist of water globules encased in a continuous phase of oil, are considered to be the most suitable for general use. They should be used as sparingly as possible, and then yield concrete of uniform colour, good durability, resistant to efflorescence, and with few mechanical defects such as blowholes. Staining is minimal. Some types are, however, unsuitable for use on steel formwork, since they have poor anti-corrosion qualities.

Water-phased emulsions consist of oil globules encased in a continuous phase of water. They are not generally satisfactory.

Oils can be purchased containing emulsifying additives, but no water, with the intention that this shall be added on site. Mixing with water on site, however, may not be sufficiently uniform to avoid variations in colour and finish on the concrete surface. Without water they may be used on steel formwork where their anti-corrosive properties may be advantageous.

Similar oils may also be used successfully on concrete moulds. Plaster-of-Paris mouldings should be given one or two coats of shellac before oiling in order to reduce absorption.

The surfaces to which the release agent is applied should be clean and smooth. The oil may be applied by means of an oily rag, a brush, or a spray, but whichever method is adopted, the forms must be covered fully but thinly. A satisfactory rate is 6 to 8 m^2/l. The important requirement that no oil should be allowed to get on to construction joints or reinforcing bars limits the use of a spray once the formwork has been fixed in place.

Surplus oil should not be left on formwork, especially if the concrete is to be plastered or rendered. Not only is there a risk of formation of a soft surface skin, but there is the probability that the surface will be water-repellent, both of which reduce the adhesion of plaster or rendering mortar, if these are to be applied.

A further choice, in addition to the various types of oils, are the chemical release agents which have been developed and used increasingly in recent years. They consist of a chemical, suspended in a low viscosity oil distillate, which reacts with the cement to produce a form of soap at the interface. Provided they are applied lightly by spray, most types can be used in the production of high quality surfaces. They are particularly suitable when a good release from awkward shaped concrete members is required. Another advantage is that they can be applied to the formwork two to three weeks in advance of concreting without incurring any loss in their release performance.

Although chemical release agents are relatively expensive, they are economical to use because they give a better coverage than the oil-type of release agent, and, since they are applied by a spray, there is less wastage. Over-application, however, will result in dusty surfaces due to the retardation of the setting of the cement in the surface skin. In certain cases, over-application has been associated with excessive efflorescence on the surface of the concrete.

Chemical release agents must not be allowed to come into contact with the reinforcement bars because of the resulting adverse effect on the bond with the concrete. Since they are normally applied by spray, it is advisable to treat the formwork before it is fixed in position. Alternatively the reinforcement should not be fixed until after the formwork has been erected and treated.

REFERENCES

1 *Formwork*, Report of the joint committee of the Concrete Society and the Institution of Structural Engineers. Technical Report, No. 13, March 1977, pp. 74.
2 *The Pressure of Concrete on Formwork*, CERA Research Report, No. 1, April 1965. Civil Engineering Research Association (now CIRIA).
3 WYNN, A. E. and MANNING, G. P., *Design and Construction of Formwork for Concrete Structures*, 6th edn, 1974. Concrete Publications, London.
4 SNOW, SIR FREDERICK, *Formwork for Modern Structures*, 1965. Chapman & Hall, London.
5 HURD, M. K., *Formwork for Concrete*. Prepared with the assistance of R. C. Baldwin under the direction ACI Committee 622. 2nd edn, 1969. American Concrete Institute.
6 RICHARDSON, JOHN G., *Formwork Construction and Practice*. Oct. 1977. Cement and Concrete Association.
7 BRITISH STANDARDS INSTITUTION. PD 6440 : Part 2 : 1969, *Accuracy in Building*.
8 CP 112 : 1971, *The Structural Use of Timber in Buildings*. British Standards Institution.
9 O'BRIEN, J., *Principles and Practice of Slipform*. Technical Report TR 33, May 1973. Cement and Concrete Association of Australia.
10 STEIN, J. and DONALDSON, P. K., *Techniques and Formwork for Continuous Vertical Construction*. Technical Report TRC 1, 1966. Concrete Society.
11 *Falsework*, Report of the joint committee of the Concrete Society and the Institution of Structural Engineers. Technical Report TRCS 4, 1971, pp. 52.
12 BIRCH, N., BOOTH, J. G. and WALKER, M. B. A., *Effect of Site Factors on the Load Capacities of Adjustable Steel Props*. CIRA Report 27, Construction Industry Research and Information Association, January 1971, pp. 48.
13 ACI 347–68, *Recommended Practice for Concrete Formwork*. ACI Manual of Concrete Practice, Part 1, 1974. American Concrete Institute.

19
Reinforcement

The design of the various parts of a concrete structure must be related to all the requirements necessary for satisfactory service. In some structures the only consideration may be that the members should be capable of carrying the loads imposed. In others there may be a necessity to limit the deflection of certain parts. In sea structures particular attention must be given to ensuring that adequate cover is given to steel reinforcement to prevent corrosion and subsequent spalling. Similar considerations apply to reinforced concrete in factories where corrosive gases are present, and generally to buildings in industrial areas where the atmosphere is polluted by industrial processes.

Cracking should be controlled to reduce the spacing and size of cracks and thus avoid an unsightly appearance, as well as to reduce the risk of corrosion of the steel. Again, cracking must be kept to a minimum in water retaining structures and oil tanks.

Various requirements concerning design considerations, including strengths of concrete and steel reinforcement, size of aggregate in heavily reinforced sections, distances between bars, cover, bond, and anchorage, and also concerning workmanship, inspection, and testing, are given in CP 110:1972 [(1)].

The basic principles involved in designing simple concrete members to satisfy the requirements for load-bearing capacity are outlined below. Reinforced concrete design is a subject in itself and reference should be made to standard textbooks [(2),(3)].

Elementary principles of reinforced concrete design

The necessity for the use of concrete and steel together arises from the nature of the stresses which result from loading. These are compressive, tensile and shear stresses, and they may occur in different members singly or all together.

Dense concrete has a high compressive strength, but its tensile strength is low. Tests have shown that the tensile strength is, in fact, only about one-tenth of its compressive strength and for the purpose of reinforced concrete design is ignored. On the other hand, steel has a high tensile strength, as well as a high compressive strength when adequate lateral support is provided to prevent buckling. Suitable combinations of these two materials, therefore, provide resistance to both compressive and tensile forces. Shear will be mentioned later.

Fundamentally the combination of concrete with steel reinforcement is made possible by two fortunate physical properties. First, the coefficient of thermal expansion of concrete is almost identical to that of steel and secondly, unless impurities are present, concrete is alkaline, providing an environment

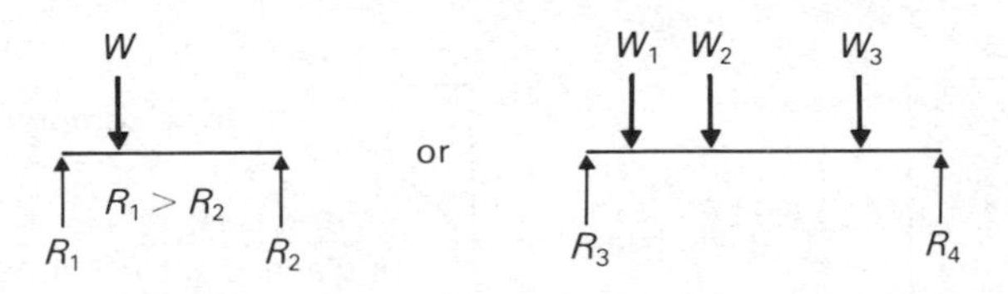

(a) Point loads on a beam

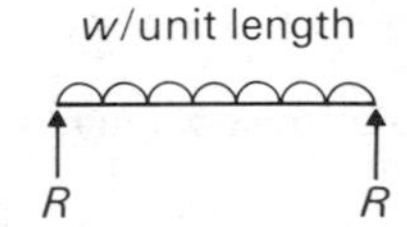

(b) Equally distributed load

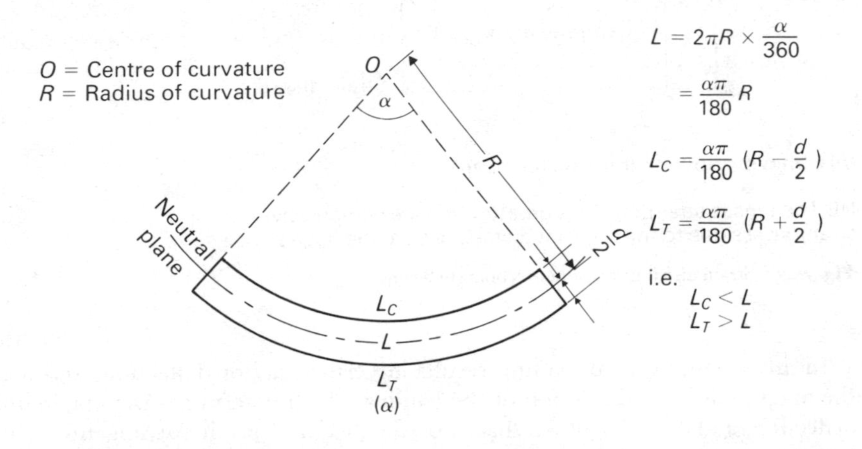

(c) Deflection of beam under load

Fig. 19.1 Loading on a beam

which inhibits the rusting of the reinforcement. A further necessity for the efficient use of steel reinforcement in concrete is that the concrete grips, or bonds well with the concrete, and this will be considered after an explanation of the development of stresses in simple members has been given.

Beams

Loading a beam gives rise to tensile, compressive, and shear stresses, so that a consideration of the design of a beam provides a convenient introduction to the elementary principles of reinforced concrete design. Loading on a beam may be a single, or a number of point loads (see Fig. 19.1 (a)); or otherwise be equally distributed along its length (see Fig. 19.1 (b)); or any combination of these, perhaps, with lateral or dynamic loading due to vibration.

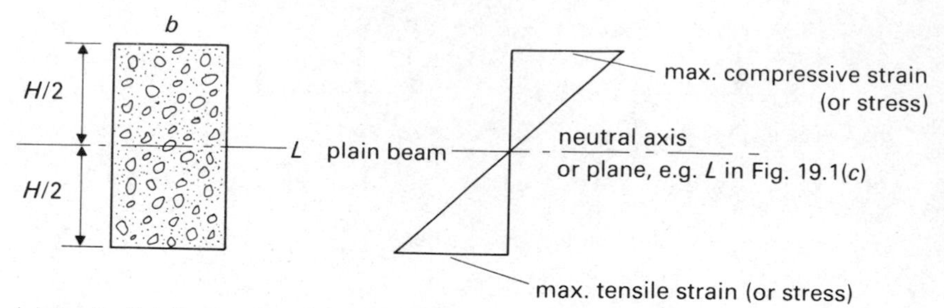

(*a*) Strain distribution for plain rectangular beam

NB Although the strain diagram and the stress diagram are the same shape, the scales will be different

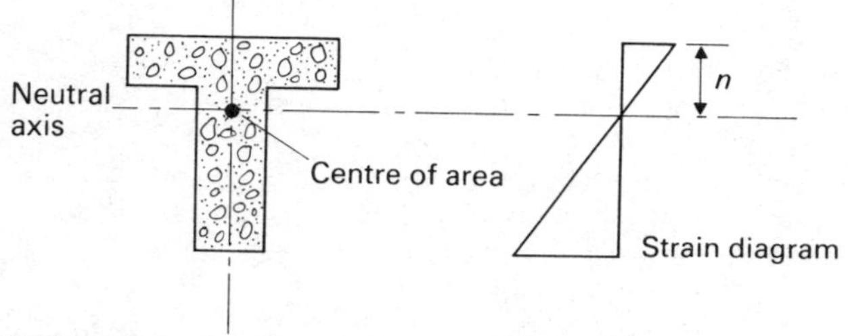

(*b*) Strain distribution for plain tee beam

NB The tensile strain/stress is greater than the compressive strain/stress due to the relevant distances from the neutral axis

Fig. 19.2 Strain distribution in plain concrete beams

In any event, vertical loading results in vertical sag or deflection, and it is the magnitude and disposition of the loading which determines the maximum deflection and the deflected shape of the beam. Certain arrangements of loading result in the deflected shape of a beam being part of the arc of a circle, and for ease of illustration this has been assumed in the following analysis.

When a simply supported beam is loaded to deflect as in Fig. 19.1 (c) the upper layers tend to reduce in length whilst the bottom layers tend to increase in length, i.e. the top is put into compression and the bottom into tension. Furthermore, it may be shown that the change in length, usually evaluated as strain, is proportional to the distance from the centre of the beam, or for a homogeneous (i.e. not reinforced) beam, what is known as the neutral plane or axis (see Fig. 19.2 (a)). For a plain beam the neutral axis will coincide with the geometric axis, i.e. it will pass through the centre of area of the cross section. Fig. 19.2 (b) shows the situation for a tee beam.

A plain beam, of rectangular section, has equal values for compressive and tension stresses, these being at a maximum at the upper and lower faces respectively. The ideal material for such a beam would be equally strong in tension as in compression, giving a balanced economic design. Since concrete is much weaker in tension than compression, it is obvious that failure under a comparatively small load will occur in the outer tension fibres, whilst the

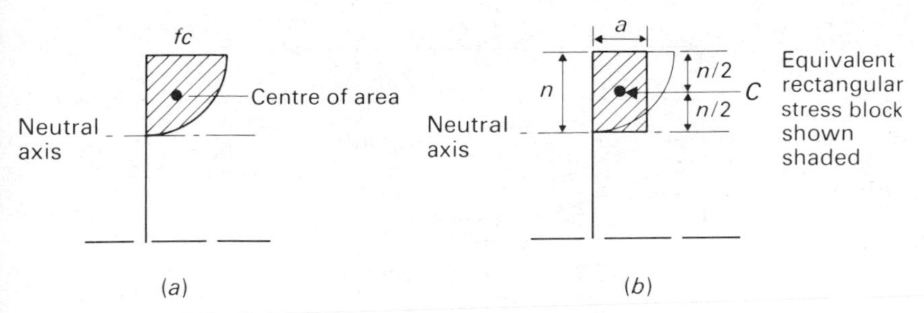

Fig. 19.3 Distribution of compressive stress in a concrete beam

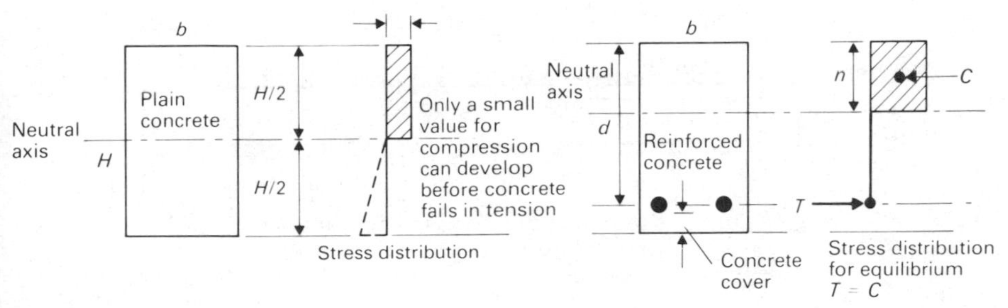

Fig. 19.4 Stress distribution in a concrete beam with bottom reinforcement

compression fibres would be capable of sustaining a much greater load. The beam would thus be uneconomic in design and of little practical use.

For design purposes the small tensile strength of concrete is ignored. A concrete beam therefore, whilst being capable of resisting compression above the neutral axis, has no significant means of resisting tension below that level. In order that the beam functions correctly, resistance must be provided in the tension zone, or putting it another way, that area of the beam must be reinforced.

The strain/stress diagrams shown in Fig. 19.2 are true for an elastic material where stress is proportional to strain. When considering concrete, whereas the shape of the strain diagram is reasonably correct, research has shown that the distribution of compressive stress is not linear, but relates somewhat to that shown in Fig. 19.3 (a) where the curve is either elliptical, parabolic or a hyperbola, depending upon the research worker. This makes the evaluation of any design extremely tedious as the geometry is complicated, and it is usual to work to an equivalent rectangular stress block (see Fig. 19.3 (b)), where the distances of centres of forces from various axes are easily determined. The value of a in Fig. 19.3 (b) depends upon the design philosophy, and reference should be made to standard works on the subject for further information.

If steel bars are placed in the lower half of the beam section near the outer tension face, the strength of the beam is no longer limited by the tensile strength of the concrete and, provided the bars are of the right sectional area and the concrete grips the bars to prevent horizontal slip, the beam can be as strong in tension as in compression (Fig. 19.4). In order to obtain the greatest

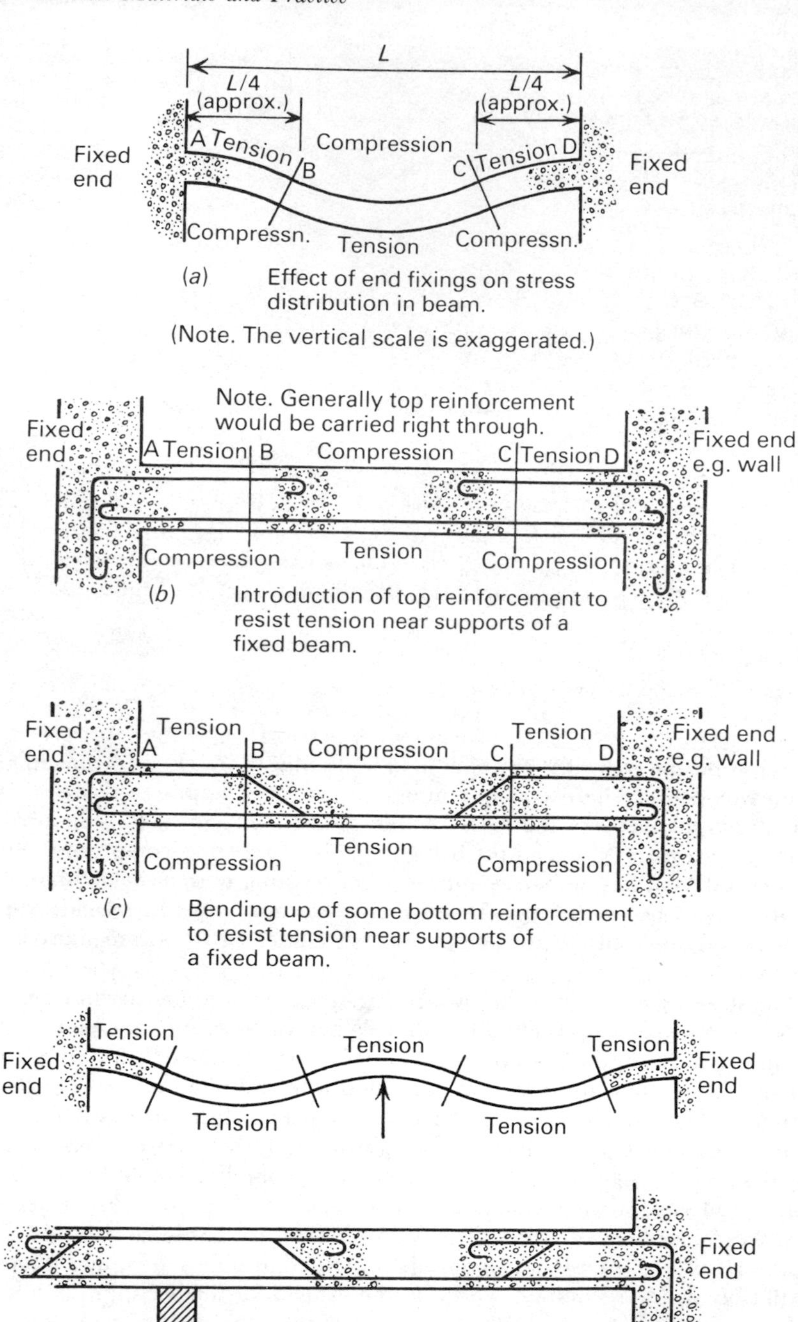

Fig. 19.5 Longitudinal reinforcement in fixed and continuous beams

advantage from the steel reinforcement bars they should be placed as close to the bottom of the beam as possible. Some cover is, however, necessary to prevent corrosion and to provide the necessary fire resistance.

The relative displacement of the neutral axis should be noted in Fig. 19.4. Steel is a very must stronger material than concrete and so only a relatively small area is normally provided, say up to 4% of the beam cross-section area. Nevertheless the area of concrete in compression required to balance the tensile force in the steel is less than that given by the breadth of the beam and the distance to the geometric axis ($b \times H/2$), and so the neutral axis moves away from the geometric axis towards the top face of the beam. This is true for normal reinforced concrete beams. However, for some prestressed beams where the tendon stress is very high the neutral axis can be below the geometric axis.

The area of steel required to give the necessary tensile resistance in a singly reinforced rectangular beam is generally not more than 4% of the area of the concrete. Ideally beam sections are designed so that the design stresses in the concrete and steel are reached simultaneously. It often happens, however, that owing to limitations of space, the size of a beam may be restricted so that additional steel must be placed in the compression zone to avoid overstressing the concrete.

Referring to Fig. 19.1(c), it will be noted that whereas a beam in its unloaded condition is level throughout its length (ignoring at present its self weight), the application of loading causes the two free ends of the beam to become inclined to the horizontal due to rotation. If this rotation is prevented by clamping or fixing the ends of the beam before loading is applied, end fixing moments are introduced and the beam takes the form shown in Fig. 19.5(a). The top surface of the beam is seen to be convex from A to B and C to D, indicating it is extended over these portions and therefore in a state of tension. Likewise, for the bottom surface of the beam, A to B is in compression, B to C is in tension, and C to D is in compression. Two theoretical diagrams for reinforcement are shown in Figs. 19.5(b) and 19.5(c), more practical applications being shown later in this chapter. It becomes necessary to introduce top reinforcement near the supports to take care of the tensile forces there. This may be achieved as shown in Fig. 19.5(b) or by bending up some of the bottom bars as is shown in Fig. 19.5(c).

Continuous beams, i.e. beams which continue over more than one span without joints at each support, deform in the manner indicated in Fig. 19.5(d) and should, therefore, be provided with top reinforcement at each end, as in Fig. 19.5(b), and also over each intermediate support.

If any beam is supported and clamped at one end only, it becomes a cantilever, and is stressed similar to the end portion of a fixed beam as shown in Fig. 19.6(a). The position of the main reinforcement is shown diagrammatically in Figs. 19.6(b) and (c). Mistakes are more often made in the positioning of reinforcement in cantilevers than in beams and slabs due to a misunderstanding of their functional behaviour. Special care should be taken to see that the reinforcement is near the top and not half way down or near the bottom of the beam.

Shear in a beam comprises two components acting simultaneously; horizon-

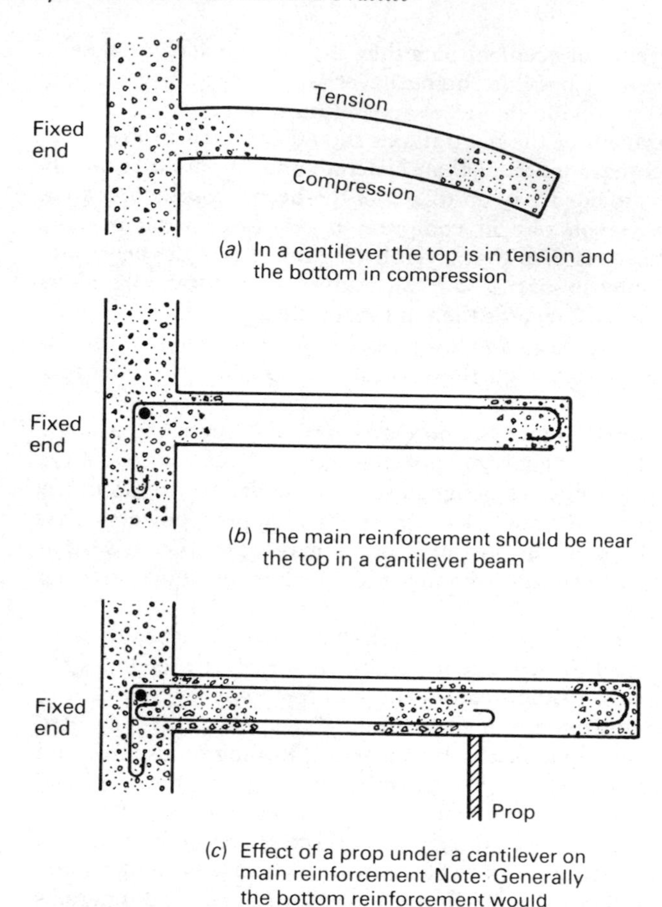

Fig. 19.6 Reinforcement in cantilever beams

tal shear due to changes in length (Fig. 19.1(c) and Fig. 19.7(a)) and vertical shear producing a vertical slicing or guillotine action (Fig. 19.7(b)). Fig. 19.7(a) has been drawn as though the beam comprises a stack of scaffold boards; in a concrete beam this does not happen, and there is a shortening and lengthening, previously mentioned, which appears as a horizontal shear force. Furthermore, it may be shown that for any point along the beam the value of the horizontal shear is equal to that of the vertical shear, the resultant therefore acting at 45° to the beam axis. Normally, the effects of shear are most severe at the ends of a beam or on either side of an intermediate support.

The shear force at any point along a beam gives rise to pure tensile and compressive forces in the concrete, again inclined at 45° to the axis of the beam, at the neutral axis level; away from this level the angle does change due to the combination of other stresses.

The concrete is normally capable of resisting the compression set up by the shear forces, but has very little ability to resist the corresponding tensile forces. Thus, under certain conditions of loading, shear cracks may occur in the lower

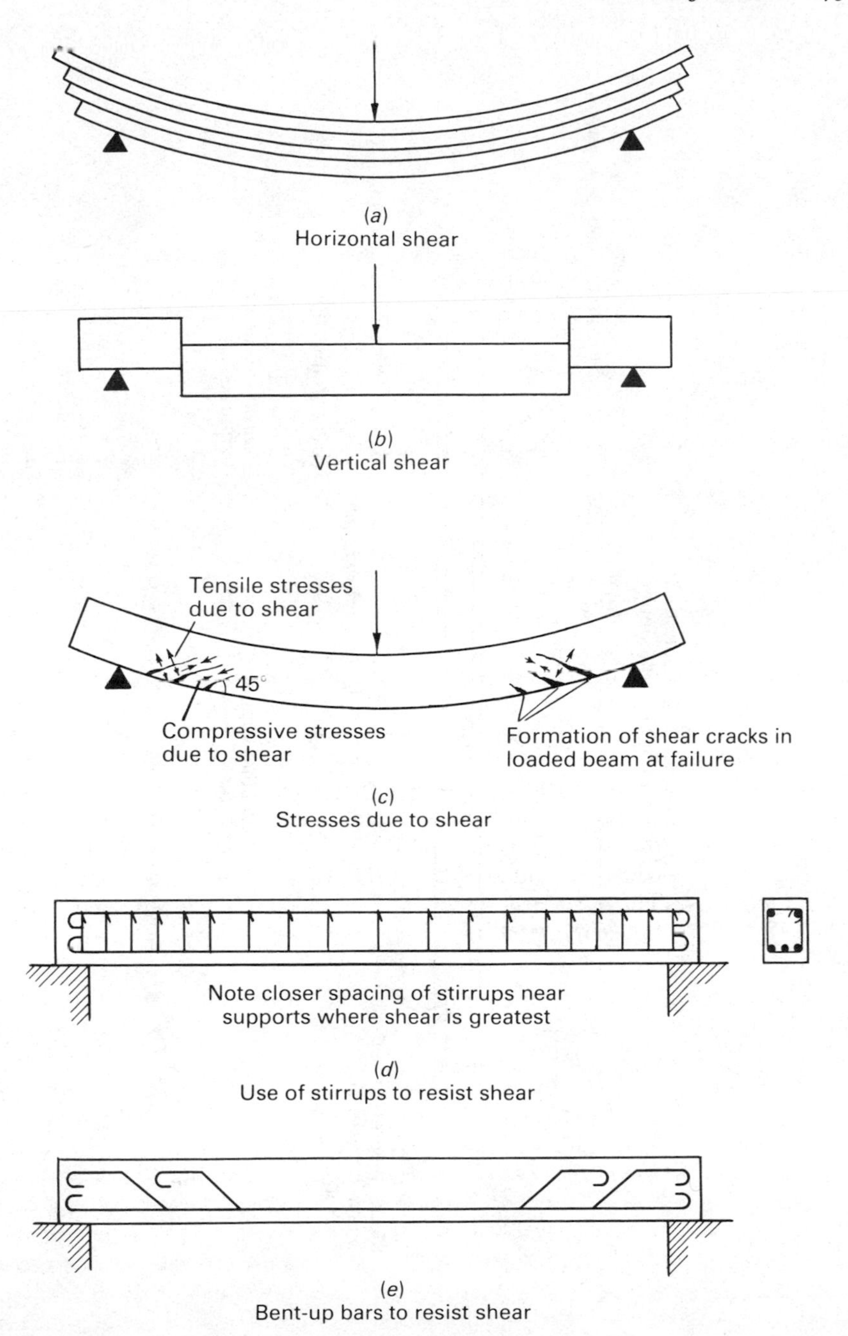

Fig. 19.7 Shearing action in beams. Note that top layers in (a) are actually compressed. In this diagram in which no resistance is assumed between layers (as in a pack of cards) each layer should be of equal length

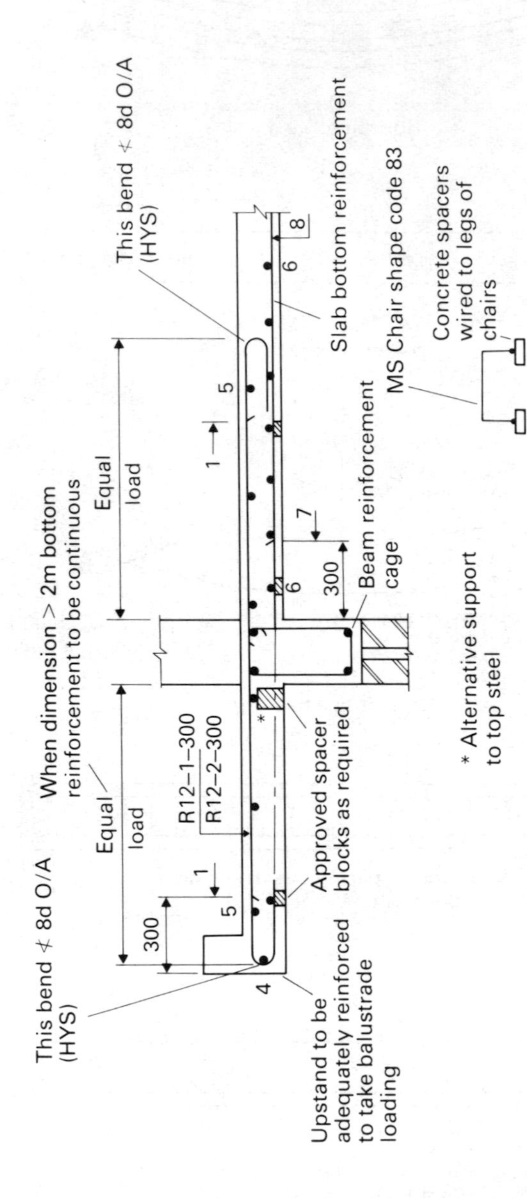

Fig. 19.8 Reinforcement detailing for cantilever over external beam

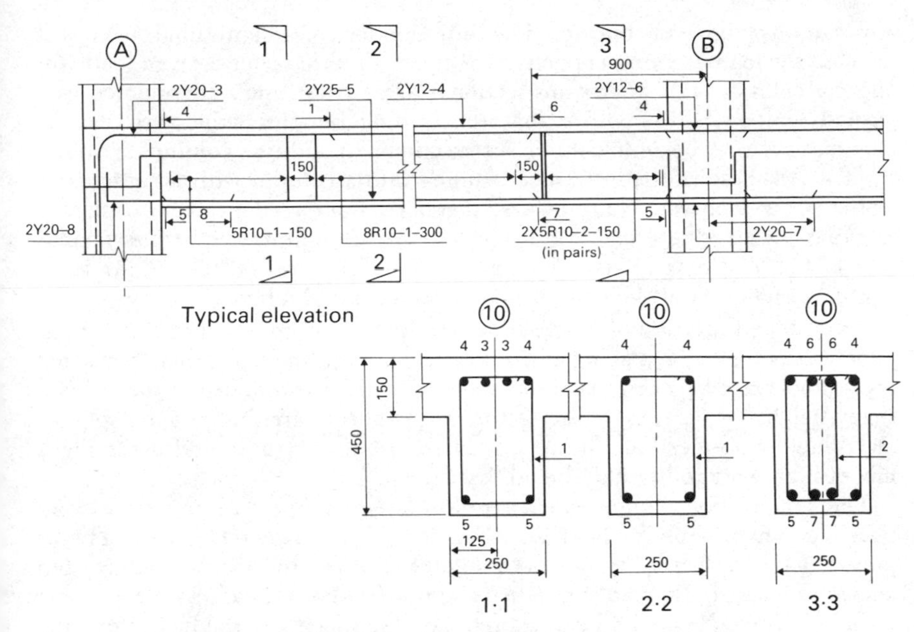

Fig. 19.9 Reinforcement detailing for continuous rectangular beam

half of the beam near the supports. These cracks will be perpendicular to the tensile plane, i.e. again at 45° to the long axis of the beam. A classic shear crack pattern is illustrated in Fig. 19.7(c). Shear reinforcement must then be provided in addition to, and to act with, the longitudinal reinforcement. Ideally such reinforcement is most effective if it is placed in the plane of tension and may be provided in the form of bent up bars as in Fig. 19.7(e). In practice the angle of these bars to the horizontal may vary from 30° to 60°.

Another method of providing shear reinforcement, giving easier steel bending and fixing, is in the form of vertical stirrups as in 19.7(d). This also gives the designer the facility of varying the spacing of the stirrups to satisfy the shear force at any point along the beam. For practical and design reasons, bent up bars are always used in conjunction with vertical stirrups.

In structural beams, shear reinforcement is always provided even though the design shear stresses are very low and may be thought to be adequately resisted by the concrete. CP 110 : 1972[1] states that a minimum, nominal amount of shear reinforcement must be provided. In the event, the provision of vertical stirrups allows beam cages to be prefabricated and facilitates the fixing of the reinforcement on site, with consequent savings in overall cost. Stirrups must pass around the main tensile reinforcing bars and must be anchored at both ends in order to develop the full working stress.

The points enumerated above are illustrated in Figs. 19.8 and 19.9, which are practical examples of reinforcement detailing.

Columns

Short plain concrete columns are capable of carrying considerable axial load, as the stresses produced are compressive only and the short length avoids any sideways bow or buckle. The introduction of longitudinal bars will increase the load carrying capacity of a given column, or for a given load will allow a column of smaller cross-sectional area to be used. In practice it is virtually impossible to ensure that the loading remains axial and that no moments occur at beam to column intersections. In addition columns are often slender. All these effects introduce bending into a column with a consequent tensile stress. Concrete columns are, therefore, reinforced to avoid unsightly cracking or a collapse condition. CP 110 : 1972[1] limits the cross-sectional area of the longitudinal reinforcement to between 1% and 6% of the gross cross-sectional area of the column. Bar sizes are also limited.

When the proportion of longitudinal reinforcement approaches 6%, serious difficulties arise in the placing of the concrete around the bars due to restricted access and space between bars (concrete is normally introduced at the top of a storey height shutter), especially if there is a zone where the bars are spliced. Detailing continuous bars, using a smaller size of aggregate and designing a mix of high workability may be advisable.

The compression reinforcement requires lateral support to prevent it from buckling outward under load and bursting off the concrete cover. This is provided by the use of transverse reinforcement, in the form of lateral ties known as links, so disposed that every longitudinal bar is held against outward buckling. The ends should be anchored, and the spacing of the links along the length of the column should be 12 times the diameter of the smallest longitudinal bars.

In order to resist buckling of a reinforced concrete column it is necessary to place the bars as near to the outer surface as possible, consistent with adequate cover.

Slabs

Along with beams and columns, slabs complete the three basic elements of most structures. Slabs may consist of precast units, or be of in-situ concrete cast, as far as possible, monolithic with the supporting beams. Only the latter will be considered here.

The behaviour of slabs under load, in respect of end conditions and over intermediate supports, is similar to that of beams. Depending upon the shape of the slab panel under consideration, i.e. the ratio of length to breadth, it may be designed as spanning one way only, or in two directions usually perpendicular to each other. In practice a slab designed to span one way only will try to some degree to span in the other direction owing to the monolithic nature of the construction. A minimum amount of reinforcement is, therefore, provided at right angles to the main reinforcement, and is described as secondary or distribution steel. The disposition of reinforcement between the top and bottom of a slab is similar to that for beams.

Slabs may be reinforced using a prewelded fabric mesh of small diameter wires, usually supplied as standard sheets or rolls, or by using rod reinforcement.

General

It is not possible to deal here with other reinforced concrete sections, but the brief description of the elementary principles of reinforced concrete design should be sufficient, not only to indicate the necessity of fixing reinforcement accurately in the positions determined by the designer, but also to enable the concrete engineer to appreciate the possible serious consequences of any errors. It is most important that there should be full compliance with the requirements given on the working drawings by the designer.

Bond and anchorage

An effective bond between the concrete and the reinforcement is essential since the efficient use of combinations of steel and concrete is dependent upon the transfer of stresses from the concrete to the steel. The bond strength, or the measure of the effectiveness of the grip between the concrete and the steel, is best defined as the stress at which a very small slip occurs. Bond is due initially to adhesion and frictional resistance, but as soon as slipping begins the adhesion fails, and the subsequent bond is due to friction and mechanical resistance.

The bond strength is improved by increasing the crushing strength of the concrete but is not directly related to it. The development of bond strength is most rapid under moist curing conditions.

Deformed bars, i.e. with a pattern of raised ribs or square twisted, permit higher bond stresses than plain round bars. Furthermore, deformed bars are classified in CP 110 : 1972 as Type 1 or Type 2 in respect of bond characteristics. Type 1 bars are generally square twisted, whilst Type 2 bars are generally those which are rolled to give a pattern of raised transverse ribs. For a given set of conditions Type 2 bars permit the highest value of bond stress in design. The decision as to whether any bar is a Type 1 or a Type 2 bar, depends upon the results of tests carried out by the manufacturers for their standard products. The accepted ratio for practical design purposes in the bond strength of plain bars, square twisted bars and ribbed bars is 1 : 1·4 : 1·8.

Bars which are stressed must extend for a sufficient distance from any stressed section to develop a satisfactory resistance to failure of the bond between the bar and the concrete. Values for permissible bond stress and methods of calculating the bond stress are given in CP 110 : 1972.

Bending dimensions for reinforcement are given in BS 4466 : 1969. In general, hooks and bends are used at the ends of bars to reduce the length of straight bar that would otherwise be required for anchorage. Standard hooks for mild steel bars should have an inner diameter of at least four times the diameter of the bar, whilst if high yield steel is being used the inner diameter should be at least six times the diameter of the bar. For both types of steel the length of the straight portion of the bar beyond the end of the curve to the end of the hook should be at least four times the diameter of the bar.

Loose or flaking rust, paint or oil on a reinforcing bar is detrimental to the bond and must be removed before fabrication of the reinforcement. Cement grout on a bar is acceptable provided it firmly adheres to the bar.

Fig. 19.10 Shrinkage cracks in a reinforced concrete plinth. Note the darkened patch above the grill and the cracking of the trowelled-on sill

Special precautions may be necessary when the temperature of the steel is below 5°C, such as reducing the rate of bending. The reinforcement may be warmed to a temperature not exceeding 100°C. Reshaping of steel, previously bent, should be avoided if possible because of the risk of fracture. Where necessary the engineer's approval should be obtained.

Control of cracking

The tensile strength of the concrete is generally ignored in normal reinforced concrete design and it is assumed that the concrete is cracked at the working load. A certain amount of cracking is, therefore, normal in the parts of reinforced concrete members in tension, but if the cracks are distributed along the length of the member they can be made small enough to be of no account. On the other hand, if the cracking is not controlled, one or more large cracks may occur, causing unsightliness and a route for the penetration of water to the reinforcement, with consequent corrosion. Cracking can also occur due to tensile stresses set up in restrained members by shrinkage, Fig. 19.10. This shrinkage may be the normal drying shrinkage of the concrete and there may also be thermal movements. Cracking from these causes, in such parts of a structure as wall panels, is unavoiable if there is an external restraint.

The formation of cracks is accompanied by a local failure of bond between the steel and concrete, so that a high total bond resistance is more likely to impede the development in width of cracks than a low total bond resistance. From this it may be concluded that the development in crack width is retarded by the use of bars of small diameter (i.e. with a large relative surface area), a high percentage of steel and a close spacing.

In general, the probability of cracks forming as a result of bending of concrete members is greater as the stress in the tension steel is increased. Reinforcement may, however, be used to control the width of cracks, the greater the quantity of reinforcement that is provided and the closer it is spaced, the smaller being the width of individual cracks, while the total width of cracking remains constant. Rules for the maximum distance between bars in tension to avoid undue cracking are given in BS Code of Practice CP 110 : Part 1 : 1972. Calculations may also be made of probable crack widths, and if they show a greater distance between bars to be acceptable, this greater distance may be adopted.

The use of higher stresses with high tensile steel has led to a closer study of crack formation. High yield steel is usually produced as deformed bars, giving increased bond resistance, which also has the merit of distributing cracks. These are in consequence smaller, although greater in number. It is, however, necessary to limit tensile stresses in the steel to prevent unduly large cracks forming, and a number of crack formulae have been developed for this purpose.

One such formula, included in CP 110, is as follows. The strain in the tension reinforcement is limited to $0.8\, f_y/E_s$.

$$\text{Design surface crack width} = \frac{3 a_{cr} \varepsilon_m}{1 + 2\left(\dfrac{a_{cr} - c_{min}}{h - x}\right)} \qquad (1)$$

where a_{cr} is the distance from the point considered to the surface of the nearest longitudinal bar,

ε_m is the average strain at the level where cracking is being considered, calculated allowing for the stiffening effect of the concrete in the tension zone, and is obtained from equation (2),

c_{min} is the minimum cover to the tension steel,

h is the overall depth of the member,

and x is the depth of the neutral axis found from the analysis to determine ε_1 (see below).

$$\varepsilon_m = \varepsilon_1 - \frac{1{\cdot}2 b_t h (a' - x)}{A_s (h - x) f_y} \times 10^{-3} \qquad (2)$$

where ε_1 is the strain at the level considered calculated ignoring the stiffening effect of the concrete in the tension zone,

b_t is the width of the section at the centroid of the tension steel,

a' is the distance from the compression face to the point at which the crack width is being calculated,

and A_s is the area of tension reinforcement.

For other details see CP 110 : 1972.

Limits regarded as reasonable for most purposes are:

(1) normal exposure: 0·3 mm

(2) very severe exposure, e.g. exposed to sea water or moorland water and with abrasion: 0·004 × nominal cover to the main reinforcement.

When tensile stresses are allowed in prestressed concrete the calculated crack widths should be limited to 0·2 mm for normal exposure and 0·1 mm for very severe exposure.

In a reinforced concrete structure many factors influence cracking, so that crack widths calculated by the formula above give only an approximate assessment on which to base design. The actual widths will vary within wide limits.

It is usual to provide both vertical and horizontal reinforcement near the outer and inner faces of wall panels. Extra shrinkage reinforcement should be provided at sudden changes of section as, for example, corners of doors and window openings.

Types of reinforcement

Mild steel bars

Mild steel bars both plain and deformed manufactured in accordance with BS 4449 : 1969 are employed for concrete reinforcement and are required to have a specified characteristic strength in tension of 250 N/mm² with a minimum elongation of 22%. The characteristic strength is the value of the yield stress below which should fall not more than 5% of the test results. The bars can be readily bent without fracture provided that the radii of the bend or hooks are not less than the standard appropriate to the particular bar diameter. Heating should not normally be necessary, but if it is employed to aid bending of the larger diameter bars, cooling by water quenching should be avoided. Permissible stresses are given in BS Code of Practice CP 110 : 1972.

High yield deformed bars

Hot rolled high yield deformed bars are included in BS 4449 : 1969. They are required to have a specified characteristic strength, as defined under mild steel bars, of 410 N/mm² with a minimum elongation of 14%.

For the purpose of identifying the grade of steel of any bar rolled from other than mild steel, the manufacturer is required to roll on to the surface of such bars at regular intervals a legible mark in one of two styles defined in BS 4449 : 1969. Bars should always be examined for such marks to avoid the possibility of mild steel being included in the structure when high yield steel is specified. Errors in delivery are not unknown.

Heating of bars to aid bending, or welding at splices and junctions up to 100°C is permissible.

Cold-worked steel bars

The manufacturing process for this type of bar involves the twisting or stretching of square or deformed section steel rods made in accordance with BS 4461 : 1969. The bars may generally be recognized by ribs or other features forming a pattern along their length. The 'cold working' given to the steel bars raises their tensile strength and yield strength. Cold-worked steel bars are required by BS 4461 : 1969 to have a specified characteristic strength, which is the value of the 0·2% proof stress below which not more than 5% of the test results fall. The specified characteristic strengths are 460 N/mm² for bars up to and including 16 mm diameter, and 425 N/mm² for bars over 16 mm diameter. The respective minimum elongations are 12% and 14%.

It is important to avoid, as far as it is practicable, rebending or reversed bending. For example, cranked bars projecting from a partly completed structure should not be temporarily bent one way, to give clearance for a construction operation and then bent back again to their final position, unless a larger radius than normally used for bending is provided. This operation would involve bending the bars at least twice and possibly three times, and fracture would then be likely.

Heating above 100°C should not, in any circumstances, be employed during the bending of high-tensile cold-worked bars. If this is done the tensile strength of the steel will revert to the unworked quality. Similarly heat produced in welding bars at junctions or splices can weaken the steel; therefore welding should only be carried out under skilled technical supervision.

Steel fabric

Steel fabric reinforcement to BS 4483 : 1969 is fabricated from plain round, indented or other deformed wires or welded or interwoven cold-worked square or other cold-worked deformed steel bars. The wires or bars should comply with BS 4482 : 1969 or BS 4461 : 1969, as appropriate. The coiled hard-drawn wire is straightened mechanically, and the straightened lengths are cut and either electrically welded or interwoven to produce the various sizes of mesh. In other processes, square section rods are twisted and cut to length for the fabric. The straightening and welding processes can alter the mechanical properties of the wire; therefore the British Standard requires the purchaser to include in his order, besides a reference number identifying the size of wires and spacing, the type of material and process of manufacture.

The purchaser may also ask for certification by the manufacturer that all welded intersections are capable of withstanding a load in shear of not less than one-quarter of that necessary to develop the load, calculated from the specified characteristic proof stress in tension, of the smaller of the intersecting wires. For tensile tests the sample is required to contain one or more cross-welds in its length. For bend tests the test piece is cut from the wires between the welds. The results of tests should comply with the requirements of the relative British Standards.

The cold working given to the steel in the wire drawing and twisting (if any) processes raises the tensile strength at the expense of some ductility. However, because steel fabric reinforcement is normally fabricated from small diameter bars there is not usually the same risk of fracture on reverse bending as occurs with the larger cold-worked high-tensile bars. There is, however, a risk of fracture at carelessly made bends where damage has occurred to the steel forming a notch or a sharp kink.

Steel fabric reinforcement is used mainly as reinforcement for slabs and roads.

Working drawings for reinforced concrete members

Working drawings should provide information concerning the reinforcement to be placed in the various members and should be sufficiently detailed to enable the reinforcement to be correctly scheduled in accordance with BS 4466 : 1969. Reference may also be made to a publication 'Standard Method of Detailing Reinforced Concrete' by the joint committee of the Concrete Society and the Institution of Structural Engineers [4]. The method for preparing such details cannot be discussed fully in this book but the following points may be used as a guide.

All the reinforcement in a beam should be shown on the elevation together with an indication of the number and size of bars in each layer, and any other necessary dimensions. It is wise also to make a large scale drawing of the reinforcement at laps, intersections, etc., to ensure that there is room for the bars at these points, and that the intended positioning can be maintained.

The size of stirrups should be specified and the spacings shown.

Cross-sections of the beam should show the overall size, the various bars, and the cover of concrete to be provided.

Columns should be shown in elevation in a similar manner to beams giving overall sizes and cross-sections at various points. The drawings should show the size and spacing of the main bars and the shape and distances between the links.

In the case of slabs and wall panels it is usual to draw in only a few of the bars of each type in full, together with their sizes and spacing. An indication should be given as to whether the bars are top bars or bottom bars. Sections should also be given showing the thickness of the slab, the concrete cover and the laps, bends and cranks in the bars. In all cases, the positions and lengths of laps and other such details should be given.

Requirements for the cover to the reinforcement are given on p. 295.

Bending schedules

Bending schedules are prepared for all the bar reinforcement in order to show the size, number, total length and bending dimensions. Each bar should be identified on the drawing, the bending schedule, and the labels attached to the bars, by the same mark which should be a symbol or number and as simple as possible. The bar bender should be given all the dimensions necessary to enable him to bend the bar as accurately as possible to the shape required. Important items are the sizes of cranks and the dimensions of hooks. It should be noted that the dimensions of the parts to be cranked are measured to the outside of the bar, whereas the dimensions of stirrups and binders are given from inside to inside and not from centre to centre. Examples of bending schedules using coded shapes and shape codes are given in BS 4466 : 1969 and in 'Standard Method of Detailing Reinforced Concrete' [4].

Inspection of reinforcement on arrival

Reinforcing bars should be inspected on delivery for damage in transit or heavy rusting. Some definition is required of the term heavy rusting, since ordinary rust which adheres tightly to the bars has been found to have no detrimental effect on the bond strength. Objectionable rust and scale may be defined as that which can be removed by firm rubbing with a piece of sacking or a hard brush.

The diameters of the bars should be checked to ensure that they are correct, and that corrosion has not caused a material reduction in the cross-sectional area.

On a site near a chemical works, a reduction of bar diameter of nearly 1 mm was measured after a period of storage of six months in the open.

Checking diameter of deformed bars

The effective cross-sectional area of deformed bars is obtained by weighing lengths of at least 0·5 m. For bars uniform along their length:

$$\text{Gross cross-sectional area in mm}^2 = \frac{m}{0{\cdot}007\,85L}$$

where m = mass in kg
L = length in m

The procedure is slightly different for deformed bars which vary along their length (see BS 4449 : 1969).

Tests on reinforcement

Occasional laboratory tests should be made on samples of reinforcement to ensure that it complies with the relevant standards. These tests should be made more frequently in the case of bars or fabric manufactured from high-yield steels, since the risks of failure, should the steel be of an inferior grade or quality, are higher than for mild steel.

Certificates should be obtained from manufacturers. This may not be easy to arrange because of the number of hands through which the steel may pass from the original manufacturer to the final purchaser. For example, billets of steel may be purchased by a rolling mill for the production of hot-rolled high-yield steel. The bars so produced may then pass into the hands of a steel stockist before it is purchased by the contractor. In the case of hard-drawn wire mesh fabric, the steel may pass from the rolling mills in rod form to a wire drawer, after which the coils are straightened and resistance welded to produce the fabric reinforcement. Further complications arise when fabricators carry out their rolling and re-rolling and welding processes on imported Continental steel.

Because of the above circumstances, the identity of the originally manufactured steel billet can easily be lost, and in the event of failure of test specimens there is none of the basic data on which the results of mechanical tests can be judged. For this reason steel suppliers should be required to furnish information about the original steel manufacturer, the steelmaking process employed, and the analysis of the original cast. Steel fabricators do not always seem to be aware of the effects of their processes on the mechanical properties of the steel.

Fig. 19.11 Bending steel reinforcement

For example, the tensile strength of a hard-drawn wire may be lowered quite appreciably in the process of straightening the wire from coils.

Bending reinforcement

Bends and hooks in reinforcing bars should be made in accordance with the requirements of BS 4466 : 1969. The radius of the bend is dependent on the steel stress at the point of the bend, on the permissible concrete stress and the position of the bend. The danger of splitting of the concrete due to compression induced by the bend is dependent on its position. Splitting is unlikely if the bend is protected by other reinforcement, but care should be taken in narrow beams and similar members.

On works contracts and in steel yards where large quantities of reinforcement are dealt with, mechanical bar-benders of the type shown in Fig. 19.11 are used. Inefficient equipment is liable to result in inaccurate bending and consequent difficulty in fixing. Skilled bar-benders should be employed on this work in order to avoid inaccuracies. Stretching of bars normally occurs to some extent during bending, and after completing the first bar it is usually necessary to make minor alterations to the positions of the bends of subsequent bars in order to obtain closer agreement with the intended dimensions. This applies particularly if there are several bends in the same bar. The extent of the adjustment necessary depends on the type of machine and the method adopted to position the bars in relation to the mandrel of the machine.

Fig. 19.12 Fixing steel reinforcement to a beam

Bars should be bent in a slow and regular movement. Rapid movement is likely to lead to fractures which may not be detected at the time the bars are bent. Heating up to 100°C may be permitted to aid bending. In order to avoid fractures during cold weather, say below 5°C, special precautions such as reducing the speed of bending or increasing the radius of bending may be necessary.

A second bending of reinforcement bars is best avoided, but when they have to be bent aside at construction joints and afterwards bent back into their original position, care should be taken to ensure that the radius of the bend is larger than that normally accepted.

Some form of marking should be adopted to enable the steel fixers to identify bars without difficulty. The bars should be stacked so that the various lengths and sizes can be found easily, and in such a way that damage is avoided.

Reference may be made to 'Steel Reinforcement. Production and Practice'[5] for other information.

Fixing reinforcement

It is very important that reinforcement should be fixed in the correct position and properly supported to ensure that displacement will not occur when the

Fig. 19.13 Reinforcement supported on 'chairs' (Shape code 83, BS 4466 : 1969). The supporting units are tied to the main reinforcement with wire

concrete is placed. Skilled steel fixers are essential if the work is to be carried out efficiently, especially with the complicated reinforcement which is required in many cases.

Reinforcement is, in some cases, built up on supporting frames and then hoisted into position in the moulds, the erection being completed after it has been placed in position. This often facilitates the work and leads to greater accuracy, although some tolerance must be allowed as suggested under 'Cover', p. 295. It is not always realized just how difficult it is to align a mass of reinforcement when efforts to straighten a bar at one point are quite likely to result in further disturbance elsewhere.

Supports of various types are obtainable which prevent displacement during concreting. Commonly used materials are concrete (preferably) and plastic spacing blocks or rings, asbestos cement rings, metallic supports and other suitable devices, see Figs. 19.13 and 19.14. It is important that some such means is adopted to prevent movement of the steel which might reduce the cover and lead to spalling after exposure to the weather. Examples of displacement of reinforcement revealed during the reconstruction of gas industry plants[6] are shown in Fig. 19.15.

Reinforcing rods are usually held together by soft iron wire ties. The tying wire is ordinarily between 16 and 18 Imperial Standard Wire Gauge. Spring wire fixings are available for use instead of tying wires. They have the advantage that they can be clipped on quickly.

Distance between bars

In order to facilitate the placing of the concrete and its compaction, whether by hand or by internal vibrator, adequate space should be provided between bars. This will usually mean a minimum between individual bars or

Fig. 19.14 Use of concrete spacer rings for location of reinforcement in the shuttering

groups of bars of about 75 mm. Smaller spacings will usually be required in places and recommendations are made in BS Code of Practice CP 110 : Part 1 : 1972 for minimum recommended spacings. Generally a spacing of less than the bar size should be avoided if the bar is larger than the maximum size of aggregate (h_{agg}) by more than 5 mm. This also applies to bundles of three or four bars in bundles. Other recommendations for pairs is a minimum horizontal distance of h_{agg} + 5 mm, and a vertical distance of 2/3 h_{agg} when a pair are vertically above one another. For pairs side by side the vertical distance becomes h_{agg} + 5 mm.

Corrosion of reinforcement

Reinforcing bars are normally protected from corrosion by the impermeability of the surrounding concrete, and its alkalinity. The latter property has probably saved many reinforced concrete members, in which the concrete has not been sufficiently impermeable to prevent the entry of moisture, or where fine cracks have occurred.

The object of specifying a minimum cover is to provide a practical guide based on general experience. However, such cover may not give sufficient protection in the presence of salts. Sea salts present in the aggregate, or the use of brackish water or sea water for mixing or curing, increase the tendency for corrosion to occur. Overdoses of calcium chloride have the same effect. Similarly in the Middle East the presence of salts, particularly chlorides, in some aggregates has caused serious cracking in a very short period, sometimes a matter of months. Sulphates present in these aggregates have caused expansion in the concrete and so caused cracking allowing moisture to penetrate to

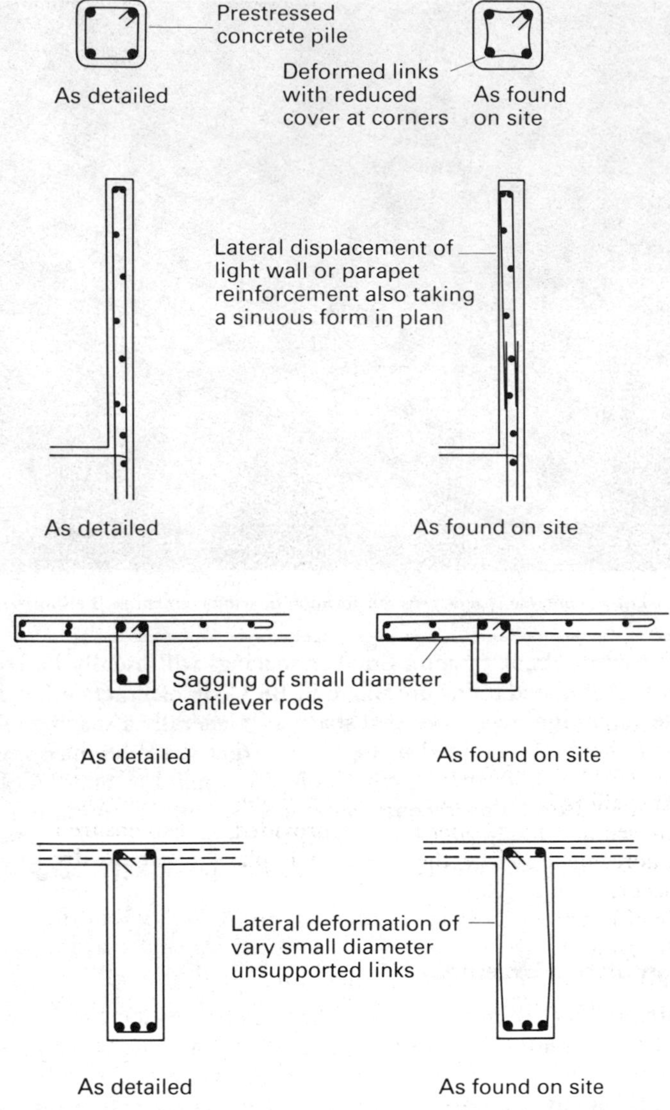

Fig. 19.15 Typical examples of defects exposed in concrete work during reconstruction

the reinforcement, thus accelerating the attack. Serious failures have been recorded, even in arid climates, where the only moisture present is likely to have been that due to condensation.

Maritime structures, exposed to alternate wetting and drying, should be constructed with strong, dense concrete, and adequate cover given to the reinforcement, if failures such as that illustrated in Fig. 19.16 are to be avoided.

Fig. 19.16 Failure of a beam and corbel supporting a seaside promenade. Failure was caused by penetration of sea salts and lack of adequate cover to the reinforcement, resulting in rusting, expansion and spalling of the concrete

Cover to bars

Adequate cover to reinforcing bars is necessary if trouble from corrosion and consequent spalling of the concrete is to be avoided. It is therefore particularly necessary to see that the proper cover is provided, and to ensure by inspection of the reinforcement before the concrete is placed that no displacement is likely to occur.

BS Code of Practice CP 110 : Part 1 : 1972 recommends that the cover to be used for design and to be shown on the drawings should be as given in Table 19.1. This is the cover to all reinforcement including links.

Table 19.1 is for dense concrete made with good natural aggregate. If lightweight aggregate is used the cover should be increased by 10 mm, except for internal non-corrosive conditions where a cover of 25 mm is generally acceptable.

The nominal cover should always be at least equal to the size of the bar, and in the case of bundles of three or more, equal to the size of a single bar of equivalent cross-sectional area. When reinforced concrete is in a situation where resistance to fire is of particular importance the nominal cover may need to be increased, or other measures taken, to prevent spalling.

Protective coatings on the concrete as, for instance, paints of all types including bituminous preparations, and sprayed plastic skin types, including epoxy resin based applications, have not as yet proved to be permanent. If the cover is reduced, regular maintenance of protective coatings of these types becomes necessary. Below ground, asphalt tanking, and similar relatively thick

Table 19.1 Nominal Cover to Reinforcement

Condition of exposure	Nominal cover				
	Concrete grade				
	20	25	30	40	50 and over
	mm	mm	mm	mm	mm
Mild: e.g. completely protected against weather, or aggressive conditions, except for brief period of exposure to normal weather conditions during construction	25	20	15	15	15
Moderate: e.g. sheltered from severe rain and against freezing whilst saturated with water. Buried concrete and concrete continuously under water	—	40	30	25	20
Severe: e.g. exposed to driving rain, alternate wetting and drying and to freezing whilst wet. Subject to heavy condensation or corrosive fumes	—	50	40	30	25
Very severe: e.g. exposed to sea water or moorland water and with abrasion	—	—	—	60	50
Subject to salt used for de-icing	—	—	50*	40*	25

* Only applicable if the concrete has entrained air.

coatings give adequate protection, provided leaks do not develop where services pass through; if this happens water may enter any porous areas of the concrete.

The need for adequate cover, wherever concrete is exposed to the weather, cannot be too strongly emphasized. In providing such cover, it must be remembered that reinforcement cannot be bent and fixed with great accuracy, except at an unjustifiable cost, particularly when the larger sizes of bar are used. It is suggested that a working tolerance of 25 per cent of the sizes of the bars is a reasonable provision. The recommendation in CP 110 is that in all normal cases the design may be based on the assumption that the reinforcement is in its nominal position. Under working conditions the actual cover is likely to vary by up to say 25 per cent from the nominal cover and this needs to be borne in mind. For concrete likely to be produced under difficult conditions, whether due to climate, materials, or shape of member, it may be wise to increase the cover by 50 per cent or more.

Failure to provide adequate cover has resulted in the corrosion of the reinforcement, and the consequent spalling of the concrete on many structures. Examples of such failures are shown in Figs. 19.16 and 19.17. Reference may also be made to Fig. 19.15, which gives examples of misplacement of reinforcement recorded in concrete work for the gas industry.

If abrasion or other damage is expected, sufficient cover should be provided to give a reasonably long life. This may be difficult to assess, and it may be necessary to take other measures either to protect the concrete at vulnerable points, or to allow repair when necessary.

Fig. 19.17 Insufficient cover to reinforcement on concrete parapet

Welding

Welding is now used fairly extensively in the precast concrete industry, but is not seen to the same extent on site work, except on the Continent. The use of welded fabric for roads, floor slabs, and similar purposes is widespread.

Welding is of two kinds:

(1) Tack or positional welding between rods crossing more or less at right angles so as to fix them in position, and for which no special precautions in regard to stress are necessary.

(2) Butt welds between rods in line, stresses being transferred across the section.

Gas, electric resistance, and metal arc welding are the alternative methods available, but the last-named is the best and the most economical for mild steel, being used almost exclusively in practice nowadays. Cold-worked high-yield bars can be resistance welded without loss of strength. Metal arc butt welding of cold-worked high-yield steels is also possible without loss of strength but it is necessary to allow the weld to cool between runs in order to control the heat input, and the area of metal must be increased proportionally when mild steel electrodes are used[7].

With suitable combinations of pressure and current it is possible to flash butt weld all types of steel without loss of strength but the process is essentially one for the factory[7]. Skilled welders are essential.

BS Code of Practice CP 110 : Part 1 : 1972 recommends that welded joints in principal tensile reinforcement should be staggered, the longitudinal distance between joints being not less than the appropriate end anchorage length for the bar. Where tests prove the strength of the joint to be at least as strong as the bar itself, the design strength may be taken as 80% of the characteristic strength for a joint in tension, and 100% in compression. The strength of welded joints in tension may be increased to 100% if they are made under strict supervision and not more than 20% of the tensile reinforcement is welded at any cross section.

Butt welding of main bars is of little advantage as regards any direct saving in cost, but may lead to a saving due to the elimination of laps, additional bars, hooks, etc. and to reductions in the cross sections of some members where the area is determined by the spacing of the reinforcement rather than strength requirements. Congestion caused by hooks and lapped bars may be relieved. A further advantage lies in the rigidity of welded reinforcing frameworks consisting of intersecting bars, stirrups, etc.

In order to take advantage of the use of welding, the design must be adjusted to exploit every possible reduction in the size of members, and the elimination of steel included solely to provide rigidity to the reinforcement framework.

Final inspection

The concrete engineer should see that all reinforcement is fixed in the correct position, and properly supported in the forms by means of built-in concrete spacing blocks or rings, asbestos–cement rings, steel chairs, or other suitable devices.

He should also check the position of splice lengths, and the lengths and diameters of the hooked ends.

The rods should be rigidly held by the tying wires. Welded joints should be carefully inspected to see that they are properly made.

Perhaps more for the psychological effect than the actual direct benefit, occasional welds should be cut out and tested to destruction. The introduction of gamma-ray equipment has given a valuable non-destructive method of examining welds.

Embedded metalwork

It is not always appreciated that embedded metal work apart from steel may require special protection to prevent corrosion. Moisture is necessary for corrosion to proceed, so that parts embedded in impervious concrete with a reasonable cover are not likely to be attacked. Some parts may, however, be embedded in the surface, and these should be protected. Copper is not normally affected but may be corroded if calcium chloride is present. Lead, zinc, aluminium, and cadmium plated parts are susceptible to corrosion in fresh concrete and in moist concrete at all ages. They should therefore be protected by suitable paint.

Corrosion of metals in concrete is accelerated by the presence of stray electric currents, but again moisture must be present if any appreciable action is to occur.

REFERENCES

1 CP 110 : Part 1 : 1972, *The Structural Use of Concrete: Design, Materials and Workmanship*. British Standards Institution.

2 REYNOLDS, CHAS E. and STEEDMAN, J. C., *Reinforced Designer's Handbook*, ISBN 0 7210 0923g. Cement & Concrete Association.

3 MANNING, G. P., *Reinforced Concrete Design*. Longman's Green & Co., London.

4 *Standard Method of Detailing Reinforced Concrete*. Report of the joint committee of the Concrete Society and the Institution of Structural Engineers, London.

5 LANCASTER, R. I., Steel reinforcement. Production and practice. Current Practice Sheet 2P/20/1, *Concrete*, No. 7, June 1973.

6 GARDNER, S. V., *The Use of Concrete as an Engineering Material in the Gas Industry*. Oct. 1954. Reinforced Concrete Association.

7 LANCASTER, R. I., Steel reinforcement. *Structural Concrete*, **3**, No. 4, July/Aug. 1966, p. 171.

20

Prestressed Concrete

In a normal simply supported reinforced concrete beam the reinforcement is designed to take the tensile forces in the bottom of the beam which, together with the compressive forces in the concrete in the top of the beam, resist the bending caused by the load. When the reinforcement is in tension the concrete which is bonded to it in the bottom of the beam is also in tension. As the tensile strength of concrete is low, fine cracks form in the bottom of the beam before its safe working load is reached and when the load is removed, there is only partial recovery, that is to say, cracks once formed do not disappear and, apart from being unsightly, they reduce the protection given to the reinforcement by the concrete cover. If the working load is exceeded, but not so much that the steel stress reaches the yield point, the lack of recovery and consequent widening of the cracks will be very much more noticeable, although the structure may still be able to carry the working load with safety.

When a beam is prestressed (i.e. stressed before the working loads are applied) generally the whole section is put into compression. In one of the earliest known prestressed beams this compression was obtained by jacking both ends of the beam horizontally off firm anchorages at each end. Nowadays, although the principle is the same, this compressive force is obtained by tensioned wires within the beam which are either bonded to the concrete throughout the length of the beam, or are held between firm anchorages embedded in each end of the beam. This then is the starting point comparable with the no-stress condition of the unloaded normal reinforced concrete beam. When the load is applied to the prestressed beam the force which would cause tension and cracking in the bottom of a reinforced concrete beam only reduces the compression which the prestress has already provided. At the same time the compressive force at the top of the beam caused by the bending moment, is added to that caused by the prestress. The effect of this principle is that full advantage can be taken of the potential compressive strength of good quality concrete, and its low tensile strength is not of any consequence. The resistance to shear is greatly increased, and shear reinforcement, so common in normal reinforced concrete in the form of stirrups, is seldom required. Lighter members can be used than with normal reinforced concrete, and this is a particular advantage with bridges, long span roofs and similar structures, in which the dead weight of the members themselves is a major part of the total load which the structure is designed to carry.

A prestressed member is generally designed to carry the working loads without tension in the concrete occurring. However, if the member is overloaded it has the capacity of almost complete recovery, provided that the yield point of the steel has not been reached, and tension cracks caused by the overload virtually disappear.

Table 20.1 Preferred Sizes and Characteristic Strengths of Prestressing Wire

Nominal wire diameter mm	Specified characteristic strength N/mm²
7	1570
5	1570
4	1720
3	1720

Note the specified characteristic strength of the wire is a guaranteed value of the tensile strength below which not more than 5 per cent of the test results shall fall.

Circular structures such as liquid retaining tanks usually require prestressing in two directions, circumferentially to resist hoop tension and vertically to resist the moments imposed by the various loads including prestress.

The use of precast, prestressed units in conjunction with in-situ concrete for floor units and bridge beams has increased in recent years, mainly because of economic advantages. These advantages are due to the fact that only part of the structural unit has to be prestressed and that the precast element usually acts as the formwork for the in-situ concrete. Moreover, small prestressed units are ideally suited to factory production.

Prestressing steel

The prestressing steel must be capable of resisting high tensile forces without excessive 'creep' occurring, i.e. without stretching when such forces are maintained. Any appreciable yielding of the steel would relieve the prestress and the beam would then behave as a normal reinforced concrete beam. As such, it would be very much weaker, and quite unable to carry the loads for which it was originally designed. Ordinary mild steel does not meet these requirements and 'high-tensile' steel is therefore necessary in prestressed concrete. High-tensile steel does not show a definite yield, as in the case of mild steel, and when held under constant stress it will show creep, with time, similar to that of concrete. However, the loss of stress due to creep of high-tensile steel is relatively small and is catered for in the design.

Prestressing steel may be hard-drawn wire, strand, twisted wires, or alloy steel rods similar in size to normal mild steel reinforcement. In order that the high-tensile properties are fully maintained, prestressing steel should not be heated for bending or die drawing purposes. The preferred sizes of plain hard-drawn steel wire together with the corresponding tensile strength (taken from BS 2691) are given in Table 20.1.

Ordinary mild steel reinforcement is used to resist the bursting forces induced by loads at the anchorages; such reinforcement is usually spirally wound round the line of action of the tendons. Steel reinforcement is also used to control early thermal shrinkage in the case of post-tensioned concrete.

Prestressing steel and anchorages should be free from loose mill scale rust, oil, grease or any other harmful matter. The steel should not be kinked or pitted. Cleaning of the steel when necessary should be done by brushing, immersion in suitable solvent solutions, or any other means which will not adversely affect the tendons.

All prestressing steel should be stored in a ventilated weather-proof shed, protected from ground damp. Steel should always be carefully inspected before use to ensure that it is not pitted.

The cost of a prestressing cable complete and in place in a concrete member may be taken as roughly 3 to 4 times the cost of the wire as delivered to the site.

Quality of concrete

In order to obtain the maximum advantage from prestressing, it is necessary to use high strength concrete, and specification requirements of compressive strength of 40 N/mm² or more, at 28 days are not unusual. The production of such high quality concrete requires the adoption of strict control measures in order to reduce variations in strength to a minimum.

Cement used for prestressed concrete should normally comply with the requirements of BS 12 Portland Cement (Ordinary and Rapid Hardening). Other cements may be used in special circumstances but high-alumina cement is inadvisable.

From the data on mix design, Chapter 7, it will be seen that rich, relatively stiff mixes are necessary in order to obtain satisfactory strengths. Vibratory methods of placing are usual. A congestion of reinforcement and ducts frequently occurs at the ends of members cast in situ, as for instance in bridge construction. The webs are often narrow and also contain ducts varying in their vertical positions. Often, little room is left for getting in the concrete. Sometimes it is necessary to reduce the maximum size of aggregate to 15 mm or even 10 mm although this is inadvisable because of the high cement content which is entailed.

Three properties of concrete must be borne in mind in applying prestressing methods, namely, the drying shrinkage which may continue for some time and the creep under load, which have been described in Chapter 2, and the elastic deformation of the concrete at load transference.

The use of a high-strength concrete accompanied by a low water/cement ratio reduces the movements due to these factors and increases the efficiency of the prestressing system. These movements result in a shortening of the concrete which tends to relieve the prestressing effect of the wires or cables with time and it is necessary to increase the tension beyond that considered necessary for design purposes, in order to provide for this gradual loss.

The loss in prestress is greatest at early ages, but the rate of loss gradually diminishes with time. From tests which have been made over periods of two years and more, it appears that the limit of loss of prestressing due to the above factors will be of the order of 16 per cent.

Because of the dangers associated with the corrosion of steel, particularly of thin wires, calcium chloride should never be used in prestressed concrete or in grout used to fill ducts enclosing prestressing cables.

Fig. 20.1 Prestressed beams being manufactured by the long line process. The stressing gear is shown in the foreground

Pretensioning

In pretensioning, the steel tendons, which usually consists of high tensile steel wire or 7-wire stress-relieved strand, are stressed and then the concrete is cast round them. The tension is maintained until the concrete has attained sufficient strength to resist the compression imposed on it, and to transfer by bond the prestress into the concrete. The wires are then cut from their anchorages.

Pretensioning is seldom applicable to structural members cast on the site, but lends itself to the factory production of precast units, ranging from small floor beams and railways sleepers, to hollow piles which may be over 30 m in length. For factory production, the most usual method of pretensioning is known as the long-line system, in which wires are stretched between anchorages at opposite ends of a long stretching bed on which forms are constructed. Stop ends fitted round the wires form small gaps at distances apart which give the desired length for the individual beams. The stressing bed is of concrete or steel and may be provided with heating ducts in order to increase the rate of hardening of the concrete and so allow the wires to be cut between each unit at as early an age as possible. A factory operating on the long-line system is seen in Fig. 20.1.

Fig. 20.2 Post-tensioned roof beams. The end of the beam on the left has holes ready to receive the cables; in the centre beam the cables are threaded through ready for stressing; on the right the cables have been stressed and the wires trimmed off

When producing individual units by the pretensioning method, the resistance to the tension in the wires is taken by the moulds themselves, which must be designed and built to withstand the prestressing loads without distortion. Special measures are necessary to ensure that the formwork does not restrain shrinkage of the concrete, otherwise cracking can occur before the formwork is removed.

The use of individual moulds has two advantages, namely, greater flexibility in size of unit produced, and the loss of only a single unit, if the anchorage should slip, as compared with a whole line of units in the long line process. Slipping of one or more prestressing wires and unequal stressing can occur, and are difficult to detect in prestressed units. Tests should be made on a proportion of the units manufactured.

Post-tensioning

In contrast to the pretensioning method, the steel tendons to be used for post-tensioning are initially prevented from bonding with the concrete by some form of sheath, or by other means, or are threaded through holes cast in the concrete, as seen in Fig. 20.2. The wires or cables or steel bars are stressed by jacking directly off the ends of the structural member after the concrete has hardened sufficiently. Various patent methods such as those illustrated in Fig. 20.3 are used for anchoring the wires within each end of the member before releasing the jacks, but the essential in each is a device which grips them by a

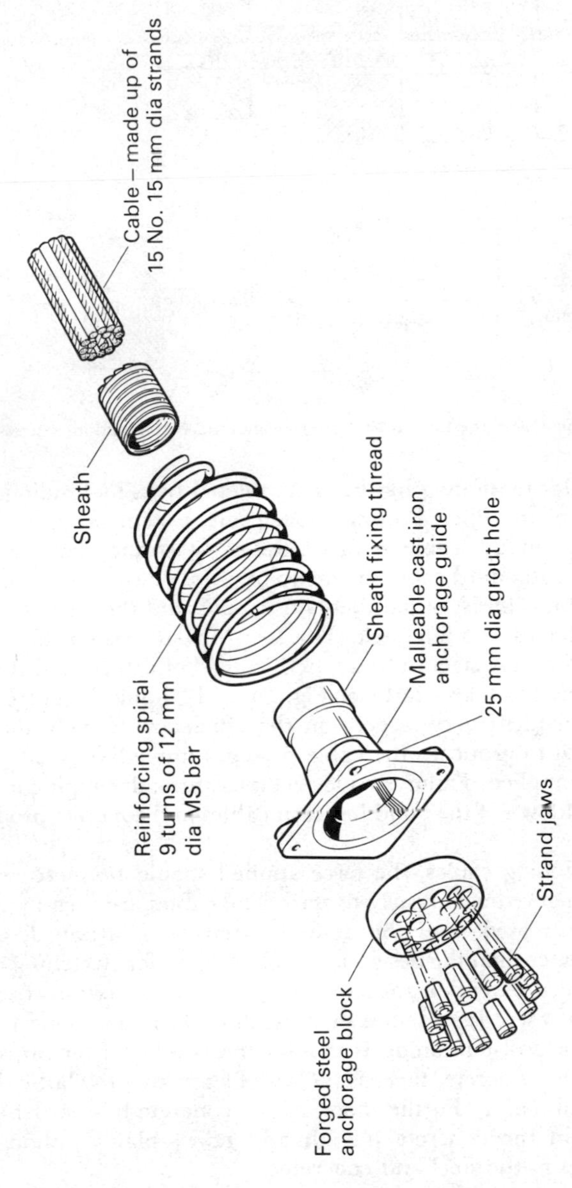

fig. 20.3 An exploded view of the PSC 'Freyssi' monogroup 15/15 mm normal strand system

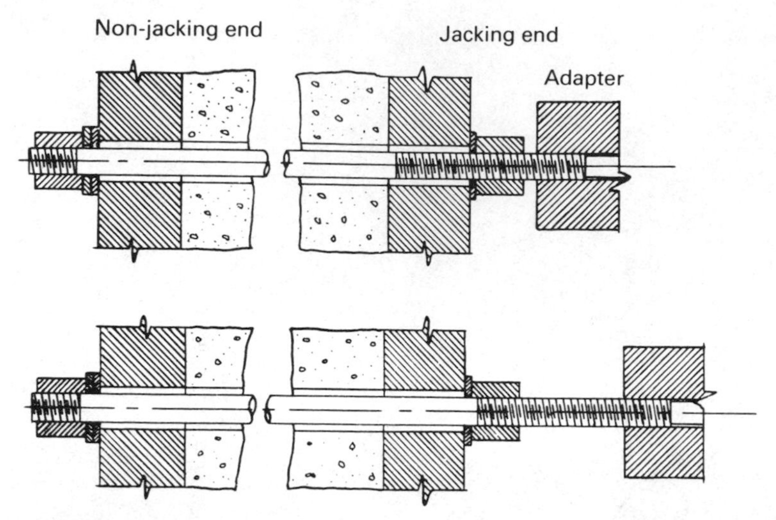

Fig. 20.4 The arrangement of the Lee McCall prestressing units before and after stressing

wedging action after tensioning has been completed by a hydraulic jack. The component parts of the PSC 'Freyssi' monogroup system are shown in Fig. 20.3. In this diagram the cable is made up of 15 strands each of 15 mm diameter, but other standard cables made up of 7, 13, or 19 strands can also be used. All these cables consist of parallel-laid strands and they are stressed in a single pull with the appropriate jack. The Lee McCall system differs in that 20–40 mm diameter alloy steel rods are used instead of cables, and the ends of the rods are threaded to take a nut (see Fig. 20.4). It should be noted that the prestressing is dependent very largely on the efficiency of such anchorages, although it is usual to grout up the cable passages immediately after the full prestress has been applied. Failures have occurred through slipping at anchorages and the breakdown of the bond between cables and concrete produced by grouting.

When post-tensioning cables, the force applied should be increased to the design load and then reduced nearly to zero. The cables are then marked and the force again increased until the desired extension is attained; this may require a force greater than the design load. The reason for stretching cables to give a predetermined extension rather than to achieve a certain stress at the jacking end, is that there is some loss of stress due to friction along the cable.

An advantage of post-tensioning is that as the reaction from stressing the wires is taken on the concrete, there is no loss of stress due to elastic deformation, as with pretensioning. Furthermore, as the concrete has hardened, some of the shrinkage in the concrete has already taken place, although there remain creep losses in the steel and concrete.

The ducts for post-tensioning systems must, of course, be maintained in their correct positions and not be displaced or damaged during concreting. Usually they are firmly fixed at intervals to mild steel reinforcement put in specially for this purpose. Poker vibrators should be operated with care and concrete should not be dropped from too great a height in the location of the ducts.

Whereas the capital investment for pretensioning is generally high, it is much lower for post-tensioning equipment.

Other methods of prestressing are described in the literature on the subject. Reference should also be made to CP 110:1972 'The structural use of concrete' and BS 5337:1976 'Code of Practice for the Structural Use of Concrete for Retaining Aqueous Liquids (formerly CP 2007).

Grouting and sealing

Post-tensioned cable passages are usually grouted up immediately after stressing and anchoring. The grout, consisting of neat cement and water, should be of thin consistency at first in order to clear out air. It should then be thickened as much as the pressure requirement to push it through the cable passage will allow. Grout is forced in at one end until thick grout issues from the other; the ends are then sealed off to ensure that none of the grout flows out. The total quantity is small and even a small loss of grout may mean a long ungrouted passage.

In long cable ducts an adequate number of injection points should be provided for the grout. Apart from each anchorage, it is advisable to provide grout injection points at every low point and every high point in the duct; they should in any case be not more than 20 m apart. A grout injection point should have an internal diameter of at least 10–15 mm; the smaller entry points in some prestressing anchorages are too small for satisfactory injection.

The water in the grout cannot evaporate, and under winter conditions, bursting has occurred caused by freezing of water in cable passages. Rusting is unlikely because the passageway is virtually sealed to air which is also required for rusting to proceed. The inclusion of a water-reducing admixture in the grout will reduce the amount of water required and thereby decrease the risk of water separating out. An expanding agent is another admixture which is sometimes used. This produces slight expansion of the grout just before it sets, giving a tight fit in the duct. An expanding agent is useful for vertical ducts since the expansion of the grout displaces water which may be formed at the top of the duct due to bleeding.

It is also necessary to patch up the recess in the end of the structure where the anchorages are fixed. This can easily be a source of weakness if moisture can gain access to the ends of the cables, and particular attention should be given to the prevention of this possibility. An ordinary mortar patch is unsuitable and is certain to be pushed off by corrosion of the reinforcement ends.

A useful advisory note on grouting ducts in prestressed concrete has been issued by the Cement and Concrete Association [1].

Safety

All the equipment for prestressing operations should be in a good and safe condition. The accuracy of the equipment used for measuring the prestressing force should be checked before its first use and subsequently at intervals of a month or so. If any doubts arise a check should be made immediately. Calibration errors should not exceed 5 per cent.

Wires do break occasionally and safety precautions against injury should be rigorously maintained. Failure of the wires or of the anchorages are of an almost explosive character and parts such as anchorage wedges can fly outwards like bullets. No person should be allowed within a 45° angle of either end of a tendon from the time the jack picks up the first load until the full load has been transferred to the permanent anchorage.

REFERENCES

1 *Notes on the Preparation and Grouting of Ducts in Prestressed Concrete Members*, Publication 47.012. Reprinted Nov. 1972. Cement and Concrete Association 1971.

21
Road Construction

WELL-designed and well-constructed concrete roads have a long life and low maintenance costs. The riding quality has at times been compared unfavourably with the best asphalt or tarmacadam roads, mainly because of unevenness at the joints. However, with the introduction of cut joints, and improved methods of 'planing' the surface, the riding quality has been greatly improved.

In this chapter it is only possible to survey the more important aspects of concrete road construction. More detailed information may be obtained from publications dealing more specifically with the subject[(1),(2)].

Road base

The main concrete slab is generally laid on a base, sometimes also a sub-base, consisting of such material as compacted hoggin, sand, slag or crushed rock. In recent years increasing use has been made of dry lean concrete (Fig. 21.1) and stabilized soil. Base materials are generally compacted by rolling, but vibrating plate compactors are sometimes used for loose granular materials which have no cement binder.

On all base materials it is advisable to lay waterproof paper or plastic sheeting as an underlay for the concrete. A waterproof underlay prevents the escape of cement paste into the permeable material of the base; it reduces the friction between the concrete slab and base, and thereby reduces the tendency for cracks to form in the concrete slab during movements caused by temperature or moisture changes in the concrete; it prevents any loose material on the surface of the base from becoming mixed with the lower layer of freshly placed concrete; and it eliminates the need to wet the base immediately before placing the concrete. It also prevents sulphates and other deleterious salts from moving up into the concrete with their detrimental effects. Sometimes the road base is given a bituminous surface dressing as soon as it is laid, in order to protect it from the ingress of water or, where the base has a cement binder, to act as a curing membrane. Although this dressing has limited effect in reducing friction between the slab and the base, a subsequent underlay is not generally considered necessary.

Concrete for road construction

Concrete used in pavement construction needs to be of high strength and low water content in order to carry the imposed loads without cracking, to resist abrasion, to have a low drying shrinkage, and to be frost-resistant even when it has been treated with salt for de-icing purposes. Air entrainment is

Fig. 21.1 General view of dry lean concrete being laid in 8·4 m width using box spreading machines and compacting with an 8 tonne vibrating tandem roller and with a small vibratory roller for edge rolling

usually recommended to improve the resistance to frost and salt.

The compressive strength (by 150 mm cubes) of concrete used in road pavements and bridges, exposed to frost and salt action, should not be less than 28 N/mm² at 28 days. (See p. 122 for recommended minimum cement contents, and p. 380 for the effect of the water/cement ratio on durability.) The entrained air should be 5–6 per cent for aggregates of maximum size, 37·5 to 20 mm.

The general principles involved in good concrete production described in other chapters apply equally well in road construction.

Slab thickness and reinforcement

The main concrete slab may be from 125 mm (100 mm on very stable subgrades) to 300 mm (325 mm on weak subgrades) in thickness, depending upon the intensity of the traffic, the quality of the subgrade and the type or thickness of the base. The thickness also depends to some extent upon whether or not the slab is reinforced. In most countries, including the United Kingdom where the traffic density is high, reinforcement is considered to be desirable in order to prevent the opening of cracks, and to allow wider spacing of joints, but in some European countries concrete roads have been constructed unreinforced. The general requirement in reinforced roads is a single layer of reinforcement placed near the top of the slab, generally with 60 mm cover in slabs over 150 mm thick. The cover may be reduced to 50 mm in slabs less than 150 mm thick.

The road thicknesses required and the quantity of reinforcement recommended are given in Road Note No. 29[1]. Reference may also be made to ACI Standard 325–58[3], which gives recommended practice in America for the design of concrete pavements.

Joints

High stresses or uncontrolled cracking can result from the warping and changes in length which accompany changes in the moisture contents and temperatures of concrete slabs. Joints are provided to reduce these stresses and to control the cracking. The different types of joint are shown in Fig. 21.2. The requirements of a good joint include the following:

(1) There should be good alignment of the surface on each side of the joint.

(2) The joint should be waterproof and should exclude small stones and dirt. The edges of the concrete should preferably be slightly rounded to reduce fretting.

(3) Jointing material should allow contraction and expansion without itself being extruded or permanently compressed.

(4) The joint should be capable of distributing the traffic load between adjacent slabs, but should not unduly restrict any curling of the slabs.

(5) The formation of the joint should not be difficult to carry out in practice, and it should not easily be displaced by the passage of vibratory equipment.

(6) The formation of the joint should interfere as little as possible with the continuity of the work.

Longitudinal joints

Concrete roads are often constructed in widths of 3 m to 4·5 m which have the advantage that they correspond with the widths of traffic lanes and are free from longitudinal cracking, but with present day plant, slabs may be constructed in single widths of up to about 11 m. In any slabs more than 4·5 m wide, whether or not they are reinforced, it is advisable to provide a central longitudinal joint. Some authorities recommend increasing the amount of transverse reinforcement over the centre third in order to control longitudinal cracking. Longitudinal joints are formed in a similar manner to contraction joints (see paragraphs to follow). Tie bars consisting of 12 mm diameter rods, each 1 m long and spaced at 600 mm centres, should be provided at mid-depth of the slab across longitudinal joints. For further details reference may be made to Road Note No. 29[1].

Joint spacing

The spacing of joints is influenced by the thickness of the slab and the weight of reinforcement, the temperature during the setting and initial hardening of the concrete, and the restraint provided by the subbase or subgrade. In unreinforced concrete the spacing should be close enough to minimize the risk of uncontrolled cracking. In reinforced slabs the joints assist in controlling

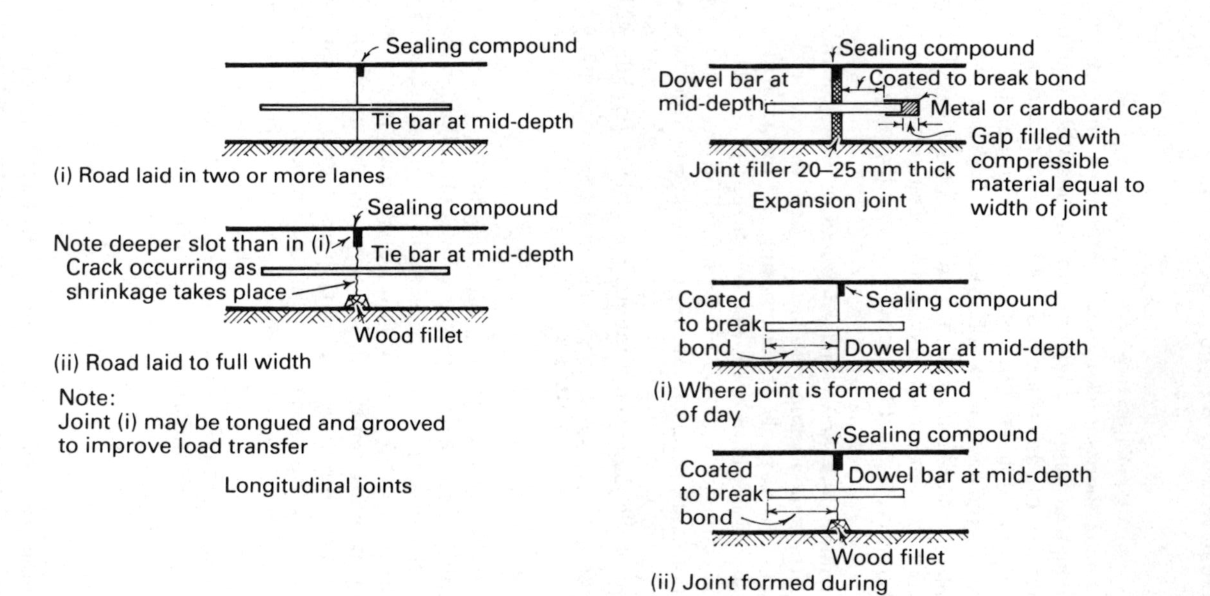

Fig. 21.2 Examples of joints used in road construction. The reinforcement should stop short of joints in all cases

the opening of any cracks which form, and the expansion joints permit thermal expansion to occur without undue compressive stress in the concrete.

Recommendations for the spacing of transverse joints are given in Road Note No. 29, 1970[1]. These are related to the traffic intensity which also affects slab thickness.

Joints in reinforced concrete roads are spaced according to the weight of reinforcement. The spacing may vary from 16·5 m for 2·61 kg/m² of reinforcement (standard long mesh) to 35 m for 5·55 kg/m² of reinforcement[1]. Every third joint should be an expansion joint, the others being contraction joints. The maximum joint spacing may be increased by 20 per cent if limestone aggregate is used throughout the slab.

The maximum expansion joint spacings recommended[1] for unreinforced slabs are 60 m for slabs of 200 mm or greater thickness, and 40 m for slabs of lesser thickness. Intermediate contraction joints are required at 5 m intervals. Again, these spacings may be increased by 20 per cent if limestone aggregate is used throughout the slab. This is because concrete made with limestone expands less than that made with other types of aggregate.

Expansion joints may be replaced by contraction joints during construction in the summer months, provided fixed structures are isolated by suitable means.

Expansion joints

An expansion joint provides a complete separation between the concrete slabs on either side; the gap formed should ideally be of sufficient width to allow compression of the infilling material without crushing it sufficiently to destroy its elasticity, and to avoid permanent exudation of the sealing material at the surface. This ideal is not achieved with materials so far developed, and indeed continual movement involving an elastic compression of 50 per cent or more, continuing over several years, coupled with other requirements in the case of the sealing compound, such as the exclusion of stones, and strong adhesion to the concrete, may prove difficult to achieve. A compromise must therefore be made, and regular maintenance of the joints is necessary.

Expansion joints are usually formed by the inclusion in the slab of a filling material extending from the bottom to within 25 mm to 30 mm of the surface (more in thick airfield slabs). The joint filler should be 25 mm thick if the maximum spacings given above are adopted (see Road Note No. 29) but may be reduced in proportion to the expansion joint spacing if a closer spacing is adopted. Materials commonly used are impregnated fibre board, soft knot-free wood, and cork base boards.

The gap above the filler is sometimes formed after the passage of the compacting machinery, by driving a tee iron into the surface. While this method is acceptable for estate roads, it is not suitable for main trunk roads as it causes a slight irregularity at the joint which affects the riding quality of the surface. A better method, suitable for large roadworks, is to use a machine fitted with a long vertical vibrating plate which is lowered into the concrete before it has hardened, to form a slot or groove of any depth required. Another method is to saw a slot in the concrete after it has hardened. In Fig. 21.3 a longitudinal

Fig. 21.3 A joint sawing machine in operation

joint is being formed in this way. This method has the merit of avoiding any ridge being formed near the joint. Unless the sawing is completed while the concrete is relatively soft, the cost is high; if the concrete is too soft incipient cracks are formed and the corners break away under traffic. Hard aggregates such as flint are also costly to cut. Careful control is necessary.

The gap above the filler should be made 5 mm wider than the filler itself, and a sealing compound poured in after hardening of the concrete. Bitumen has been used in the past but is not elastic and exudes in hot weather. Newer materials have been developed which provide greater elasticity, provided they are not overheated when being prepared for pouring. Other materials have been developed to give greater resistance to the hot blast from jet aircraft. Strong adhesion is difficult to achieve in wet winter weather conditions, and it may be necessary to fill the joint temporarily, until drying out occurs. Another method is to press neoprene rubber strips into the joint with a simple machine. If this method is adopted special care is necessary in forming the groove.

Failures of roads at expansion joints may occur because the joint is not vertical, because the gap for the sealing compound is not made vertically above the filling material, or because of insufficient care in cleaning out concrete above the jointing material before filling with the sealing compound. In the latter case, as soon as expansion occurs, the concrete adjacent to the joint spalls off, as shown in Fig. 21.4. Diagrams showing faulty constructions are also given in Fig. 21.5.

Fig. 21.4 Failure of the joint in a concrete road, caused by not cleaning out properly before pouring the sealing compound. See Fig 21.5 (ii)

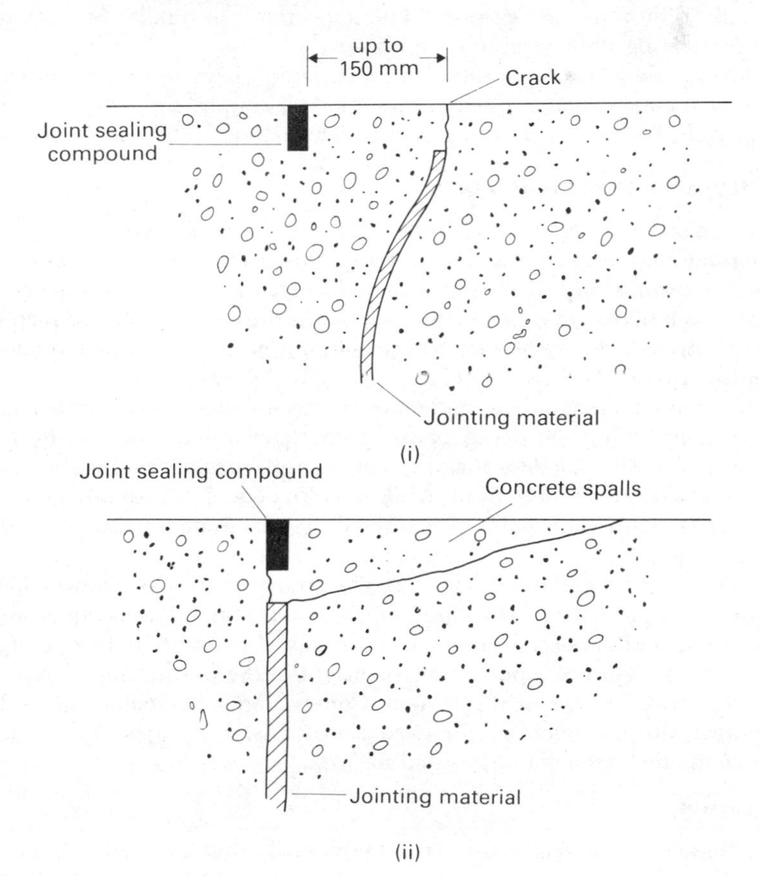

Fig. 21.5 Two common causes of failure at expansion joints in roads

For joints in unreinforced slabs where the anticipated traffic during the life of the road is less than 0·15 million axles and, for all joints in slabs less than 150 mm thick, no dowel bars are required. For roads carrying heavier traffic and more than 150 mm thick, dowel bars should be provided at 300 mm centres along expansion joints. They should be fixed straight and parallel with one another and with the centre line and surface of the road. Suitable sizes and lengths are given in Road Note No. 29 [1]. Failures brought about by misalignment of dowel bars occur more often than need be if sufficient care were exercised. They should be rigidly held in place so that they are not disturbed during concreting. In major roadworks it is now common to use wire cradles to support the joint filler and the dowel bars. The sliding half of the dowel bar should be coated with a thin film of bitumen, or covered with a close fitting plastic sheet to prevent bond with the concrete; on the end it should be provided with a steel or cardboard cap, the end 25 mm of which should be filled with soft material to allow closure of the joint.

Experiments have shown [4] that the base fixing for expansion joint assemblies should not be more than 600 mm apart. The dowel bars should fit tightly in the joint filler, the stiffness of which has a considerable influence on the overall stiffness of the assembly. The importance of quality of workmanship in fabricating the assembly is emphasized.

On small scale works, alternate bay construction is sometimes adopted, in which case the construction of expansion joints is simple, since the transverse forms provide a vertical surface against which the joint filler can be fixed.

Contraction or dummy joints

Contraction joints, or dummy joints relieve warping stresses, and localize crack formation due to contraction, whether due to drying shrinkage or temperature changes. Contraction joints should be provided with dowel bars, as with expansion joints. The only difference is that the cap at the debonded end is omitted. Reinforcement and tie bars should not be carried through contraction joints.

Contraction joints should be made one-quarter to one-third of the thickness of the slab, including the depth of the groove and fillet. The width of the groove will vary with the joint spacing. For example, a groove of width 15 mm may be used at spacings of 8–15 m, while a width of 20 mm is suitable for joint spacings of 15–20 m. The depth of seal should be 20–30 mm, again depending on joint spacing.

The ACI recommendation [3] is that the groove should be between one-sixth and one-quarter the thickness of the pavement. If a parting slip is installed at the bottom of the pavement the depth of groove may be reduced to 40 mm and the depth of groove plus the height of the parting slip should not exceed one-third the pavement thickness. Contraction joints may be formed by the methods already described for expansion joints. The gap is filled with a sealing compound similar to that used for expansion joints.

Road forms

Steel forms are generally used for the construction of roads. They are usually about 3 m long and so designed that they can be held to the sub-base or subgrade of the road by pins driven in at 1 m to 1·5 m centres. For minor roads, for which the concrete is spread by hand and compacted with a small manually operated tamping beam, the forms can be of relatively light construction (Fig. 21.6). For major roads and airfield pavements, heavy vibrating and finishing machines are used and these require forms of relatively heavy construction. Generally, these forms have a wide flat base and, as in the case of those in Figs. 21.8 and 21.9, have provision for attaching a rail on which the spreading and compacting machines can travel. There are notable advantages in using rails instead of allowing the machines to run on the edge of the forms. These include the prevention of distortion of the forms, more even spreading of the weight of the machinery over the bed supporting the forms, and the provision of a surface for the machines to run on which is free from the concrete droppings generally to be found on the top edge of the form.

It is important that road forms are set accurately to line, as this directly affects the riding quality of the road surface. Modern motorway specifications require that forms with rails shall be laid to within 3 mm of the level shown on the drawings, and that the rails themselves shall be set so as to ensure an accuracy of finished surface to within 3 mm under a 3 m straight edge. A less strict tolerance than this is permissible on minor roads.

It is also important that forms be uniformly and firmly supported to prevent them deflecting under the loads involved in concreting the slab. They should be laid on a continuous bed of cement mortar or concrete; when heavy machines are used, a substantial concrete foundation may be needed for the forms.

An interesting development for large scale works is that of the 'slip-form paver'. This is a spreading and compacting machine which requires no fixed forms because it trails a suitable length of forms behind itself. It nevertheless requires a uniform running surface on each side of the construction width.

Spreading and compaction

For small roads works it is usual to mount vibrating equipment on a wooden or metal beam, which serves to screed off, compact, and finally level off the surface. The uncompacted concrete should be levelled evenly and to sufficient height above the forms to allow for the reduction in volume from the loose state as deposited. On small jobs, the concrete is usually spread by hand and is preferably levelled with the aid of a template moved along the forms. The reinforcement is laid in place after a first layer of concrete has been spread. Vibratory equipment for compacting concrete on small scale works usually consists of a wooden or metal beam on which is mounted electric, pneumatic or petrol driven vibrators. Fig. 21.7 shows a wooden tamping beam, to which is attached a petrol driven vibrating unit, being used in the construction of a minor road. Two such vibrating units are sometimes used on a long beam. With these types of compactor it is necessary to leave the uncompacted concrete about 2·5 cms above the side forms and then to lower the beam vertically until the concrete has been compacted. The beam is then raised and lowered vertically on the adjacent strip, and so on, working in strips the width of the compacting beam. With both these types of equipment the final

Fig. 21.6 Steel road form for minor road construction. The driving home of double wedges to secure the form to the pins prevents movement during construction

levelling is obtained by tilting the beam slightly and pulling it slowly along the forms. Two extra men are usually required for this operation. Any lifting and replacing of the beam during final levelling, concrete on top of the forms, screed pins insufficiently hammered in, or uneven joints between forms, cause ridges, and careful attention should therefore be given to avoid these things happening.

On large jobs a high standard of riding quality is generally required. If the concrete is dumped in heaps between the forms, the position of the heaps is reflected in the level of the finished surface whatever method of spreading and compacting is used. Modern specifications therefore require that the concrete be spread to the required surcharge level, by a method which prevents the concrete being compacted while being spread. The most common type of spreader now used is known as the 'box spreader' seen in Figs 21.8 and 21.9. It has its own motive power to move along the forms. The box, a long

Fig. 21.7 Manually operated petrol driven vibrating beam compacting a mix of stiff consistency

trough parallel with the forms, is filled at one side of the road, generally from side tipping lorries; it then moves transversely and, as necessary, longitudinally, to distribute the concrete evenly to approximately the required surcharge.

Reinforcement may be laid after a first layer of concrete has been spread, or may be supported on 'chairs' resting on the base. The spreader in the latter case deals with the whole depth of concrete in one layer.

Modern compacting machines such as that shown on the right of Fig. 21.8 and to the left of the train in Fig. 21.9, generally incorporate strike-off paddles to trim the uncompacted concrete to just the surcharge required, a heavy vibrating beam to compact the concrete, and an oscillating finishing screed to provide a smooth regular surface.

An improvement in riding quality can be achieved by finally 'planing' the surface with straight edges, operated by hand.

Fig. 21.8 The main units in a concreting train on the construction of a modern road. On the left is the box spreader, and on the right the compacting and finishing machine followed by a wet surface profileometer recording the profile of the finished surface

It is necessary when vibrating concrete in the construction of roads, to use a mix consistency which will be plastic enough to enable either the manually operated or the mechanical compacting machine to compact fully, without leaving honeycombed patches along the forms and at the bottom of the slab. The large vibrating and finishing machines are particularly sensitive to the consistency of the mix, and it is necessary to adjust this to suit the machine. With the type of machine shown in Figs. 21.8 and 21.9, a workability of barely 12 mm slump and a compacting factor of about 0·82 gives the best results; for hand manipulated vibrating beams a higher workability is needed (see Table 7.5). Mixes that are too dry for the method of compaction leave the concrete honeycombed at the bottom and are extremely difficult to finish true to level and free from ridges and other blemishes. On the other hand, wetter mixes are apt to sag and overflow the forms, part of the camber thus being lost. Wet mixes are, of course, a disadvantage also on account of the lower strength of the concrete and the higher drying shrinkage, and the increased risk of frost damage.

Curing

In paving work, it is necessary to adopt a form of treatment which does not cause any damage to the newly finished surface, especially during the period before appreciable drying out has taken place. In summer weather, to keep off the wind and sun's rays, curing should commence immediately after finishing the surface, by the erection of coverings on frames, or by means of a resinous curing membrane sprayed directly on to the surface. If a bituminous surface layer is to be applied subsequently, early curing can be effected by spraying the sealing coat on to the surface. Omission of this early curing is a contributory cause of the cracking shown in Fig. 2.5 which occurs in road slabs during

Fig. 21.9 Box spreader to the right followed by the compacting machine on the left. The strike-off paddles, which leave the uncompacted concrete at the surcharge required, are seen in the front of the compacting machine

warm dry weather, even within half an hour of finishing. Protection by coverings on frames should be followed as soon as the concrete has hardened sufficiently, by other forms of treatment such as spraying water on a cover of hessian, straw mats, or sand laid directly on the concrete.

Waterproof paper, plastic sheeting and tarpaulins are effective without spraying, provided the concrete is already wet and they are overlapped and secured at the edges to prevent wind from blowing underneath.

Ponding of horizontal surfaces provides a very effective means of curing, but it is not very practicable, since the small banks of clay usually built round the edges of each bay are not easily maintained in a condition such that the area remains filled with water.

'Sprayed resinous curing membranes' have been widely used over large concrete surface areas as the sole curing agency. They have the advantage that once applied they need no further attention, and while such curing membranes are not always applied with sufficient care to be completely effective, it is widely recognized that, in practice, they are probably more effective than methods of curing which require continuous spraying with water—a process not always strictly adhered to.

On roadworks they are normally applied through a motorized pump, and spraying lance, or on large road or airfield works automatically by a machine mounted on a bridge towed by the finishing plant or having its own motive power. It cannot be emphasized too strongly that application of a curing membrane must follow as soon as possible after the final finishing operation on the concrete, i.e. as soon as free water brought to the surface by the finishing process has been reabsorbed by the concrete or has evaporated. Under windy conditions special care in spraying is necessary to ensure complete coverage of the whole area to be cured. Further comment is given in Chapter 14.

REFERENCES

1 *A Guide to the Structural Design of Pavements for New Roads*, Road Note No. 29, Department of the Environment, Road Research Laboratory, 3rd edn, 1970. HMSO.
2 *A Guide to the Structural Design of Pavements for New Roads*, Revision of Road Note No. 29, Introduced by D. Croney for informal discussion. 18 Mar. 1971. Institution of Civil Engineers.
3 ACI 325–58, *Recommended Practice of Design of Concrete Pavements*, ACI Manual of Concrete Practice, Part 1, 1974. American Concrete Institute.
4 Weaver, J., *An Examination of the Stability of Expansion Joint Assemblies under Loads applied by Concrete Road Making Machines*, Technical Report TRA/396, Aug. 1966. Cement and Concrete Association.

22

Piles, Fluid Retaining Structures, Dams and Tunnels

DIFFERENT forms of construction have their own special features and while the technique employed may vary considerably, the production of good concrete is dependent on the principles already described. Variations in method of construction and special requirements in certain circumstances make it necessary, however, to modify the proportions and workability of the mix, or to make other changes. Concrete piles, fluid retaining structures, dams and tunnel linings are examples of constructions in which special care is necessary to obtain satisfactory results.

Concrete piles

Piles are generally divided into two main types according to whether or not soil is displaced during their formation. These types are:

(1) *Displacement piles.* Conventional precast concrete piles which are driven into the ground come within this category. Also in this category are the driven cast-in-place piles which are constructed by driving a closed-ended tubular section (or sections) of steel or concrete into the ground to form a void. The void is then filled with concrete.

(2) *Non-displacement piles.* In this case, a void is formed in the ground by boring or excavation and is then filled with concrete. The most common pile type in this category is generally termed a bored pile.

Alternatively, piles may be classified according to whether they are precast or cast in situ. However, it is difficult to apply this classification rigidly, particularly when piles under consideration include a displacement pile where in-situ concrete is placed inside driven precast concrete shells.

Precast concrete piles

Although the use of precast concrete piles has declined somewhat in recent years due to the increased use of other pile types, notably bored piles, they are still widely used for marine structures such as jetties where the pile extends above soil (sea-bed) level, in the form of a structural column. It is also worth noting that the transport, handling and driving of long precast piles is easier on marine structures than on land.

Precast concrete piles are generally of square section for short and moderate lengths, but in the case of the longer piles, circular, hexagonal or octagonal shaped piles are often preferred. They are usually cast in suitable moulds on a platform on the site. In some cases, especially in the shorter lengths, they may be cast at a central depot and transported to the site after hardening, but the risk of damage in handling is thereby increased considerably.

Table 22.1 Cement Content and Cube Strength for Precast Reinforced Concrete Piles

Condition	Minimum 28 day cube strength N/mm²	Minimum cement content kg/m³
Hard and very hard driving conditions for all piles and in marine works.	25	400
Normal and easy driving conditions	20	300

The most important considerations in the design of precast concrete piles, whether using mild steel or prestressed high tensile steel reinforcement, are the magnitude of the stresses likely to arise during handling and driving. The quantity of reinforcement is usually governed by the stresses set up during handling, and it is very important that piles should be lifted at the points indicated by the designer. Care is also necessary when stacking that supports are provided at the same points. Usually these points are situated at about one-fifth of the length of the pile from each end. Holes are often formed in the pile by means of metal or cardboard tubes fixed in the mould to ensure suspension at the correct position. Care is especially necessary during the pitching of the piles because pitching stresses can be higher than those caused by handling or driving.

Driving sets up very high stresses at the head of the pile and to a somewhat less extent at the toe. It has been found that failure due to crushing at the head of the pile is minimized by (1) efficient and properly placed packing, (2) the provision of adequate transverse reinforcement near the end of the pile, (3) the use of strong concrete at the head of the pile.

It is interesting to note in this connection that the amount of reinforcement has no important influence on the stresses set up. Impact stresses vary according to the type of soil through which the concrete pile is being driven, but in hard driving they may amount to as much as 21 N/mm². Since the impact strength of concrete is not much more than half of the crushing strength under a slowly increasing load, it is necessary to provide a cube strength of as much as 42 N/mm² near the head and toe.

BS Code of Practice for foundations CP 2004 : 1972 gives recommended cement content and cube strengths for precast reinforced concrete piles, as quoted in Table 22.1.

Adequate cover to all reinforcement should be provided. The Code requires a cover over all reinforcement, including binding wire, of not less than 40mm of concrete, but where the piles are exposed to sea water or other corrosive influences, the requirement for cover is increased to not less than 50 mm.

The concrete should be thoroughly compacted in the moulds using poker vibrators or external vibrators fitted to the forms. In most cases, a slump of 50 mm will give the right level of concrete workability, but for high concrete strengths, a reduced workability may be necessary to provide the desired water/cement ratio. It must not be too low, to ensure that the compaction equipment will give a dense concrete.

Concrete piles are usually made in timber moulds set out and bedded on a

Table 22.2 Minimum Periods for Curing and Handling Piles after Casting

Type of cement	Minimum periods from time of casting			
	Strike side shutters	End wet curing	Lift from casting bed	Drive
Ordinary Portland and sulphate resisting Portland	12 hours	4 days	10 days	28 days
Rapid-hardening Portland	12 hours	3 days	7 days	10 days

level base preferably of concrete. The moulds should be tight enough to prevent the loss of cement and water and should have fillets fitted in the corners to form chamfers on the edges of the concrete.

Placing and compaction should be carried out in accordance with normal good practice taking the precautions which have already been described in Chapters 12 and 13. If form vibrators are used it is important to see that they are fixed at intervals of not more than about 1·5 m in order to obtain good compaction throughout the length of the pile.

The curing of concrete piles is particularly important from the point of view of both strength and impermeability. Curing may be effected by wet sacking, straw, or other covering. The allowable time between casting, lifting and driving may be assessed by compression tests on concrete cubes made at the time the piles are cast. In the absence of such tests CP 2004 : 1972 recommends the minimum periods shown in Table 22.2.

Prestressed concrete piles have been used to an increasing extent in recent years, especially in the longer lengths. They give savings in reinforcement and weight. They are handled and driven in a similar manner to precast piles with ordinary reinforcement.

Pile driving

Precast concrete piles are driven into the ground by one of several types of pile-driving hammer. A common type comprises a falling weight or hammer which is lifted a predetermined distance and is then tripped so that it falls and hits the top of the pile. The weight is guided by a pile frame which is also used to retain the pile in the correct position and direction during driving.

Single-acting hammers have a heavy piston which is raised by steam or compressed air and then allowed to drop by gravity. An advantage of this type of hammer is that it is much faster than the ordinary drop hammer and causes little strain on the pile frame and other gear. Double-acting hammers are similar to the single-acting type, except that in addition to being raised by steam or compressed air the piston is also pushed down under pressure. Diesel pile hammers are designed to give an extra kick as the pile moves down under the impact of the piston, by the explosion of diesel fuel in the cylinder at the bottom of the stroke of the piston. The choice of hammer is determined by circumstances and driving conditions.

The pile head is protected from the full force of the hammer blow by means of a helmet. This is usually of steel with a hardwood timber or composite plastic dolly and some type of resilient packing between the helmet itself and the pile. The packing usually consists of such material as rope, sacking, hemp, felt, asbestos fibre, or bags of sawdust. The resilience of the packing has an important bearing on the impact stresses arising during pile driving and it is important to replace the packing as soon as it becomes hard. In general, high impact stresses can best be avoided by the use of a heavy hammer, a small drop, and a soft or resilient packing.

Recommended practice is given in CP 2004 : 1972 [1], and further information may be obtained from textbooks dealing with piling [2].

Before concrete piles are driven into the ground, they should be carefully inspected to determine whether they are free from cracks. Transverse cracks can result from poor handling or lifting, particularly when the pile is moved at too early an age. These cracks may result in subsequent corrosion of the reinforcement, especially if they are used in a marine environment.

Care should be taken when using a toggle bolt to hold a pile which is driving out of plumb, as cracking may arise if the bending stresses become excessive.

Pile heads should not, on completion of driving, be stripped by automatic hammers used downwards, as this practice is liable to cause fractures below the 'cut-off' level.

Bored piles

Piles of the non-displacement type may be formed by boring a hole in the ground with the type of equipment used in well sinking, or with a mechanical auger, and filling the hole with concrete. Some of the newer mechanical augers produce shafts up to about 3 m in diameter, which are then filled with concrete. The hole is not cased unless the ground shows signs of caving in or there is a serious inflow of water from pervious strata. The larger shafts are sometimes belled out at the bottom by hand or by under-reaming cutters to form an enlarged toe.

The load-bearing capacity is often dependent largely on shaft friction, and in order to reduce softening of the soil round the hole, the concrete should be

Table 22.3 Workability for Concrete in Bored Piles

Typical conditions of use	Range of slump values mm
Poured into water-free unlined bore. Widely spaced reinforcement leaving ample room for free movement between bars	75–125
Where reinforcement is not spaced widely enough to give free movement between bars. Where cut-off level of concrete is within casing. Where pile diameter is less than 600 mm	100–175
Where concrete is to be placed by tremie under water or drilling mud	At least 150

Fig. 22.1 Driving raking hollow precast concrete piles

placed as soon as possible after completion of the boring. Concrete for piles placed in the dry should contain not less than 300 kg/m^3 of cement, and when placed under water by tremie tube a minimum cement content of 400 kg/m^3 should be used. The concrete should have a high workability to ensure that it is 'self-compacting' and flows out against the walls of the bored hole to completely fill it. Slump values, as recommended in the 'Specification for Cast in Place Piling' issued by the Federation of Piling Specialists, are given in Table 22.3. If casing is used the concrete is required to slump down without arching as the casing is withdrawn, otherwise gaps, waisting or necking will result.

If the bottom of the hole is wet a batch of dry concrete should first be placed and well rammed. This may be sufficient to allow further concrete to be placed satisfactorily 'in the dry'. In water-bearing sands and gravels it is likely that water will rise in the hole, making it necessary to place the concrete under water, and tremie methods should be used (see p. 192 for details). Water flowing in fissures or channels in the rock surface or in bouldery material is particularly troublesome and may even wash the toes of the piles away, leaving the concrete arched over. In such circumstances it may be necessary to adopt precast piles, either driven or lowered into boreholes.

Franki piles

A displacement type of pile commonly made by Frankipile Limited is constructed by driving a steel tube to the desired depth and then by ramming

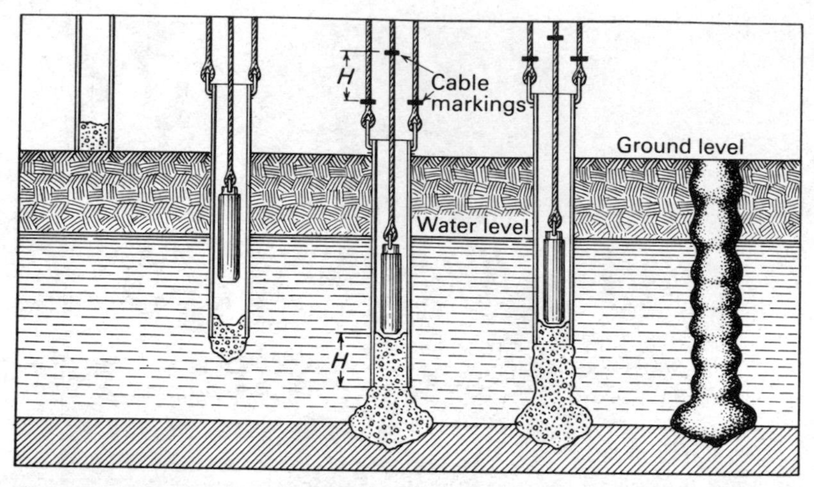

Fig. 22.2 Steps in formation of standard Franki pile

concrete down through this tube and at the same time withdrawing it slowly. The initial driving of the tube is achieved by filling the casing with a plug of semi-dry concrete to a depth of 0·5 m to 1 m and then forcing this into the ground by a suitable drop hammer. The casing follows the block of concrete down in the manner shown in Fig. 22.2. When the penetration of the tube is sufficient, further movement is prevented, and the concrete is forced downwards and outwards by means of the drop hammer to form an enlarged toe, more concrete being added as necessary. The tube is then slowly withdrawn while successive batches of concrete are added and rammed into position. The use of stiff concrete coupled with heavy ramming produces a pile of high strength providing there is no infiltration of ground water. When necessary, a reinforcing cage is lowered into the tube as soon as the enlarged toe has been formed.

Prestcore piles

Prestcore piles are a non-displacement type of pile, but differ from the usual concept of a bored cast-in-place pile. They are formed by assembling circular precast concrete units around a central steel core which is then lowered into a hole bored in the ground. When the whole pile has been assembled in this way, vertical reinforcing rods are inserted in preformed holes, the central core withdrawn and cement grout introduced under pressure. This is forced out between the joints of the precast units and fills the space previously occupied by the tube lining the hole. In this way a solid pile is built up, as shown in Fig. 22.3.

The Prestcore system is designed mainly for circumstances in which piles have to be driven in limited headroom or without vibration. They also have advantages in squeezing ground and where chemical attack may be a danger.

Prestcore piles are generally available in standard diameters of 400 mm, 450 mm and 500 mm.

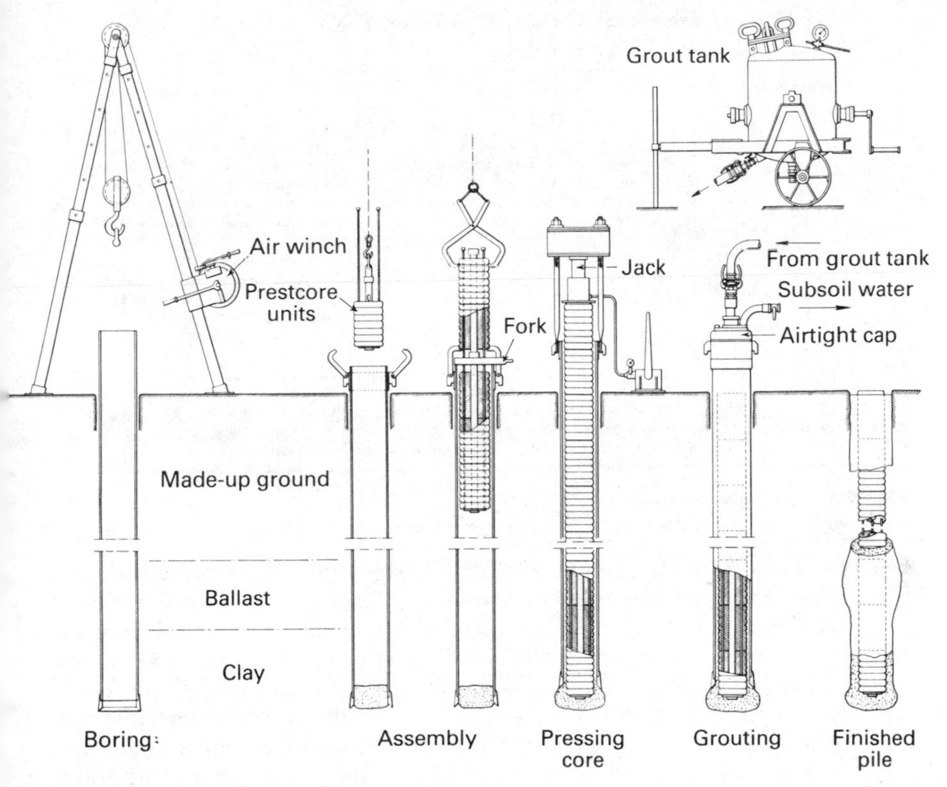

Fig. 22.3 Method of forming Prestcore piles

Comparison of different types of concrete pile

There are many other types of piles used in addition to those described above. The majority of them are proprietary systems employed by specialist companies. The choice of a particular system is dependent on many factors and in most cases must be based on informed engineering judgement.

For purposes of comparison in-situ piles may be divided roughly into three classes:

(1) Those where the lining tube or shell is left in.
(2) Piles hammered down, the tube being removed.
(3) Those which are formed by boring, the tube being removed.

There are certain advantages attached to each method, and these are summarized in Table 22.4. A comparison with precast piles is also given in the Table.

For further information, reference should be made to the specialist literature on the subject[(2),(3)].

Concrete construction below ground water level

Structures such as sewage works settling tanks, underground pumping stations, culverts, basements, and the like, are very often built in reinforced

Table 22.4 Comparison of Different Types of Pile

Type of pile	Advantages	Disadvantages
Precast concrete	The reinforcement can be placed accurately and carried right to the toe of the pile. The concrete can be placed under good conditions, a stiff mix being placed by vibration to provide high strength and density. They can be driven in water-bearing ground. They are suitable for use as columns above ground, and alternatively are easily extended.	They must be properly matured before driving, thus occasioning some delay. Space must be reserved, and a bed prepared for their manufacture, a point of some importance on congested sites. Pile driving presents difficulties in case of limited head room. The guiding and driving plant is heavy. The top of the pile has often to be cut away when the cast pile is too long, or when damage occurs due to heavy hammering. Possible damage during driving.
In-situ piles: (1) General. (Applies to all types included below.)	They are cast in their final position. No cutting is necessary at the head. The piling plant is lighter than that necessary for precast piles. Certain types have special advantages in cases of limited head room, and where vibration might affect adjoining buildings.	The concrete is not cast under the best conditions. The reinforcement is in many cases difficult to place accurately. Possibility of gaps in pile. Difficulty in cleaning out bottom of hole before concreting.
(2*a*) Bored piles, lining tube left in.	The concrete cannot become contaminated with earth or water. There is no risk of trouble during removal of the tube. Moisture and soluble sulphates are excluded which might otherwise cause corrosion of the steel or disintegration of the concrete.	
(2*b*) Bored piles, lining tube withdrawn	The cost of the tube is saved. The concrete surface provides more frictional resistance.	Water and, to a less extent, earth, may cause weakness in the concrete. The tube sometimes presents difficulty in removal. If the concrete is not properly consolidated there is a danger of corrosion of the reinforcement. The reinforcement is difficult to position correctly.
(3) In-situ piles hammered down, tube withdrawn.	The concrete must be of relatively stiff consistency, and this coupled with the heavy hammering gives a high strength. The dry concrete plug usually prevents the ingress of water and slurry. When in contact with clay it tends to absorb water and so improve the skin friction. The concrete is hammered out into any loose strata of soil, thus helping to consolidate the ground and increasing the resistance. The enlargement thus formed, increases the bearing capacity. The rough surface produced, gives increased skin friction.	Water and, to a less extent, earth, may cause weakness in the concrete if workmanship is poor. The tube sometimes presents difficulty in withdrawal. If withdrawn too quickly, waisting, necking or even gaps may occur. There is a danger of corrosion of any reinforcement if the concrete is not properly consolidated or, for some reason, the cover is insufficient. Heavy hammering may, under certain circumstances, cause vibrations detrimental to adjoining buildings and to piles already cast.

concrete. When they are constructed below the ground water table, special precautions are necessary, mainly in connection with watertightness, percolation damage to setting concrete, and buoyancy. Methods of placing concrete under water are dealt with in Chapter 12.

(1) *Watertightness.* If the structure is designed to be tanked in asphalt, no special precautions are necessary in the concrete work other than to avoid damage to the asphalt layer, which can occur either during construction or by movement due to shrinkage of the concrete walls in contact with the asphalt.

The effects of shrinkage due to moisture and thermal changes can be minimized by careful attention to the design of the concrete mix and to protection and curing of the concrete during its early life. On some large works, gaps have been left between adjacent sections and concreted in at a later time; usually two or more weeks after the early shrinkage has taken place in the sections.

Where no asphalt tanking is provided, more than usual care is necessary in the production and placing of the concrete. Methods of producing good concrete, and of avoiding weakness at construction joints have been fully described in the preceding Chapters. Reference should also be made to 'Fluid-retaining Structures', below.

(2) *Percolation damage to setting concrete.* When constructing below the ground water table, the water level is lowered by pumping from open sumps, or by well-points. The walls are very often concreted in direct contact with the ground (i.e. outside formwork is not provided). In these cases, pumping from the sump or well-points must be maintained until the concrete has attained a sufficient degree of hardness and strength to resist the external water pressure caused by the rise in head of the ground water after pumping has stopped. A head of only 300 mm or so is sufficient to burst green concrete, and if this has occurred the only remedy is to break out all the damaged concrete and start again.

In rock excavation, ground water often issues from fissures in the form of springs, after the water level has been lowered by pumping from a sump, and although the pumping is continued after concreting, the flow of water from the fissures over the rock face is liable to cause damage to the setting concrete. If the springs are few in number, they can be dealt with by cutting a chase in the rock surface into which are placed agricultural drain pipes, surrounded by chippings and led to the sump. Numerous springs over the whole excavated face can only be effectively stopped by pressure grouting with cement or cement: sand mixes, or in some cases, by plastering up crevices in the rock face with quick setting cement.

(3) *Buoyancy.* After completion of a watertight concrete structure below ground there is a risk of the whole structure floating to the surface after the water table has been allowed to rise to its normal level.

Calculations should be made to ascertain whether the dead weight of the structure exceeds the uplift due to buoyancy. If flotation is liable to occur, the interior may be kept flooded until superimposed loads are added to provide negative buoyancy. Cases of flotation of structures are by no means rare, and may be disastrous, since it is usually impossible to get them back to their original position.

Fluid-retaining structures

It is particularly important in fluid-retaining structures that cracking from any cause is reduced to a minimum. The causes of cracking are likely to be:

(1) Thermal shrinkage at an early age as the heat of hydration of the cement is dissipated. Later thermal movements due to climatic and other changes may contribute (see also Chapter 2).

(2) Drying shrinkage as the concrete dries out followed by moisture movements as conditions change (see Chapter 2).

(3) Settlement of the foundations which may impose strains on the structure.

It is presumed that cracking due to loading will be controlled within tolerable limits by the design. Reference should be made to the BS Code of Practice 'The Structural Use of Concrete for Retaining Aqueous Liquids' [4].

Cracking is controlled by a combination of movement joints and suitable reinforcement. Types of movement joint which experience suggests as the most suitable are shown in the Code of Practice [4], but there are numerous arrangements which the designer may prefer in particular circumstances. Possible variations are shown in Fig. 11.11.

Shrinkage can normally be reduced by using a lean mix and by reducing the water/cement ratio to a minimum consistent with the requirements for adequate compaction. In water-retaining structures, however, strength and permeability considerations make it necessary to use a fairly rich mix, and it is therefore important that the water/cement ratio should be reduced to a minimum. All construction joints should be given particular attention. It is, in fact, generally considered better to pour continuously whenever possible in order to avoid the formation of lift planes altogether. Compaction at movement joints should aim particularly at ensuring that the concrete penetrates into the small spaces around the waterstops.

Adequate cover to the reinforcement and a thorough inspection to ensure that none of the reinforcing bars is likely to become displaced during concreting, is essential.

Dams

There are several features in the design and production of concrete for use in dams which require special consideration. These arise from the mass of concrete involved, with resultant temperature differentials arising from the heat of hydration of the cement, and shrinkage which may lead to internal or surface cracking, or cracking of the whole structure.

In many cases, the location of dams makes it necessary to utilize local aggregates which are not always of high quality. However, the limits suggested in Chapter 3 for amounts of salts and other deleterious materials should be observed. In the larger masses of concrete it is usual to use aggregates up to a maximum size of 150 mm. The US Bureau of Reclamation [5] recommends that the grading of the coarse aggregate should be within the approximate ranges shown in Table 22.5.

Low-heat cement is often chosen in order to slow down the rate of heat evolution, as described in Chapter 4. Limits are specified for the heat of hydration of these cements. If pozzolanic materials, such as volcanic pozzolanas, groundblast furnace slag and pulverized fuel ash, are available within a reasonable distance, they are commonly used to replace up to about one-third of the cement. A further slowing down in heat evolution and hardening results, but the pozzolans usually provide better resistance to chemical attack, and often give an improvement in the workability of the mix. Air-

Table 22.5 US Bureau of Reclamation Recommendations for Approximate Ranges of Gradings for Coarse Aggregates for Dams

Screen size or number (American sizes—see (Table 3.3)	Percentages of coarse aggregate fractions (clean separation)	
	Maximum aggregate size 6 in	Maximum aggregate size 3 in
6–3 in	20–35	0
3–1½ in	20–32	20–40
1½–¾ in	20–30	20–40
¾–$\frac{3}{16}$ in*	20–35	25–40
¾–⅜ in	12–20	15–25
⅜–$\frac{3}{16}$ in*	8–15	10–15

* $\frac{3}{16}$ in sieve equivalent to No. 4 American.

entraining agents are often used to improve the workability of the mix, and to improve the resistance to freezing and thawing.

Experience, mainly in America, has indicated that cement contents as low as 150 kg/m³ of concrete give satisfactory results in the interior of the dam. For exposed faces the cement content is usually increased by 60 to 100 kg/m³. French experience suggests that a higher cement content, i.e. 250 kg/m³ and more, is required to provide sufficient frost resistance, and going further north to Scotland a minimum cement content of 300 kg/m³ with a low water/cement ratio is advisable for the exterior face. The leaner concretes are not likely to be impervious and the use of a strong facing does give some risk of crack formation or other deterioration due to shrinkage differentials. There may also be a build up of water pressures behind the downstream facing. The spillway section requires special consideration because of the abrasion problems, and dense concrete with a low water/cement ratio is necessary. This will usually lead to the necessity for a higher cement content, probably 90–100 kg/m³ more than for the other exposed faces. Erosion, frost action and chemical attack are discussed in more detail in Chapter 26.

Mixers of at least 3 m³ capacity are required to deal with the 150 mm cobbles, and sturdy transport and placing equipment must also be provided. It is usual to limit the height of lifts to 1·5 m, although lifts up to 2·5 m are allowed when measures are taken to cool the concrete after compaction. The rate of concreting is usually controlled in order to take advantage of the cooling which occurs from the exposed surface of the concrete.

In cooler climates, metal forms have an advantage in conducting away heat more quickly than wooden forms, but may be a disadvantage in hot climates unless shading is provided.

Construction and contraction joints are provided at intervals in order to reduce the risk of random cracking. These joints may take a number of forms and various measures are adopted to prevent leakage through them. In dams it is not usual to provide a water bar at all horizontal joints, reliance being placed on the massiveness of the structure preventing undue seepage, but a water bar is used at all vertical joints. Leakage at horizontal joints is not unknown, however. Another method of sealing joints, sometimes used, more perhaps in cut-off walls through permeable ground than in the dams themselves, is the formation of a series of horizontal and vertical channels at the joints, which are filled with a soft bitumen. Provision is usually made for topping up at a number of points along the top of the wall. In the larger structures expansion joints should be provided at intervals. The subject is, however, too specialized for adequate treatment in this book, and is covered in various publications, including those by the US Bureau of Reclamation[(6)]. Contraction joints are often grouted with cement some months after construction has been completed, through pipes embedded in the concrete. If the grouting is performed when the concrete has cooled, and preferably in cool weather, such grouted joints are usually effective.

Measures to reduce the setting and hardening temperature of concrete in dams include both precooling of the aggregates and the concrete mix and postcooling of the compacted concrete by a piped cooling system. Precooling is unnecessary if the air temperature at the time of placing is below 10°C. However, in many climates precooling is necessary and may be achieved by sprinkling the coarse aggregate with chilled water, by the use of refrigerated air blasts over the aggregate or in the mixer, by cooling the mixing water, by additions of chip ice to the mix, and by refrigerating the mixed concrete just before placing[(7),(8)]. Often a combination of these methods is used. It is usual to aim at a temperature of not more than 10°C in the concrete as it leaves the mixer. At the Chief Joseph and Folsom dams, further cooling ensured that the concrete was placed at a temperature below 10°C, and under those conditions postcooling was considered unnecessary, although air temperatures were 25°C and more.

Postcooling is usually achieved by pumping cool water through a grid of pipes embedded in the concrete. Pipes of 25 mm diameter are commonly used, being placed on the top of each lift of concrete after the concrete has hardened. The horizontal spacing of the tubes depends on temperature conditions, and is usually between 0·5m and 2m. Water is pumped through these pipes at about 18 litres per minute, this having been found to be the most economical rate of flow for most purposes. This artificial cooling is started immediately after placing the concrete and is continued until a stable temperature condition is achieved. Subcooling or other control of the temperature of various parts of the structure may be accomplished and, if the cement grouting of the joints is undertaken under these subcool conditions, it is possible to achieve a more favourable stress distribution under the load as a result of the subsequent temperature rise.

As in other mass concrete structures, there is some risk of surface cracking when the concrete is placed in warm conditions, and cooling occurs rapidly during falling temperatures at certain times of the year. In climates where hot sunshine may be followed by clear, cool, or cold nights, measures should therefore be taken to prevent the surface from either becoming hot in the day time or cooling at night.

Large mass concrete structures

With proper control of the temperature it is possible to induce temperature stresses which will be beneficial to the structure rather than otherwise. In order to prevent uncontrolled cracking during the cooling of mass concrete structures, which contain little or no steel reinforcement, it is necessary to provide contraction joints at intervals. The spacing of these joints is important since it is necessary to grout them up at a later stage when the major part of the contraction has taken place. An opening of 0·5 mm or more is necessary for successful grouting with cement so that the joints should be spaced at distances which will provide openings of sufficient width, but which are not so large as to create the danger of intermediate uncontrolled cracking (see also p. 245 on the sealing of cracks).

Tunnels

Particular care is necessary when placing concrete tunnel linings if cold work planes, honeycombing and porous areas are to be avoided. The concrete may be mixed at the surface or near the point of placement. If at the surface, a long haul to the heading may result in a time lapse between mixing and placing of an hour or more. Placing is usually done with a pneumatic placer which shoots the concrete between the shuttering and the rock near the crown of the tunnel. Two placers may be used near each other if the amount of concrete required is sufficiently large. Diagonal flow is unavoidable, and is conducive to segregation and the production of honeycombed areas. Pneumatic placers need a highly workable concrete of 100–150 mm slump, and a cohesive mix which will not segregate easily. A richer concrete than usual, subject to consideration of the increased shrinkage, is preferable, with an adequate rounded sand content, say 5 per cent more than usually results from mix design calculations.

Air-entraining agents are sometimes used to aid workability but a check should be made of their efficacy, since much of the air entrainment may be lost during mixing, transport, transference to the placer and, particularly, in the violent thrust of the placer itself. Retarders are used if long delays are expected between mixing and placing, especially if temperatures are high.

The pneumatic placer pipe should preferably be kept buried in the concrete. The placer pipe should certainly not be allowed to be more than about 1 m outside the concrete face at any time, otherwise splatter is likely to cause separation of the coarse aggregate resulting in porous areas in the crown and honeycombed patches in the sidewalls and invert. The invert is sometimes cast separately, in which case the concrete needs to be of stiffer consistency to allow it to be trowelled to true line and level on the sloping surfaces.

A continuous supply of concrete to the placer is essential and should be sufficient to maintain progress of at least 3 m of tunnel per hour and preferably more, in order to avoid disturbance to concrete already setting and the formation of cold joints. Internal vibrators are used through suitably placed access doors to aid the flow of the concrete and its compaction.

The broken rock at the top of the tunnel is unlikely to allow complete filling in the recesses, and settlement of the concrete itself before it sets leaves gaps which have to be filled with cement grout under a low pressure through holes drilled in the hardened concrete.

The shuttering is usually hinged to allow withdrawal through the narrow space available. It should be strong and adequately braced to ensure a good alignment. All joints should be tight to avoid loss of cement mortar, or the formation of lips. Access doors should fit exactly. Tunnel linings are usually required to have a high standard of finish since flowing water, especially at the higher velocities, may cause serious erosion at surface irregularities. Patching should follow the lines described in Chapter 17. Rather more rubbing down or grinding off of irregularities in the surface may be necesaary than is usual for most structures.

Curing of the concrete is usually unnecessary, since the atmosphere remains humid due to the amount of water which accumulates from the various constructional operations. Curing compounds may be sprayed on to the lining if necessary.

REFERENCES

1 CP 2004 : 1972, *Foundations*. British Standards Institution.

2 TOMLINSON, M. J., *Pile Design and Construction Practice*. Publication 12.070, 1977. Cement & Concrete Association.

3 WELTMAN, A. J. and LITTLE, J. A., *A Review of Bearing Pile Types*. DOE and CIRIA Piling Development Group Report PG1, January 1977. Department of the Environment and Construction Industry Research & Information Association.

4 BS 5337 : 1976, *The Structural Use of Concrete for Retaining Aqueous Liquids* (formerly CP 2007). British Standards Institution.

5 *Concrete Manual*, 1st ed. 1938, 8th ed. 1975. US Bureau of Reclamation.

6 *Treatise on Dams*, Various design standards published by US Department of the Interior, Bureau of Reclamation.

7 *Symposium on Mass Concrete*, Special Publication No. 6, 1963, American Concrete Institute.

8 STODOLA, P. R., O'ROURKE, J. E. and SCHOON, H. G., Concrete core block for Oroville Dam, *Proc. Amer. Concr. Inst.*, **62**, No. 6, June 1965.

ADDITIONAL REFERENCE

DEACON, R. Colin, *Watertight Concrete Construction*, 1973. Cement and Concrete Association.

23

Precast Concrete Products

THE term 'precast concrete product' is used to describe the many types and varieties of concrete unit which are cast in moulds either in a factory or on the site, and are not built into the structure until they have fully hardened. Precast concrete products therefore include various structural members and such items as cast stone, concrete lintels, staircase units, fence posts, kerbs, lamp standards, railway sleepers, concrete blocks and bricks.

The factory production of concrete should offer certain advantages over site manufacture, chiefly in regard to ease of supervision and control. Compaction can be carried out more efficiently, as also can moist curing. However, the benefit of these advantages is often lost, and the user should check and test precast concrete products with just as much care as he gives to in-situ concrete. Against the advantages there is also the possibility of damage to the finished products during transport and the need for lifting equipment on site of sufficient capacity and range to lift units into position.

Precast concrete products may be divided roughly into five groups:

(1) Architectural cast stone and other types of ornamental concrete.

(2) Concrete steps, paving flags, etc., which have to be wear-resisting. Kerbs may be placed in the same class.

(3) Load-bearing structural units, and other concrete units such as piles (see Chapter 22), lintels, beams, posts, railway sleepers, pipes, water tanks, and troughs. Many of these units are prestressed and in some cases the appearance is of secondary importance.

(4) Roofing tiles, which must be impermeable and weather resistant.

(5) Concrete blocks, slabs, and bricks.

Architectural cast stone

Architectural cast stone is used mainly in place of natural stone dressings on buildings, and is usually made to match a sample of natural stone or other material submitted by the architect. An example of cast stone masonry is given in Fig. 23.1. Well-made cast stone is dense and impermeable and is often indistinguishable from natural stone except by close examination. The various colours and textures are obtained by blending white and normal Portland cements with suitable inert pigments, and by the use of aggregates of the correct colour and texture. The material employed for the facing is expensive and it is the normal practice to use only a thickness of 12 to 25 mm, with a backing of normal Portland cement concrete; but this depends on the size and shape of the unit.

Aggregates for cast stone should comply with BS 882. They will not usually be larger than 20 mm and for the facing may be limited to a 5 mm maximum

Fig. 23.1 Church front at Banbridge, Co. Down, in architectural cast stone

size. Materials commonly used are: Crushed stone, such as Portland, York, Blue Penant, quartzite, whinstone and granites, Derbyshire spar; Crushed marble; Leighton Buzzard and other light-coloured natural sands.

Two methods of making cast stone are in use, namely, the process by which an earth-moist mix is heavily rammed into the moulds, and the alternative of using a plastic mix with subsequent surface treatment to obtain the required appearance. The proportions of the mix depend on the maximum size of the aggregate, since it is necessary to use a richer mix with small aggregates in order to obtain a comparable strength and a dense impermeable concrete, especially if the unit contains reinforcement. A typical facing mix is: 1 part cement, 3 parts fine aggregate 5 mm or smaller. A richer mix is undesirable owing to the increased shrinkage and tendency to craze. The backing will generally be Grade 30 (CP 110 : 1972) concrete, with a minimum cement content of 300 kg/m^3 and 20 mm maximum aggregate size. Smaller aggregate may be necessary for small units.

The advantages of using an earth-moist mix are that (1) a stonelike texture is obtained straight from the mould, (2) the casting can be removed immediately from the mould and any blemishes or loose corners attended to before the concrete sets, (3) the surface is free from blowholes, (4) the shrinkage is reduced, and (5) there is less tendency to craze.

A disadvantage of using earth-moist mixes is that if the water/cement ratio is even slightly too low, they tend to be more permeable than plastic mixes. It is advisable to contain any reinforcement within the backing concrete, ignoring the facing thickness for cover.

Early removal of the moulds is especially useful if they can be used over and over again on repetition work. Care is necessary in the use of earth-moist mixes, however, to ensure that they are the wettest which will give a satisfactory surface texture and can be removed from the mould without sagging. Overdry mixes give weak corners and permeable concrete which soon blackens in town atmospheres.

Mixers of the open pan type with a rotating star of paddles are the most suitable for earth-moist mixes. Rotating-drum and tilting-drum mixers tend to cause 'balling', a condition which renders the mix unsuitable for the production of dense, close-textured cast stone.

Plastic mixes can be placed in accordance with the normally accepted principles of concrete making. Some method of removing the cement skin, such as scrubbing before the concrete has fully hardened, rubbing down with a carborundum stone or a flexible carborundum disc, or tooling, is necessary to obtain a satisfactory architectural finish. Blowholes must also be filled with a cement mortar grout. Thin skins of stone dust and cement rubbed on to the surface with a wooden float are unsatisfactory and prone to crazing.

Moulds for cast stone may be of timber, steel or partly steel, plaster of Paris, gelatine, concrete, glass fibre reinforced plastic, or sand, according to the number of repetitions to be made of any particular unit and the amount of ornamentation. Timber is the most commonly used material owing to the ease with which it can be worked. The moulds must be strong and rigid, especially for earth-moist mixes, and must retain their shape after repeated usage. Steel may be used for plain ashlar facing, and for reinforcing timber moulds, especially at corners likely to be damaged during ramming. Concrete moulds are very useful for some classes of repetition work, but are heavy to handle. Plaster of Paris and gelatine moulds are used for ornamental work. Sand moulds are sometimes used and, although they require a very wet mix, satisfactory cast stone can be produced from them.

Earth-moist mixes are usually rammed into the moulds by hand tampers or by pneumatic and electric tampers. Plastic mixes require less vigorous treatment, while the fluid mixes used in plaster of Paris, gelatine, and sand moulds need only judicious stirring with a rod or spade, taking care that the concrete is pushed into every detail. Stiff mixes are often compacted by holding the moulds down on electric or pneumatic vibrating tables, or by attaching vibrating units to the moulds themselves.

Fig. 23.2 Cast stone as turned out of the wooden mould to the right

Many different surface textures may be obtained on cast stone by suitable treatment. Earth-moist mixes can be left as cast against the face of the mould, apart from final cleaning, and give a smooth and uniform texture. Otherwise, they can be wire brushed before they have fully hardened or etched to give a rough surface, bush hammered, combed, or rough tooled in a similar manner to natural stone. Plastic mixes can be wire brushed as soon as the mould is removed, giving an exposed aggregate texture. A surface retarder is sometimes used to facilitate the removal of the cement skin. Alternatively they can be tooled or rubbed down with carborundum stones or discs, and if necessary given a dull polish by rubbing with successively smoother stones, ending up with a soapstone.

Exposed aggregate finishes are also obtained on slabs by casting face upwards, and trowelling in extra coarse aggregate of whatever size is required to give the desired texture. As soon as the concrete has stiffened a light spray of water, aided as necessary by a soft brush, is used to remove the cement skin. The slab is tilted during this washing off. Except where this treatment is to be adopted it is advisable to form the facing of cast stone against the mould. Sometimes, however, casting is only practicable with the facing uppermost. In such cases it is important to keep the amount of trowelling to a minimum.

Cast stone should be well matured before being built into the structure in order to avoid cracking at the joints due to shrinkage. Most manufacturers coat light-coloured stones with a lime slurry before delivery in order to protect the surface and keep it clean. This should not be cleaned off until the stone is fixed in place. There should be no difficulty in removing it with a stiff brush or by a light rubbing with a carborundum stone. The joints should never be made in rich cement mortar; a mix of 1 : 1 : 6, cement : lime : sand gives satisfactory results.

Many cast stone units, particularly slender units such as cills and coping stones, contain reinforcement to withstand stresses due to handling. Such units are usually exposed to the weather and it is essential for the concrete mix used to be one which permits full compaction, otherwise a porous concrete will result and the unit will have a short life before spalling is caused by corrosion of the steel. Many instances are known of units which have deteriorated within a few months. These units are preferably made using a plastic mix rather than one having an earth-dry consistency.

Concrete units to resist wear

The principal requirement of concrete units required to resist wear by foot and other traffic is that they should have a strong dense wearing surface; this can best be achieved by good trowelling techniques to avoid surface laitance which wears off with time. See also Chapter 25, Floors and floor surfaces.

Concrete steps, floor tiles and other similar units are usually made with such aggregates as marble, granite, basalt, quartzite, and Hopton Wood stone, having a maximum size of 10 mm. In order to obtain a satisfactory grading, 2 parts 10 mm to 5 mm stone may be used either with 1 part of 5 mm to dust stone or 5 mm down washed sand. If 5 mm to dust stone is used, care should be taken to ensure that the supplies do not contain an excess of crusher dust.

A rich mix of the order of 1 : 3 cement : aggregate is used for the facing, but a leaner mix may be used as a backing. The mix must be as dry as possible, consistent with the production of dense concrete. Casting may be done with the tread either upwards or downwards; the former gives the densest and strongest surface if experienced trowel hands are employed. Too much emphasis cannot be placed on the necessity for good workmanship in producing trowelled wear-resisting concrete surfaces. As contrasted with the trowelling of plaster, considerable pressure must be employed to compact the surface without bringing up more fat than is necessary to fill the interstices between the larger pieces of aggregate. Any fat which is brought to the surface should be removed. Two or three trowellings may be necessary at intervals, allowing the concrete to stiffen and dry out to some extent between each. The period from casting to the final trowelling should not exceed about four hours; the actual time will depend on the ambient conditions. In order to increase the proportion of the larger pieces of aggregate in the surface, extra stones are sometimes trowelled into the surface. These are later exposed by polishing.

Steps and floor tiles cast face downwards are certain to have a number of pinholes in the surface which must be filled with neat cement. There is a tendency to use a wetter mix, and this added to the fact that the surface is not compacted to the same degree as a properly trowelled surface, accounts for the poorer wear resistance. Floor tiles can be made very successfully by means of hydraulic presses similar in principle to those used for paving flags (Fig. 23.3).

Steps and tiles are often given a dull polish by means of carborundum stones, beginning with a coarse stone and ending with a fine one. Various types of machine are available for this work, all having rotating carborundum

Fig. 23.3 Manufacturing paving flags by hydraulic pressure

stones to perform the grinding. Terrazzo paving is made in a similar manner.

Non-slip finishes to concrete steps and tiles are obtained by the use of carborundum grit or by inserts of material similar to grindstones. The carborundum grit may be trowelled into the surface or sprinkled in the bottom of the mould in the case of units cast face downwards. The former method is more effective. Small non-slip inserts about 25 mm square and 5 mm thick can be obtained in various colours for use in concrete steps and tiles. They are usually inserted to form a mosaic pattern 50 or 75 mm wide, and at a distance of about 50 mm from the nosing.

Paving flags

Paving flags are nowadays produced in large quantities in hydraulic presses. A mixture of 1 : 3 cement : granite chippings, 5 mm to dust, is normally used, and is mixed in a pan-type mixer situated above the hydraulic press. Steel moulds are used, perforated gauze and a new sheet of paper being put in the bottom before filling. Similarly, a sheet of paper and a zinc plate are placed on the top of the filled mould. This arrangement allows drainage from both top and bottom, and the squeezing out of water increases the density of the pressed flag, so giving a high wear and frost resistance.

The moulds are mounted on a rotating table so that each one is filled from a measuring apparatus situated below the mixer. The slab then moves round into position for pressing at a pressure exceeding 7 N/mm^2. After pressing it moves round to a third position, where it is raised in the mould by means of a hydraulically operated discharge ram. A vacuum plate is then lowered into position over the slab, which is lifted by suction, and swung round on to a stillage, usually mounted on a trolley for convenience in handling. The top perforated zinc plate and the top paper are stripped immediately, but the bottom paper is left in place, and is stripped after the paving flag has hardened.

Paving flags are made in a number of standard sizes given in BS 368 : 1971 [1]. Details are given of tests which the flags are required to pass, covering transverse strength and absorption. The test for wear resistance has been dropped, but the 10 per cent fines test, BS 812, on the aggregate has been introduced to ensure that this is of adequate mechanical strength. Paving flags should not be delivered until they have been properly matured.

Special sizes of paving flags, or those required in special colours, and in small quantities only, are made by hand in wooden or metal moulds. If wooden moulds are employed it is preferable to use a steel sheet as a facing in order to get a good finish. The mix used should be as dry as possible consistent with the production of dense concrete. The use of a vibrating table is a distinct advantage in this respect.

Kerbs

Kerbs are usually made to comply with BS 340:1963 [2], for kerbs, quadrants and channels.

Concrete for kerbs must be of good quality, since kerbs are particularly liable to attack by frost if they are at all porous, and a low water/cement ratio should, therefore, be used, even if this results in a rather less smooth finish than that obtainable with a highly workable mix. The water/cement ratio should not exceed 0·55 and preferably be less than 0·5. BS 340 states 'the water/cement ratio should not exceed 0·55 by weight immediately after moulding or pressing'.

Air entrainment is an advantage in resisting frost attack in the presence of salt solutions, but reduces the transverse strength.

BS 340 requires that the aggregate, when tested in accordance with BS 812 shall not have an aggregate crushing value exceeding 30 per cent, nor a flakiness exceeding 35 per cent.

Standard sizes are given in the specification. Tests are given for transverse strength and absorption. Kerbs may be made by hand, by vibration, or by hydraulic pressure, a pressure of not less than 7 N/mm^2 being used. The moulds must have an internal surface of steel or other material which will give a smooth uniform surface.

Load-bearing stuctural units

Precast load-bearing structural units have been used increasingly in recent years, either to form the concrete structure, or in combination with in-situ concrete. The main advantage to be gained from the use of precast units is a saving in formwork and supports, so that it is those parts of the structure for which the formwork is most expensive, that the precast unit is most likely to be

Fig. 23.4 Construction of multistorey flats entirely in precast concrete

used. The advantage to be gained depends, within certain limits, on the amount of repetition; as a rough guide it may be assumed that the use of 50 times or more of moulds for precast concrete units gives reasonable economy. The saving on formwork must be sufficient to offset the fact that it costs much less to produce and place concrete on the site than it does to take the raw materials to a factory and then to have the precast unit transported to the site, together with attendant handling costs.

Other advantages of precast units accrue from the possibilities of reducing the formwork by casting them on their side, or the possibility of saving in concrete quantity and, consequently, the weight of the unit, by the use of more complicated shapes as, for instance, beams of 'I' shape.

Accuracy of dimension is important in precast load-bearing units and, if a large number of repetitions is expected, soft wood is unsatisfactory for the moulds, and it is necessary to revert to hard wood or metal. Synthetic resin finishes are helping to protect mould surfaces and so prolong their life.

For load-bearing units, a high concrete strength is often required, necessitating the adoption of weigh batching of materials, low workability, and good control and workmanship to ensure uniformity, as described in detail in other chapters.

It is usually most economical to use the largest precast units which can be handled and transported without difficulty, and with which the erection tackle provided on the site can cope. As erection proceeds joints are made between units by bolting or by interlocking the reinforcement and infilling with concrete or cement mortar. In countries where infilling with concrete would be impracticable in long cold periods, steel plates are cast into the concrete units, and these are welded together. Jointing arrangements should be as simple as possible, and should not result in projecting concrete ribs or other details sensitive to damage during transport and handling. The infilling concrete or mortar must be well compacted and properly cured, otherwise ingress of water might lead to corrosion of the steelwork.

When precast units are used in conjunction with in-situ concrete, as for instance, to form the soffit of a floor, it is usual to form a mechanical key in the top of the unit in order to provide a bond between the two. Reinforcement is also left projecting from the tops of units for the same purpose and to resist shear stresses.

Various concrete units

Such units as lintels, beams, columns, posts, water-tanks and troughs are made in accordance with normal concrete practice. The maximum size of aggregate may be limited by the size of the mould or by close spacing of reinforcement. Care should be taken to ensure that any reinforcement is placed correctly and is not displaced during casting, since, in the small sections often made in precast concrete, any movement may result in a serious reduction in cover. In such units as fence-posts the cover is sometimes reduced to 12 mm or less. Cases of spalling of fence-posts due to insufficient cover are common everywhere, and it would appear advisable to limit the minimum size of post by the requirements of both strength and the minimum cover specified in CP 110: Part 1: 1972[3]. The minimum cover would then be 30 mm for moderate exposure. The concrete should be Grade 30 or richer. In smaller posts it is better to use a central steel rod or tube, than a cage of three or four small rods.

It is an advantage to reinforce all precast lintels and beams so that they can be used either way up. Some top reinforcement would probably be necessary in any case for handling, so that the extra cost is not likely to be serious, and provides the only practical insurance against failure resulting from fixing upside down.

Railway sleepers must be of high-grade concrete and adequately reinforced if satisfactory service is to be obtained. A minimum crushing strength for the concrete of 35–40 N/mm^2 at 28 days is normally required. In order to obtain this strength a mix containing about 370 kg of cement per m^3 of concrete is normally used, with a low water/cement ratio. The sleepers are usually made in steel moulds, compacted on a vibrating table.

Concrete pipes may be made in a variety of ways. They may be cast in steel or wooden moulds, either horizontal or on end, being compacted by hand or mechanical tampers, or they can be made in special machines. One such machine utilizes a revolving core which is slowly withdrawn as the concrete, which is of stiff consistency, is tamped into place. The revolving core brings sufficient fat to the surface to form an internal skin on the pipe and makes it more impermeable. Another method is to spin the mould so that the concrete is spread and consolidated against the mould by centrifugal force. The speed of spinning is critical, and must not be too fast, otherwise the centrifugal force may be sufficient to cause separation of the heavier particles. The coarse aggregate collects at the outer surface leaving cement mortar on the inside,

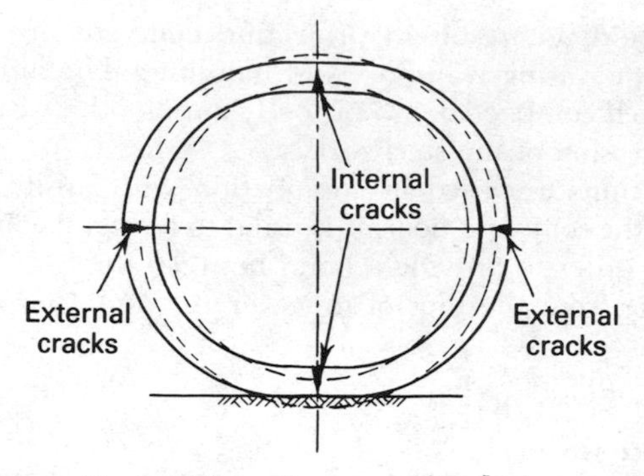

Fig. 23.5 Typical failure of a concrete pipe without haunching

and the pipe is weakened. The interior of the pipe is not formed against a mould by this method, and a round iron bar is used to smooth the surface while the pipe is still spinning. Concrete pipes, the larger sizes in particular, should be carefully handled to avoid fracture.

The method of haunching concrete pipes is important since, except in the smaller sizes, they are not capable in themselves of sustaining a load from the overburden when given point support only. An illustration of the type of failure likely to occur is given in Fig. 23.5. Concrete pipes should therefore be adequately haunched if not completely surrounded with concrete. When the fall is small or the foundation is doubtful a concrete base should be used.

The most important point to remember is that the support given to the pipe should be continuous and uniform throughout its length.

Care is necessary to see that the concrete is worked well under the pipe, often a somewhat difficult procedure. Uneven support results in unequal settlement along the length of the pipe with consequent failures.

Concrete pipes and haunching may suffer serious attack in sulphate-bearing or other aggressive soils, unless adequate precautions are taken, either by using a sulphate-resisting cement (possibly high-alumina cement) in the manufacture of the pipes themselves, or by providing a protective haunching in dense resistant concrete. Too much emphasis cannot be place on the need for dense concrete, containing adequate cement to resist sulphate attack. No cement is effective in permeable concrete. External coatings of bitumen on the pipes are also useful.

Roofing tiles

Roofing tiles are made in a range of standard sizes given in BS 473, 550 : Part 2 : 1971 [4]. Limits of variation in size and shape are given. Other requirements of a good roofing tile are that it shall be impermeable, weather resisting, and shall not fade; it should also be strong, in order to minimize breakages both in transit and after being placed in position. Tests designed to cover these points are given in the British Standard.

Roofing tiles are made in a variety of machines, designed to press the tile in a mould formed in the machine. Each tile is made on a separate pallet (shaped

Fig. 23.6 Automatic machine for the manufacture of concrete roofing tiles

to the underside of the tile) placed in the bottom of the mould. The top of the tile is struck off by a combined screed and tamper, and a lifting device is used to push the tile out of the mould. The tile, still resting on its pallet, is then removed to curing racks.

The materials used for making roofing tiles are either normal or rapid-hardening cement and sand in the proportions of 1 : 3. The mix used is of earth-moist consistency, but must be the wettest which will allow immediate removal of the tiles from the machine. Suitable pigments are added to the dry cement and thoroughly mixed, preferably by grinding them together.

Some tiles are given a surface finish after moulding, by brushing on a coloured cement:sand grout, by sprinkling with sand, by dashing with water to give a rough finish, or by using a roughened screed.

Precautions should be taken to see that the tiles do not dry out quickly, as this would result in loss of strength and high permeability.

Concrete blocks, slabs and bricks

Concrete units for masonry wall construction are usually turned out from machines in which an earth-moist mix is heavily rammed or pressed into a collapsible steel mould. A pallet is used in the bottom of the mould, and the blocks or bricks are removed immediately, or the blocks are 'laid' straight on to the block yard floor. A machine producing blocks by this latter method is seen in Fig. 23.7. Earth-moist concrete is fed to the block making machine from a high feed truck, and is then compacted in the machine by high

Fig. 23.7 Block making machine, which produces blocks by 'laying' them on the yard floor

frequency vibration plus additional tamper vibration. The mould box is immediately lifted away, three cycles per minute being a normal production rate. Fig 23.7 also shows blocks being lifted and stacked until sufficiently mature for transport to the building site.

Care must be taken to avoid the use of too dry a mix, as this would cause a loss of strength, friable corners, and a high permeability. Premature drying out should be avoided for similar reasons. The mix used for concrete blocks and slabs is usually between 1 : 6 and 1 : 8 of cement : aggregate, the maximum size of aggregate being 20 mm. A facing of 1 : 3 cement : sand is used if a good surface is required, as on 'rock-faced' blocks. Concrete bricks are made from a mix of between 1 : 6 and 1 : 10 of cement : sand.

Blocks of odd shapes are normally made by hand either in wooden or metal-lined moulds, using the same materials as outlined above.

The following British Standards cover the manufacture of concrete bricks and blocks:

BS 1180 : 1975. Concrete Bricks and Fixing Bricks.

BS 2028, 1364 : 1968. Precast Concrete Blocks (Amendment 1970).

BS 2028, 1364 deals with solid blocks, hollow blocks, cellular blocks, and aerated concrete blocks, made with a variety of aggregates. The blocks may have plain or profiled faces, or a special facing backed with dense or lightweight concrete as an integral part of the manufacture. In another method a special facing is applied after demoulding.

Three broad categories of block are defined in BS 2028, 1364. These are:

Type A. For general use in building, including the use below the ground level damp-proof course.

Type B. As Type A but not below the ground level damp-proof course externally, except when made with dense aggregates to BS 882 or BS 1047, and having an average compressive strength of not less than 7 N/mm². They may also be used if the manufacturer produces authoritative evidence that they are suitable for the purpose. (If there is any doubt it is advisable to use Type A, and particularly if there is the possibility of sulphates in the ground water.)

Type C. Primarily for internal non-load-bearing walls. Generally these will be partitions or panels in framed construction.

The density of all three types of blocks is required to be not less than 1500 kg/m³.

As would be expected the requirements for compressive strengths, for Type A blocks are higher than for Type B. There is no compressive strength requirement for Type C blocks, but they are required to satisfy a transverse breaking load test. The maximum drying shrinkage for Type A blocks is 0·05 to 0·06% (depending on strength specified), for Type B blocks 0·07 to 0·09% and for Type C blocks 0·08 to 0·09%.

Check tests should be made by the user from time to time. The cost of testing is normally borne by the purchaser if successful, or by the manufacturer if unsuccessful in achieving the specified requirements in all respects.

In order to obtain the requisite strength and drying shrinkage values, it is usually necessary to mature concrete blocks and bricks for at least 4 weeks.

The first shrinkage should be completed before the blocks are built into the wall, and this means that they should be dried out before delivery to the building site. In winter it is necessary for the manufacturer to carry large stocks or to provide suitable dry accommodation under cover. Good air circulation is required to dry the units properly (see also Steam Curing, p. 351).

Troubles due to shrinkage cracking are not likely to be experienced if the blocks comply with the relevant British Standard, particularly the clause limiting the drying shrinkage, unless the mortar used for the blockwork is unnecessarily strong. A mix of cement : lime : sand of the order of 1 : 1 : 6 or 1 : 2 : 9 by dry volume is generally regarded as satisfactory. A mix of similar strength using a plasticizer to replace the lime is often used, but is less able to accommodate movements.

Reference is made to the manufacture and use of lightweight concrete in Chapter 24.

Clinker blocks

Precast clinker blocks are used extensively for the internal partitions of houses and other buildings, chiefly on account of their cheapness. Unfortunately good clinker is not so readily available as it was and this has reduced the total output. Type C blocks (BS 2028, 1364 : 1968) are usually used, and provided the blocks comply with the requirements for transverse breaking strength, drying shrinkage and wetting expansion, there should be little risk of subsequent cracking in walls and panels or of shelling off of plaster coatings. Wetting expansion for clinker blocks should be not more than 0·02% greater than the value for drying shrinkage, as given in the previous section on precast concrete blocks.

Cracking in clinker block partitions has often been ascribed wrongly to the use of unsatisfactory blocks. Partitions are commonly built on timber joists which are quite liable to shrink 3 mm or more in their depth, which is quite sufficient to cause cracking at junctions with more rigid supports.

Reference should be made to CP 111 : Part 2 : 1970, Structural Recommendations for Load Bearing Walls, for data on working stresses for load-bearing walls constructed with blocks.

When constructing walls using clinker or lightweight concrete blocks the following points should receive attention:

(1) Long uninterrupted lengths of lightweight concrete block walling should be avoided by providing breaks or flexible mastic-filled joints in the construction at intervals of about 6–10 m in the length, and at points of marked change of section.

(2) Relatively weak mortars should be used so that any shrinkage cracking will tend to occur in the mortar joints rather than in the blocks. Mixes of 1 : 2 : 9 cement : lime : sand are recommended or, for winter conditions or severe exposure, a 1 : 1 : 6 mix is satisfactory.

(3) When blocks are used externally and rendered, a relatively weak porous rendering should be used. This recommendation should be tempered by consideration of the exposure conditions. In Britain, where renderings may remain wet for several days and may be frozen, it is unwise to use a mix leaner than about 1 : 1 : 6 of cement : lime : sand, except in sheltered situations.

Colours

Only certain types of inert pigment give satisfactory results in concrete, the principal requirements being that the colour shall not be affected by lime, shall not fade when exposed to the weather, and shall not contain any substances such as gypsum, which are deleterious to concrete. Organic pigments do not fulful these conditions and should not be used. Mineral colours are generally satisfactory, but are sometimes diluted with other cheaper constituents. However, to obtain the same colour shade, a greater quantity of the diluted pigment must be used, so that no advantage is gained, while in fact the increased proportion of fine material causes a loss in strength and increased drying shrinkage.

For general purposes the paler tints of reds, yellows, buffs and browns are the most successful. These pigments are usually oxides of iron. Pale green shades obtained by additions of chrome oxides are also quite good. Brighter colours are obtained by using white rather than grey cement.

Deep reds, blues, greens, and black are apt to come out patchy, and in any case soon lose their brightness due to the slight efflorescence or lime bloom which forms on most concrete exposed to the weather.

Dark grey may be obtained by using small quantities of manganese black or carbon black.

Pigments are not very different from silt in their effect on the physical properties of concrete and not more than 10 per cent of colour by weight of cement should be used. Larger percentages cause serious loss of strength and increased shrinkage. Pigments are not easily mixed with cement on the site, and if large quantities of coloured cement are required it is better to mix the pigment and cement in a ball-mill at a central depot, or to buy one of the coloured cements prepared by the cement manufacturers.

The effectiveness of concrete colouring depends to a considerable extent on the colour of the aggregate used. Unless a contrast is required between the aggregate and the cement, as in terrazzo, it is advisable to select an aggregate of similar colour to that required for the finished concrete. In some cases, aggregate may be selected, which, when mixed with cement, gives a suitable colour without additions of pigment.

Crazing

Crazing of concrete products results from differences in shrinkage between the surface and the interior. As has been shown in Chapter 2, this is a combination of differences in drying shrinkage and a shrinkage due to carbonation which affects only a layer up to 10 mm or so thick near the surface. Cracks due to crazing rarely exceed 10 mm or so in depth, and are therefore not serious, apart from their unsightliness.

Several methods of overcoming crazing have been advocated, but the best insurance appears to be either to use an earth-moist mix, or if a plastic mix is necessary use as low a water/cement ratio as is practicable and remove the cement skin to expose the aggregate. Aggregates which have low shrinkage in themselves are likely to produce less crazing than highly shrinkable aggregates. Proper attention to curing, especially during the early stages, is an added safeguard.

Trowelled surfaces are particularly prone to crazing, and trowelling should therefore be avoided whenever possible. If a surface must be trowelled, the amount of trowelling should be the minimum necessary to finish the surface.

Steam curing

Steam curing is used to obtain a high strength at early ages and so release the moulds more quickly than would be the case with normal methods of curing, especially during cold weather.

Steam curing may be effected either by the use of low pressure or high pressure steam, but, while the former is the cheaper, the use of high pressure steam results in an improvement of certain of the properties of the concrete which may be considered to offset the increased cost.

When high pressure steam curing is adopted the precast concrete units are stacked in steam chambers, or autoclaves, and the steam introduced for the required period of curing, usually a matter of a few hours. It is advisable to allow a short curing period at ordinary atmospheric temperatures before introducing the steam, and an excessive rate of heating should be avoided. There is considerable controversy on the question of the maximum rate of heating which may be allowed. Figures for the safe rate of increase in temperature quoted by various investigators range from $5\frac{1}{2}$°C to 66°C per hour. Rapid cooling at the end of the curing period should also be avoided.

High pressure steam curing encourages the growth of a crystalline bond and in the presence of finely divided siliceous aggreagate, there is a reaction between the aggregate and lime from the cement which accounts very largely for the increased sulphate resistance. The shrinkage and moisture movement is also reduced.

Low pressure steam curing is effected either by stacking the precast concrete units in specially built chambers, or by using light portable boxings which are placed over the units wherever they are cast, the steam being supplied through flexible connections. This latter method has the advantage that small individual chambers take less time to warm up than large chambers containing a large number of units, and the amount of handling is reduced. Low pressure steam curing speeds up the rate of hardening of concrete units, but does not improve sulphate resistance or reduce shrinkage and moisture movements.

As with high pressure steam curing, excessive rates of heating and cooling must be avoided.

If hot concrete units dry out after having hardened in a steam curing chamber for only a few hours, sufficient to facilitate demoulding, the loss in ultimate strength can be serious. Both with high pressure and low pressure steam curing, it appears that the ultimate strength is not fundamentally different from that of concrete cured at atmospheric temperatures, unless the rate of heating is too high. Under these conditions, further gain in strength on subsequent moist curing appears to be arrested.

Efflorescence

The chief cause of efflorescence on cast stone is the formation of crystals of calcium carbonate and calcium sulphate on the surface due to leaching and chemical reactions of the free lime (calcium hydroxide) with rain-water. This action is unavoidable in porous concrete, but is unlikely to be serious if the concrete is dense and impermeable.

Efflorescence is more apparent on dark or deeply coloured surfaces than on those which are white or tinted.

Treatment with waterproofers containing soluble salts, or which react with the formation of soluble salts, should be avoided. Aggregates containing alkalies or sulphates are liable to cause efflorescence, as also are aggregates containing sea salts.

REFERENCES

1 BS 368 : 1971, *Precast Concrete Flags*. British Standards Institution.

2 BS 340 : 1963, *Precast Concrete Kerbs, Channels, Edgings, and Quadrants*. British Standards Institution.

3 CP 110 : Part 1 : 1972, *The Structural Use of Concrete: Design Materials and Workmanship*. British Standards Institution.

4 BS 473, 550 : Part 2 : 1971, *Concrete Roofing Tiles and Fittings*. Metric units, Amended June 1976. British Standards Institution.

24
Lightweight Concrete

THERE are various methods of producing lightweight concrete but they all depend on either the presence of air voids in the aggregate; or the formation of air voids in the concrete by omitting fine aggregate; or the formation of air voids in a cement paste by the addition of some substance which causes a foam. In some types of lightweight concrete, the first two methods may be combined.

No-fines concrete made with gravel aggregate is not strictly a lightweight concrete, although its weight is only two-thirds that of dense concrete, but it is convenient to consider it with concrete made with other lighter aggregates.

Lightweight concrete is used not only on account of its light weight, but also because of the high thermal insulation compared with normal concrete. Generally, a decrease in density is accompanied by an increase in thermal insulation, although there is a decrease in strength. The relationship between density and thermal conductivity is shown in Fig. 24.1, which is derived from results obtained on various types of lightweight concrete at the Building Research Establishment[1].

It is important to realize that the high thermal insulation is only obtained while a lightweight concrete remains dry. It is, therefore, not always satisfactory for roof insulation because of practical difficulties in drying out before applying the waterproof finish, and because condensation may cause saturation later. This condensed water sometimes shows as damp patches on ceilings and is mistaken for rain penetration. Similarly lightweight concrete is of no use for thermal insulation in any position in which it may become saturated, for example underground duct work.

Structural 'lightweight' aggregate concrete can be produced with a strength in excess of 30 N/mm^2 and even higher strengths have been attained in certain cases, although at the expense of increased densities. The aggregates used include sintered pulverized fuel ash, expanded shales, clays and slates, foamed slag, and pumice or scoria. As most of these aggregates absorb considerable quantities of water (up to 80 per cent by volume), the effect on workability within a few minutes of mixing is such that a wet mix can become too dry to work. It is, therefore, necessary to wet, but not saturate, the aggregate before mixing. A good portion of the mixing water is also best added before introducing the cement. Rich mixes containing 350 kg/m^3 cement or more are usually required to give a satisfactory strength.

The cover to reinforcement when using lightweight concrete should be 10 mm more than that used for normal dense concrete. The increased cover is necessary because, besides being more permeable, lightweight concretes carbonate more quickly than dense concrete and the protection afforded to the steel by the alkaline lime is lost (see also Expanded Minerals, p. 356).

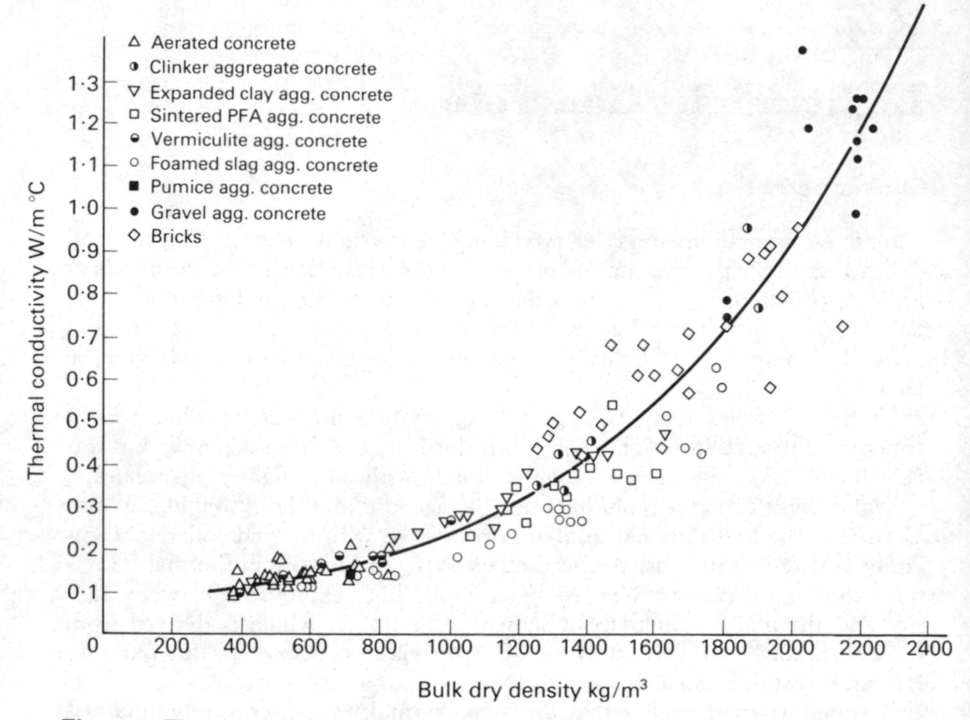

Fig. 24.1 Thermal conductivity of lightweight concrete corrected to moisture content of one per cent by volume vs. dry density. Reproduced from BRS Current Paper CP1/70 by permission of the Director, Building Research Establishment. Crown copyright, HMSO

Properties of lightweight aggregate concrete

The air-dry densities of lightweight aggregate concrete vary considerably, depending on the choice of aggregate and whether natural sand or crushed lightweight aggregate fines are used. A density of 1850 kg/m³ may be considered as the upper limit for a true lightweight concrete, although this value is sometimes exceeded. Typical density values of various lightweight concretes are given in Table 24.1. Lower densities than those stated will be obtained in cases where the concrete is only partially compacted.

Table 24.1 also includes typical values of compressive strength. In general the lower densities can only be achieved at the expense of lower strengths. The range of strengths given in each case are typical of what may be achieved in practice, but it should be noted that rather rich mixes are required for the higher strength values in the case of the lightweight aggregate concretes. It is also necessary to use natural sand instead of lightweight fines to achieve these higher strengths. The tensile and shear strengths of lightweight aggregate concrete are less than that for natural aggregate concrete of the same compressive strength. The reduction in the case of tensile strength may be as much as 30%.

The modulus of elasticity of lightweight concrete is about 0·5 to 0·75 times the value of natural aggregate concrete of the same compressive strength varying from 7 to 21 kN/mm^2 [2]. Values of elastic deformation, shrinkage and creep are, therefore, greater for lightweight concrete. Extra reinforcement is also necessary.

Clinker and breeze

Clinker and breeze aggregates have been in use for many years in the production of blocks and slabs for internal partitions and other interior walls. These aggregates are cheap and plentiful, and provide a very useful product after they have been crushed and graded.

There is no definite demarcation between materials described respectively as clinker and breeze. In general, clinker is regarded as a well-burnt fused or sintered mass containing little combustible material; whereas breeze is a more lightly sintered and less well-burnt residue and therefore contains more combustible matter.

Increasing quantities of combustible matter causes corresponding increases in moisture movement. The combustible content is determined approximately by igniting a small sample of the clinker at specified temperatures and finding the percentage loss in weight. The specified limits for combustible content are varied according to the position in which the concrete is to be used. They are given in BS 1165 : 1966 [3] as follows:

Class A1.	Not more than 10 per cent	General purposes, plain concrete.
Class A2.	Not more than 20 per cent	Interior work not exposed to damp conditions, cast in situ.
Class B.	Not more than 25 per cent	Precast clinker blocks. (Some British Standards call for a clinker aggregate which, by virtue of the test requirements in those standards, requires a lower limit on the percentage of combustible content.)

Clinker and breeze aggregate is quite unsuitable for reinforced concrete due to its porosity and absorptive properties, which maintain a more moist condition than the surroundings. The sulphur content is also a factor in accelerating the corrosion of embedded steel.

The main sources of clinker aggregates are the older power stations where the boilers are fired by solid fuel in chain-grate stokers. However, many of the more modern power stations now use pulverized fuel for firing their furnaces and consequently the availability of clinker is expected to decline. The use of pulverized fuel produces two forms of burnt product, namely pulverized fuel ash (pfa) and furnace bottom ash. The latter residue may occur partly in the form of a clinker, similar to that from the older power stations, but it is very variable in quality and is accompanied by a great deal of useless dust. The better grades can however be used as lightweight aggregate for blocks.

Pumice concrete

Pumice is the only natural lightweight aggregate in common use. Provided it is free from fine volcanic dust and materials not of volcanic origin, such as clay, pumice produces a satisfactory lightweight concrete with a density of between 720 kg/m³ and 1440 kg/m³. Other properties are given in Table 24.1.

Pumice provides better thermal insulation than other types of lightweight concrete.

Foamed slag

Foamed slag is made by rapidly quenching blastfurnace slag produced in the manufacture of pig iron. Its texture and strength are dependent on the chemical composition and the method of treatment, but in general the structure is similar to natural pumice.

Foamed slag aggregates are supplied in two sizes, coarse 15–3 mm and fine, 3 mm down. They should comply with BS 877:Part 2:1973[4]. Foamed slag may be regarded as satisfactory if it fulfills the following requirements:

(1) Freedom from contamination by heavy impurities, including air-cooled slag.

(2) Freedom from volatile impurities such as coke or coal.

(3) Freedom from an excess of available sulphate.

The properties of foamed slag concrete are given in Table 24.1. The thermal conductivity when dry can be almost as low as pumice concrete.

In common with other lightweight concretes, foamed slag has a higher moisture movement than ordinary concrete. The value is, however, as good as the best of other lightweight concretes.

Expanded minerals

Naturally occurring clays, shales, and slates may be used to produce lightweight porous material of a cellular texture by suitable treatment and heating up to temperatures of about 1000°C–1200°C. Material with similar characteristics may also be obtained from pulverized fuel ash or 'fly ash' which is a residue from the burning of powdered coal at power stations. After crushing and screening to the desired size, these processed materials form good lightweight aggregates. The properties of the lightweight concrete they are used to make are given in Table 24.1.

Expanded clay aggregate is produced on the Continent and at one plant in Britain where it is produced under the trade name of Leca (Lightweight Expanded Clay Aggregate). It is made by a patented process using a special grade of blue clay which bloats readily when heated. The resulting material consists of light, hard, rounded pellets with a dense skin and a honeycomb interior. Angular material may also be present if oversized material has been crushed and added. The material is supplied in three nominal sizes: coarse, 20–10 mm; medium, 10–3 mm; and fine, 3 mm down. Expanded shale is also produced on a relatively small scale, the manufacture being based on an American process of sintering colliery shale and clay. This material, which is known as Aglite in Britain, has similar properties to Leca except that it is crushed and is, therefore, angular. It is also slightly more porous.

A somewhat similar lightweight aggregate known as Lytag is produced by pelletizing fly ash and burning it on a sinter strand at temperatures up to

Table 24.1 Typical Properties of Different Types of Lightweight Concrete

Type of lightweight concrete	Air-dry density kg/m³	Compressive strength N/mm²	Drying shrinkage per cent	Thermal conductivity W/m°C	Working properties	Nail and screw holding
Sintered pulverized fuel ash (Lytag)	1360–1760[1]	14.0–42.0[1]	0.04–0.07	0.32–0.91	Easily worked	Satisfactory
Expanded shale or clay (Aglite & Leca)	1360–1840[1]	14.0–42.0[1]	0.04–0.07	0.24–0.91	Easily worked	Satisfactory
Foamed slag	1680–2080[1]	10.5–42.0[1]	0.03–0.07	0.24–0.93	Easily worked	Satisfactory
Pumice	720–1440	2.0–14.0	0.04–0.08	0.21–0.60	Easily worked	Satisfactory
Clinker	1040–1520	2.0–7.0	0.04–0.08	0.35–0.67	Easily worked	Satisfactory
Aerated cement mortar	400–960	1.4–4.9	0.05–0.18	0.10–0.22	Easily worked	Satisfactory
No-fines concrete						
(a) gravel aggregate 1 : 8 aggregate/cement ratio (by volume)	1600–1840	3.5–11.0	0.02–0.03	0.65–0.80	Difficult	Fixing blocks necessary
(b) lightweight aggregate 1 : 6 aggregate/cement ratio (by volume)		2.4–3.1	Depends on aggregate used	Depends on aggregate used	Easily worked	
Dense concrete containing gravel or crushed stone[2]	2240–2480	14.0–70.0	0.03–0.05	1.40–1.80	Hard	Good when plugged

[1] These higher values of density and strength are obtained by replacing the lightweight fines by a natural sand
[2] Typical properties of normal dense concrete mixes given for comparative purposes

1400°C. A bed of pellets, on a moving steel grid, passes under a furnace and the pellets become sintered into a clinker-like mass. After cooling, it is screened and graded into three sizes: coarse, 13–8 mm; medium, 8–5 mm and fine, 5 mm down.

Since these lightweight aggregates are porous, they absorb considerable quantities of water. The amount of water absorbed by Aglite in 24 hours can exeed 30% by volume and the ultimate water absorption is of the order of 50% by volume. The corresponding figures for Leca and Lytag are generally about two-thirds of these values, indicating that a considerable amount of water can be present within these aggregates. This can have a considerable influence on the thermal insulation, shrinkage, and creep of the concrete which is made with them.

Although the use of these processed aggregates produces a saving in weight of the concrete as compared with the use of natural aggregates, the cost of the lightweight concrete (of equivalent strength) is 30–40% more. Their use does not, therefore, normally result in a saving in the cost of the structure.

Exfoliated vermiculite is very light in weight, but is so soft that the action of mixing with cement and water is sufficient to cause collapse of particles. Vermiculite concrete is, therefore, likely to be of uncertain density and to have little compressive strength.

No-fines concrete

No-fines concrete is composed of cement and coarse aggregate only, the fine aggregate being omitted in order to leave uniformly distributed voids throughout the mass. The aggregate may be gravel or crushed stone, blastfurnace slag, crushed brick, or one of the light aggregates already mentioned. The quality of the aggregate should accord with the requirements already given, and the relevant British Standard.

Recommended practice for the use of no-fines concrete cast in situ for one to three storey housing was originally given in Post-war Building Studies, No. 1 [5]. The following comments are based on experience in the production of no-fines concrete on a large scale for houses and multistorey flats, both in Britain and abroad.

(1) *Grading of aggregate.* The aggregate should be as nearly single sized as practicable. In Great Britain, it is usual to use an aggregate of which not more than 5 per cent is retained on a 20 mm mesh sieve, and not more than 10 per cent passes a 10 mm mesh sieve.

(2) *Recommended proportions.* Normal dense aggregate, 50 kg cement to 0·28 m^3 aggregate; light aggregate, 50 kg cement to 0·21 m^3 aggregate.

(3) *The water/cement ratio* should be the minimum necessary to ensure that each particle of aggregate is coated with cement grout. Too little water gives a friable appearance with a large proportion of uncoated particles. Too much water causes the cement grout to run and separate from the aggregate.

(4) *Mixing.* It is most important that the aggregate is used in a damp condition or wetted before adding the cement. Mixing should continue until the aggregate is evenly coated with cement grout. With rotary-drum mixers it is generally advisable to add some of the water before raising the loading skip

and then to add the remainder of the water after the materials have entered the drum. In this way a more uniform mix is obtained. This difficulty does not occur with open-pan mixers fitted with a rotating star of blades.

(5) *Placing.* No-fines concrete will not segregate and can therefore be readily placed in deep lifts of up to three storeys high in one operation. Pressures on the formwork are not high and a comparatively light timber framework serves to support plywood, hardboard, sheet metal, or expanded metal coverings. No-fines concrete should not be rammed but careful rodding should be done wherever obstacles occur, since bridging is more likely to occur than with ordinary concrete.

Delays in placing the no-fines concrete, particularly during hot weather, must be avoided since the progressive stiffening of the cement paste around the aggregate can cause a serious reduction in the density and strength of the no-fines concrete in the structure.

(6) *Joints.* Since the bond between new and existing work is weaker than in ordinary concrete, construction joints should be horizontal and as few as possible. Vertical joints weaken the structure, especially if they occur near an external angle, and except where necessary for the provision of expansion joints they are better omitted altogether. Expansion joints should be used at suitable spacings to take care of shrinkage and thermal movements.

(7) *Cube strength.* Experience has shown that with good site control a minimum crushing strength of 2·8 N/mm^2 or more at 28 days is easily practicable with the mix given under (2), above, enabling the building of unreinforced structures five or six storeys in height. Greater strengths can be obtained by using richer mixes and by compacting to a higher density. In Germany, no-fines concrete has been used without reinforcement for structures up to about eight storeys. The total height of these structures is, in some cases, greater than eight storeys, but this is about the maximum number of storeys in no-fines, as defined above. In Great Britain, one system of construction utilizes no-fines panels to form the moulds for reinforced beams and columns which eventually provide the supporting framework. By this method multistorey structures have been produced.

The BS Code of Practice, CP 111 : Part 2 : 1970[(6)], gives minimum cube strengths and maximum permissible stresses, as in Table 24.2. These are for walls with a slenderness ratio of less than 15. For greater slenderness ratios it is necessary to apply a reduction factor.

It must of course be borne in mind that the cube strengths quoted refer to cubes made according to the standard procedure for no-fines concrete cubes. This procedure is different from that for dense concrete cubes since the compaction is limited to a specified number of blows from a specially designed rammer. Details of making no fines concrete cubes are given in BS 1881 : Part 3 : 1970.

Other characteristics. No-fines concrete presents some difficulty in the fixing of various fittings and it is necessary to embed nailing blocks of timber, sawdust-cement, or foamed slag. Cutting away for services should be avoided by forming suitable openings and chases at the time of placing.

No-fines concrete has little resistance to the penetration of water, but there is also very little capillary action. Thus, there is no tendency for the water to

Table 24.2 Recommendations for Cube Strengths and Permissible Stresses in No-Fines Concrete

Nominal mix	Volume of coarse aggregate per 50 kg of cement m^3	Cube strength * within 28 days after mixing N/mm^2	Maximum permissible stresses N/mm^2
Special mixes	0.28 or less	7.0 †‡	1·3
		3·5	0.6
1 : 8	0.28	2.8	0.5

The recommendations cover the use of natural aggregates to BS 882 and air-cooled blastfurnace slag to BS 1047, mixed with Portland cement, Portland blastfurnace, or other cement included in CP 110.

* These requirements may be deemed to be satisfied if two-thirds of the value is obtained at 7 days.

† The average cube strength for mix design purposes should be 2.10 N/mm^2 in excess of the cube strength specified unless more definite information is available.

‡ The attainment of this strength increases the density of the concrete to an extent where the thermal-insulation properties may be impaired.

be drawn into the wall; the provision of a rendering, with care in arranging and fixing flashings at various openings, is sufficient to waterproof the structure satisfactorily.

The rendering should preferably have a rough surface texture, dry dashings being particularly suitable. A 1 : 1 : 6 or 2 : 1 : 9, cement : lime : sand rendering is normally considered suitable, except perhaps in cold climates where a somewhat richer mix may be used to provide increased resistance to disintegration by frost and continually wet conditions.

The thickness of no-fines concrete is more often determined by the requirements of thermal and acoustic insulation than stability. For instance, in order to obtain insulation similar to that of a 275 mm closed-cavity brick wall, plastered internally, it is necessary to use 250–300 mm of no-fines concrete rendered externally and plastered internally.

A general indication of other important properties of no-fines concrete is given in Table 24.1.

Aerated cement mortars

Aerated cement mortars (often referred to simply as aerated concrete) are made by introducing air, or specially formed gas, into a cement slurry so that, after setting, a hardened mortar with a cellular structure is formed. The slurry usually consists of a mixture of cement and siliceous material such as sand and/or pulverized fuel ash.

There are two main methods of forming the cellular structure, namely (1) the addition of powdered aluminium or zinc, which combines with the lime in the cement to generate hydrogen gas and (2) using a foaming agent. In the first of these processes, the aluminium or zinc powder is added to the cement slurry during mixing, the quantity of powdered metal being about 0·1 to 0·2 per cent of the weight of the cement. Within a few minutes, hydrogen gas begins to evolve causing the slurry to rise, the action continuing for an hour or so. The slurry then sets to form a material consisting of a multitude of closed bubble holes surrounded by hardened cement mortar. The density of this material depends on the quantity of powder metal used and the temperature at the time of manufacture. For most purposes densities within the range 550–950 kg/m^3 are satisfactory.

While aerated cement mortar made in this way can be relatively impermeable to moisture, owing to the closed type of pores, it has the disadvantage of a high drying shrinkage and moisture movement. Precast units, which can be matured well before use, thus materially reducing effective shrinkage movements when the units are built into a structure, do not suffer from cracking to the same extent as aerated cement mortar cast in situ. Curing in an autoclave, using steam under pressure, considerably reduces drying shrinkage and moisture movements. This procedure is generally adopted for the production of blocks and precast units.

In the proprietary systems which incorporate the autoclave treatment, the cement is mixed with sand or blast-furnace slag which has been ground in a ball mill to the same fineness as the cement. Pulverized fuel ash or lime may also be added.

Aerated cement mortars with a similar texture can also be produced by the use of foaming agents such as resin soap. The foaming agent is either mixed with the cement, sand and water and the air entrainment achieved by whisking in a high speed mixer, or is turned to foam by the use of compressed air using a special foam producing apparatus, and the foam then added to a cement : sand mortar in an ordinary concrete mixer. The latter method provides a more uniform density, provided the foaming time is carefully controlled.

Foamed cement mortar can be made with a density as low as 320 kg/m^3, but then has no appreciable strength and is only of use for insulation purposes in dry situations.

Some foaming agents are sufficiently stable to permit the use of cement: sand mixes for the production of foamed concrete at densities above about 800 kg/m^3. The average crushing strength varies with the density as indicated in Fig. 24.2. As with other lightweight concretes the relatively high drying shrinkage presents problems in the use of this material in situ, and ample provision of 'expansion joints' is essential. Nevertheless, aerated mortars cast in situ have been used successfully in the construction of some thousands of dwellings in countries such as Kuwait and Borneo, where sand has been in ample supply in the vicinity of the construction site. Precast blocks must be cured for several weeks before use unless steam curing is adopted. Even then it is advisable to build in a cement : lime : sand mortar not richer in cement than 1 : 1 : 6 if visible cracking is to be avoided. The inclusion of steel reinforcement has extended the use of autoclaved aerated cement mortars for structural units such as wall units and lintels. Because of the danger of corrosion of the reinforcement, however, such units should be protected from the weather and the reinforcement should be given a protective coating of cement latex compound or similar material.

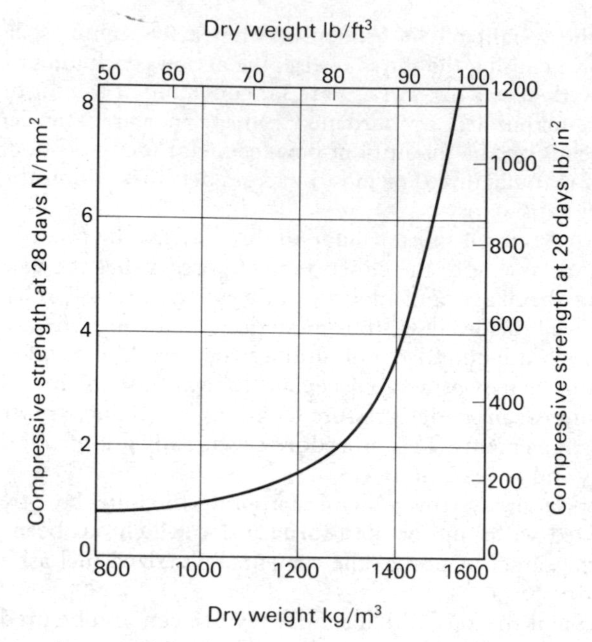

Fig. 24.2 Compressive strength-density relationship for aerated cement mortar

Aerated mortar concrete

Coarse natural or lightweight aggregate can also be used in conjunction with aerated cement mortars. Natural aggregates increase the density, and to some extent reduce the moisture movements. Lightweight aggregates do not cause the same increase in density, but the moisture movements are more nearly those of aerated mortars.

Lightweight blocks

A variety of concrete blocks are made with different types of lightweight aggregate or with various types of aerated cement mortar. Reference is made to their manufacture in Chapter 23.

Wood-cement mixes

Wood in the form of fibres, chips and sawdust is sometimes used in various types of block, partition slab and floor surfacings. The high drying-shrinkage and moisture movement is apt to cause cracking, and the use of these materials should not be considered unless freedom of movement of each individual unit can be arranged.

REFERENCES

1 ARNOLD, P. J., *Thermal Conductivity of Masonary Materials*. BRS Current Paper CP1/70. Building Research Station.

2 SPRATT, B. H., *The Structural Use of Lightweight Aggregate Concrete*, 1974. Cement and Concrete Association.
3 BS 1165 : 1966, *Clinker Aggregate for Concrete*. British Standards Institution.
4 BS 877 : Part 2 : 1973, *Foamed or Expanded Blastfurnace Slag Lightweight Aggregate for Concrete*. British Standards Institution.
5 *House Construction*, Post-war Building Studies, No. 1, Ministry of Works, 1944. HMSO.
6 CP 111 : Part 2 : 1970, *Structural Recommendations for Loadbearing Walls*. British Standards Institution.

25

Floors and Floor Surfacings

CONCRETE floor construction may conveniently be considered under two headings, floors on the ground and suspended floors. Floor surfacings may be produced by trowelling up the construction concrete or by applying a topping of granolithic concrete.

Floors on the ground

The minimum thickness for floors on the ground should be 100 mm. This is generally sufficient for residential purposes where the subgrade is good and the traffic is light. In public places this should be increased to 125–150 mm for foot traffic and light rubber-tyred vehicles. For industrial traffic and floors carrying loadings of 5 to 20 kN/m^2, a thickness of 175 mm or more will be required. These thicknesses should be increased by 25 mm or more if the subgrade is known to be weak. A sub-base of somewhat weaker concrete may be desirable, its thickness depending on the subgrade and the loading expected.

The uniformity of the subgrade and possible movements due to shrinkage and expansion in the subsoil, especially if this is a clay, are more important to the incidence of crack formation in the floor than the thickness. Reinforcement may be desirable, say 50 mm from the top, to reduce the risk of cracking and more particularly to resist the opening of those cracks which do occur.

Further information on slab thicknesses may be obtained from a CACA publication[(1)]. Reference may also be made to Road Note No. 29[(2)] for thicknesses suitable for moving vehicles.

Floors placed on the ground should be isolated by suitable movement joints from columns, beams, etc., supported on deeper foundations. Normally a vertical strip of compressible filler board with sealant or other suitable material, say, 20 mm thick, and of the type used in road joints, will be satisfactory.

Control joints are necessary in large floor areas, and it is usual to form these between columns. If such bays are more than 6 m in length they should be sub-divided by movement joints very similar to those used in road construction, see Chapter 21.

Vapour barriers are necessary in concrete floors on the ground, when a dry floor is needed. PVC membranes about 0·5 mm thick, laid on the formation before laying the concrete, have proved satisfactory and are now widely used. The formation should be finished with fine material to avoid puncturing, the PVC should be lapped and folded over to give a reasonable seal at joints, and some care should be exercised to avoid tearing during concreting operations. The PVC should be turned up at walls to meet the damp proof course.

As in other forms of construction, concrete for floors should be dense and strong. A Grade 20 (N/mm^2) concrete should be satisfactory if a further wearing course, or coverings such as tiles or carpets, is to be applied. For floors to resist wear, a Grade 30 concrete with a minimum cement content of 330 kg/m^3 should be used. If heavy trucking or moderate chemical conditions are expected, these requirements should be increased to 40 N/mm^2 and 400 kg/m^3 or more of cement. These strengths and cement contents apply to concrete to resist wear; if a further applied topping, e.g. granolithic, is to be used the grade of concrete and the minimum cement content may be decreased by 10–20 per cent. Good quality aggregate complying with the requirements of BS 882, particularly as regards hardness and abrasion resistance, should be used. Good trowelling technique is essential.

The concrete should be spread evenly, avoiding segregation, and compacted by vibrating beam as in the case of roads. A suitable beam is shown in Fig. 21.7. For slabs thicker than about 150 mm these are notched so that the beam rides below the level of the forms. This enables the beam to compact the lower part of the slab, and is also convenient if reinforcement is used. The forms are partially filled and compacted, the layer of reinforcement placed on the surface, and then the second layer of concrete is placed and compacted, with an un-notched beam, to the top of the forms. If no extra wearing course is to be applied, further levelling by means of straightedges (preferably metal) and by darbies of wood or light alloy, becomes necessary. Skip floats can also be used. (A skip float is a flat blade 600–800 mm wide on the end of a tubular handle, and is pulled lightly across the concrete when excess water has drained from the surface.) It is important that excess moisture and laitance are not brought to the surface during these operations. Joints and edges are then given attention as necessary.

The floating is delayed, depending on atmospheric conditions, until the surface is free from any water sheen and is firm enough to stand on. Disc type floats and trowelling machines provide light pressure to compact the concrete and give a dense smooth surface. If steel trowels are used by hand, considerable pressure is necessary to give a good wearing-resistant surface. A second trowelling, an hour or so later, may be necessary and beneficial. The finishing process is very similar to that employed on granolithic floor surfacings, to which reference should be made.

Further guidance may be obtained from Cement and Concrete Association publications [1].

Suspended floors

Many flooring systems are available, both in-situ and precast, which are designed to reduce the dead weight of concrete. Cavities are formed in floors by various methods, for example hollow concrete blocks and other precast units, hollow clay blocks, glass fibre domes (see Fig. 25.1) and retractable steel or wooden tubes. The use of retractable cores is restricted to precast units. Holes for services are sometimes formed by inflated rubber tubes. More often, however, services are laid in grooves formed in the upper surface, or simply laid in a cement : sand screed which is trowelled to provide a smooth level surface on which to lay the floor finish.

An effective saving in weight for heavily loaded and long span concrete

Fig. 25.1 Coffered ceiling produced by use of glass fibre reinforced polyester moulds. Reinforcement runs in the ribs

structures can be achieved by using a waffled or coffered floor or ceiling as shown in Fig. 25.1.

Suspended floors will normally be designed to comply with the relevant code of practice, e.g. CP 110 : 1972. If the concrete is to be finished to provide the wearing surface it may be necessary to increase the cement content, as for floors on the ground. If a wearing course is to be added, in certain circumstances CP 110 : 1972 allows this to be taken into account in the structural design.

Precast concrete beams and slabs (Fig. 25.2) for floors achieve their greatest economy when dimensions are standardized as far as practicable, and they can then be produced in a highly mechanized factory. A further advantage accrues when conditions are such as to render in-situ concreting impracticable or at least very expensive. In Russia this condition occurs for several months each year, and in consequence factory produced precast concrete is used extensively.

Among the advantages of the precast beam type of floor as against the in-situ type are (i) the saving in timber supports and shuttering, (ii) increased speed of construction, and (iii) the fact that the floor can be traversed immediately the beams are placed in position and grouted up. On the other hand, the in-situ floor is (i) usually cheaper to construct, (ii) better suited for distributing concentrated loads, (iii) not subject to cracking at intervals (at joints) as so often happens with a series of beams, and (iv) it provides better sound insulation than the lighter hollow floor.

Fig. 25.2 Floor of precast concrete slabs, 2 m wide, each containing four cavities, 300 mm wide

All floors formed from precast concrete units should have a topping of concrete, preferably 75 mm thick, with mesh reinforcement, if cracking due to temperature changes, drying out and other causes of movement between elements is to be minimized.

Prestressed floor units

Prestressed floor units produced on the long line system described in Chapter 20, have the advantage of a saving in weight, an especially useful feature when long spans are envisaged. Their deflection may, however, be greater than that of ordinary reinforced concrete units, a point which requires consideration in design. An example of a possible source of trouble is that of a lightweight block partition, carried on a prestressed floor which is part beam and part cantilever, in which cracking may occur over the support adjoining the cantilever unless deflections are limited.

Variation in camber between units can present a problem in finishing the soffit. This variation increases with length of span, and differences between units of up to 50 mm have been measured.

Floor surfaces

Concrete floors may be finished either by trowelling up the concrete surface or by applying a special finishing course of the granolithic type. The technique of trowelling up the surface of the construction concrete has improved considerably in recent years. Floors constructed by this method will have surfaces suitable for most purposes, but surface wearing courses of the granolithic type, which may be produced with smooth hard surfaces highly resistant to wear and easy to clean, are often preferred. In some conditions it may also be preferable to leave the completion of the floor surfacing until other building operations are complete, thus reducing the risk of damage and discoloration.

For heavy-duty floors a carefully controlled technique is essential if good results are to be achieved. The following considerations apply to granolithic and similar finishes.

Aggregates

The aggregate should consist of a hard tough natural rock, of suitable grading to comply with BS 1201 : 1973 (included with BS 882 : 1973, Aggregates from Natural Resources (including Granolithic)).

The specification limits the type of rock to one of seven trade groups. An essential property of aggregates for granolithic floors is wear resistance, for which no direct test is available, although the ten per cent fines value gives an indication of hardness; the test is made in accordance with BS 812, the limit being ten per cent fines under a load of not less than 15 tonnes.

BS 1201 also gives maximum desirable quantities of clay, silt and fine dust.

Cement

Normal Portland cement is used almost exclusively for granolithic floors, coarsely ground cements being preferred as against the more finely ground rapid-hardening cements. Experience suggests that the latter result in an increased tendency to crack.

Additions

Coloured floors are not entirely satisfactory, as variations in colour intensity are difficult to avoid, and efflorescence is often the cause of white patches. Colours have the same effect in a concrete mix, as the presence of clay and silt, reducing strength and wear resistance while increasing drying shrinkage. Large percentages of colour have been the cause of failures; in one case investigated it was found that the use of dusty granite aggregate and 10 per cent (by weight of the cement) of red pigment had increased the drying shrinkage to twice that of the undercourse.

Metallic aggregates are widely used in the belief that they improve the wear resistance of floors subjected to heavy traffic. They are mainly composed of particles of cast iron of varying degrees of fineness. The finer materials have a large portion passing a BS 150 μm sieve and their value is open to doubt. They may be either incorporated in the mix or thrown on to the surface and trowelled in.

Large proportions of metallic aggregates have caused spalling due to rusting and expansion of the iron. The safe limit depends on the density of the concrete, but about 8 kg/m^2 is suggested as a maximum.

Metallic aggregates may improve well-laid floors but are valueless in badly-laid floors. They are more evenly distributed when incorporated in the mix than when applied by throwing on the surface and trowelling in.

Discoloration by rust during damp curing of the floor is difficult to avoid.

Carborundum grit trowelled into the surface is useful for preventing slipperiness, but being brittle does not improve the wear resistance under conditions of heavy impact and abrasion by iron-tyred trucks and similar vehicles.

Calcium chloride is sometimes added to accelerate the setting and hardening. It should not be used where there is reinforcement in the main slab which may become corroded (see p. 80).

Mix

The mix used should be 1 : 3 cement : aggregate by weight. The aggregate should be measured accurately by weight or by volume, and cement by weight. The water content should be the minimum necessary to give sufficient plasticity for laying and compacting and does not generally need to exceed a water/cement ratio of 0·42. The influence of cement content and the cement/water ratio (converse of water/cement ratio) on wear resistance is indicated by Fig. 25.3 (*a*) and (*c*), which is plotted from the results of tests made by Schumann and Tucker[(3)].

In two-course work the bottom course may be 1 : $2\frac{1}{2}$ to 1 : 6 by weight and of somewhat drier consistency than the top course, which is mixed in the proportions given above. Care should be taken to ensure that the bottom course is not so dry that a porous, friable layer having little or no bond to the base concrete, results. This point cannot be emphasized too much, as many floors have failed because of the use of a dry bottom course which has failed to bond to the base concrete. The consistency should be such that a dense layer is obtainable with the compacting method employed.

Preparation of base concrete

The topping may be laid before the base concrete has completely hardened, thus forming a monolithic slab, or it may be necessary to lay the topping on set concrete. At whatever stage the topping is to be laid the surface of the base should be freed from laitance, and the best way to achieve this appears to be to brush the concrete with a stiff broom just before it sets. This method also leaves a roughened surface which forms a good key.

If measures have not been taken when laying the concrete, hardened bases should be thoroughly roughened; a few chips hacked out here and there is not sufficient and mechanical means of roughening are strongly recommended if a good bond between the hardened concrete and the topping is to be achieved. The base should then be thoroughly cleaned and wetted before commencing to lay; surplus water should, however, be removed from the surface before placing the topping.

Lifting and curling of granolithic floors due to failure to clean away dust, chippings and other loose debris from the base, has commonly occurred. Plaster droppings are specially detrimental to the bond, and even small quantities may cause sulphate expansion under damp conditions. Careful cleaning is essential, particularly in the corners.

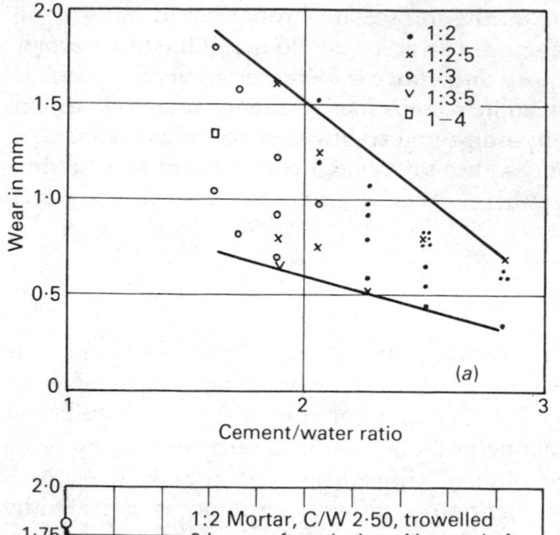

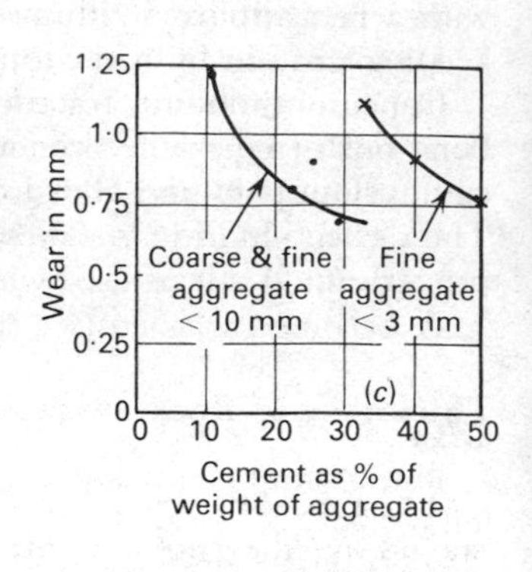

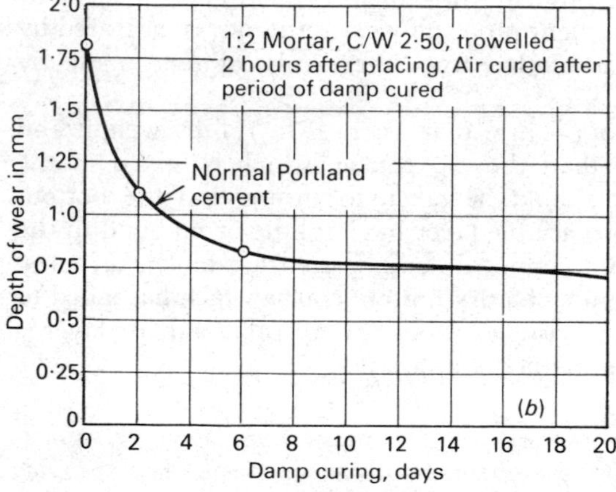

Fig. 25·3 Relation between wear resistance and cement/water ratio, length of curing, and cement content

Note. Points are average of all specimens at each mix, irrespective of water content, types of aggregate or method of finishing.

When laying on an oily or otherwise contaminated base, it becomes necessary either to chop out all the oil-soaked concrete, or, if conditions of height and level make it possible, to overlay a thickness of, say, 75 mm of topping, possibly reinforced.

American practice favours the use of a detergent on bases which have been laid a week or more previously, followed by commercial hydrochloric acid broomed over the surface. When the acid stops foaming the surface is flushed with water and all loosened material removed with a stiff broom. Rubber boots and suitable protective clothing are necessary.

Grouting base

After cleaning and wetting, the hardened base concrete should be grouted with a cement : sand (in about equal parts) grout scrubbed into the surface, keeping just ahead of the topping.

In lieu of grouting, certain epoxy and epoxy-sulphide resins may be used to bond fresh to hardened concrete. Their use is likely to be restricted to special circumstances by the added cost and the more involved technique necessary. Their effective life is unknown, but should be at least several years, if used properly.

Thickness of floor surfacing

The minimum thicknesses of floor surfacing are usually specified to be as follows:

(1) If laid before the base has hardened, minimum 20 mm.

(2) If laid on a hardened base to which good bond can be expected, minimum 40 mm.

(3) If laid on a hardened base to which little or no bond can be relied upon, due to contamination or other causes, minimum 75 mm.

Unfortunately as the thickness increases, the tendency to curl upwards at the edges, and to break the bond with the base concrete is increased. Attention to those details which improve the bond with the base concrete, is probably more important than thickness. Entirely satisfactory granolithic surfacings, 20 mm thick, have been laid on a hardened base carefully prepared. Wire fabric may be used in the thicker surfacing to resist shrinkage and to assist in reducing crack width.

Laying

Granolithic surfacing is laid in bays which may be up to 6 m square although areas about 3 m square are generally advisable, particularly for the thinner surfacings, as less likely to crack within the bay.

The bays are usually laid alternately in order to reduce the opening at joints due to shrinkage. Wooden battens are used at the edges, and act as guides to the finished level. Joints should coincide with those in the base concrete, otherwise cracks are certain to form over all joints in the base. The new concrete is butted against the old, no special jointing medium being used. It is an advantage to chamfer the edges slightly at the joints.

(1) *One-course work.* The surface mix should be placed on the grouted base, spread level, and then compacted by means of a punner, a heavy straightedge or a roller.

(2) *Two-course work.* The underlying course is usually of a somewhat drier mix than the top course, and is compacted by punning or rolling. A top course of at least 12 mm thickness of a wetter finishing mix is then spread roughly and either struck off or tamped down with a straightedge. Excess water from the top course is absorbed by the bottom course.

There is a danger that the use of too dry a bottom course may result in poor bond with the base concrete. Many failures have resulted from the use of over-dry, and consequently weak and friable, bottom courses.

Fig. 25.4 Use of power trowel in finishing granolithic floor surfacing

Trowelling

Trowelling is the most important factor in the laying of a good granolithic surfacing, and too much emphasis cannot be placed on the necessity for experienced trowel hands for this work. Trowel hands, although classed as 'plasterers', should be specially trained in floor laying and need not be experienced plasterers in the normal sense. Conversely, experienced plasterers are not necessarily good trowel hands.

Finishing is carried out with the trowel and the float, the important points being the period at which the trowelling is done, and the pressure used by the trowel hand. Immediately after screeding off, the surface should be floated just sufficiently to level it while not bringing too much laitance to the surface. The surface should then be allowed to dry and stiffen until all the excess water has disappeared before any further trowelling is attempted. The next trowelling should be accompanied by considerable pressure, just enough work being done to tighten and smooth the surface. Further trowelling should be given at intervals as the mix stiffens so that a hard compact surface is produced without bringing laitance to the surface. Experienced trowel hands obtain such a surface with no more than a thin skin of cement covering the aggregate.

Machine tools are available for floating and finishing the surfaces of concrete or granolithic concrete floors. One such tool, known as a power float, employs a heavy rotating steel disc for floating out the concrete. Power floating may be a preliminary operation to power trowelling which is carried out by the machine shown in Fig. 25.4. This power trowel is fitted with four rotating steel trowels. The basic machine can often be used for both floating and trowelling by changing over the disc and the trowels.

The power trowel is used to best advantage in finishing large surface areas such as aircraft hangars, or factory floors. It may not be economical for small isolated areas, such as floors for dwelling houses or flats. The manufacturers of this machine claim that lean mixes of low water content can be laid at a speed of 5 to 10 times faster than that of hand floating. In easing the physical effort involved in trowelling, and facilitating the use of rather less rich mixes, e.g. 1 : 4 to 5, cement : aggregate, the use of a power trowel results in lower shrinkage, thus reducing the risks of cracking and curling. It is important, however, that a good bond is obtained with the underlying concrete, and the grading of the aggregate and the cement content of the mix must be carefully adjusted to ensure that the compacted surfacing is dense and in full contact with the roughened surface beneath.

Dustings of cement intended to dry up the surface during trowelling are dangerous and liable to produce a neat cement skin which later peels off. They should therefore not be tolerated. This risk of peeling is also present when dry mixtures of cement and iron filings, with or without pigment, are trowelled in to produce a hard-wearing skin or to give a coloured surface.

Artificial drying can be effected very satisfactorily by laying cotton sheets on the floor and covering them with an inch or two of dry cement; this absorbs water and the sheets can be lifted after, say, 10 minutes, when trowelling can proceed. The damp cement can usually be used for other concrete work which may be proceeding at the time.

A more recent development is that of vacuum dewatering. A flexible vacuum mat up to 5 m square is laid on the surface. This is connected through a flexible pipe to a vacuum generator. A filter sheet prevents the loss of cement from the surface. The vacuum is applied for 3–5 minutes per 25 mm depth of concrete; thicknesses up to 300 mm may be dewatered. Power floating can commence as soon as the vacuum mat is removed, by which time the concrete is stiff enough to support a man.

All trowelling should be finished on the same day as the laying operation.

Builder's traffic should be prevented from damaging the floor soon after laying. Attempts are often made to hide such damage by trowelling in neat cement slurry, but this is useless as it invariably flakes off or becomes dusty.

Curing

Granolithic floor surfacings should be moist cured for 7 days after laying by hosing, flooding, or by covering with a PVC or similar plastic film, a suitable thickness being about 0·5 mm. The film should be laid over the surfacing while still wet, as soon as it is hard enough not to be marked. No further watering is then necessary. Other methods such as coverings of wet hessian or wet sand require frequent watering in dry weather. Wet sawdust often causes staining. The value of moist curing is not generally realized and reference should be made to Fig. 25.3, which gives the results of tests on the wear of a 1 : 2 mortar after different lengths of curing, and illustrates the large reduction in wear resulting from moist curing during the early stages of hardening. Furthermore, prolonged curing delays the start of drying shrinkage and enables the topping to develop sufficient tensile strength to withstand the shrinkage stresses.

Surface hardeners

The most commonly used surface hardener in this country is sodium silicate, but magnesium silicofluoride, other silicofluorides, and certain other salts are also in use.

A properly laid and trowelled granolithic floor surfacing will be hard wearing and durable without the use of surface hardeners. The value of surface hardeners is therefore a matter of controversy. Sodium silicate is generally believed to produce a hard skin which may reduce the wear of new floors at early ages. On old floors it is regarded as effective in reducing dusting. In either case the effect only appears to last for a limited period, perhaps six or twelve months.

Floor hardeners are not effective on weak and friable floors.

Common causes of failure

Common causes of failure in the past have been:

(1) The use of aggregates containing a large proportion of dust.

(2) Overwet mixes coupled with unsatisfactory trowelling.

(3) Lifting due to poor bond with the base concrete resulting from omission to clean and roughen it properly. When laying between previously constructed bays, failure to clean out loose dust near the edges has been a common cause of lifting.

(4) Possibly, insufficient thickness of surfacing.

(5) Use of a dry, friable mix for the bottom course in two-course work, failing in bond with the base concrete or even in itself.

(6) Unsatisfactory trowelling.

(7) Dustings of cement. A neat cement skin is formed even up to 3 mm thick, and this later flakes off.

(8) Trowelling-on of a skin of neat cement slurry to hide poor finishing or damage. This invariably flakes or dusts off.

(9) Oversized bays.

Terrazzo

Terrazzo floors are laid in much the same way as granolithic floors. The aggregate used is marble, which is obtainable in a range of sizes from 2 mm to 25 mm. White cement, often pigmented, forms the matrix. A rich mix is used, usually about 1 : 2·5 cement : marble chippings.

The thickness depends on the size of marble chippings needed to give the desired appearance. Compaction is by tamping and heavy roller. The trowelling follows the lines of that for granolithic.

Grinding-down follows after two or three days, followed by filling of blowholes and further grinding. A high-speed grinding machine is used. For final polishing a fine grade carborundum stone is used.

Specialists should be employed for this type of work.

General

Granolithic floor surfaces can be made hard, dustless, and wear-resisting provided the technique and workmanship are good. Some comment is, however, necessary concerning floors subject to exceptionally heavy abrasion and chemical attack.

Fig. 25.5 10 mm thick cast iron perforated plates laid on granolithic base course. After levelling further granolithic is placed and compacted to fill the perforations. Note the dry texture of the granolithic in the background

(1) *Dairy floors.* Rapid deterioration, particulary where heavy impact or abrasion is combined with the action of milk wastes, is liable to occur unless particular attention is paid to all the details that ensure a hard and dense surface.

Where heavy trucking or churn rolling occurs, heavy cast iron or steel grids embedded in dense granolithic concrete give good service (see Fig. 25.5). They are, however, expensive and the floor is noisy. The best size of opening seems to be about 50 mm square; grids with small openings become slippery, while grids with larger openings permit potholes to form.

(2) *Food factories.* In food factories where the floor is likely to be affected by sugar, vegetable and fruit juices, acid solutions and fats, very similar remarks apply as in the case of dairy floors. Sugar solutions are more detrimental than dry sugar.

In repairing floors in sugar factories care is necessary to prevent any contamination of the mix by traces of sugar, otherwise the concrete may be spoilt or prevented from setting.

High-alumina and super-sulphated cements are more resistant than normal Portland cement, providing attention is paid to curing at early ages.

(3) *Engineering workshops.* Lubricating and cutting oils which have a deleterious effect on concrete are usually present and soften the concrete whereever they penetrate the dense surface skin. Such penetration may occur where the floor is accidentally chipped and where there is heavy trucking by irontyred vehicles.

Again, the effective life of the floor is dependent on the hardness and denseness. Cast iron or steel grids are effective in trucking corridors.

REFERENCES

1 DEACON, R. Colin, *Concrete Ground Floors, their Design, Construction and Finish.* 1974, revised and reprinted 1976. Cement and Concrete Association.

2 *A Guide to the Structural Design of Pavements for New Roads,* Road Note No. 29. Department of the Environment, Road Research Laboratory, 3rd edn., 1970. HMSO.

3 SCHUMANN, L. and TUCKER, J., Tests of wear resistance of concrete, *Nat. Bur. Stand. J. Res.*, **25**, No. 5, 1939, pp. 549–570.

ADDITIONAL REFERENCES

SHACKLOCK, B. W., The design and construction of granolithic toppings, *Structural Engineer*, **41**, No. 9, Sept. 1963.

BARNBROOK, G., *Concrete Ground Floor Construction for the man on the Site*, Parts 1 and 2. 1974/76. Cement and Concrete Association.

26

Durability of Concrete

Atmospheric pollution

Atmospheric pollution in the neighbourhood of large towns is a major cause of the decay and discoloration of concrete. Most of the substances which pollute town atmospheres are derived from coal, gases and soot from vehicular exhausts and acid gases from factory chimneys. The introduction of the Clean Air Act (1956) has reduced pollution in some areas.

Analyses of the grime on buildings in towns have shown that it consists mainly of chimney smoke, soot from motor vehicles, and street dust. Stains in rural surroundings contain only a small quantity of soot (from motor vehicles), while the quantity of firmly adhering oily or waxy pollen and fragments of decaying vegetable matter is much larger.

Stains due to weathering cannot be entirely removed without rubbing off the stained particles from the surface, and since penetrations of 7 mm to 12 mm have frequently been noted in columns and walls, this is extremely difficult.

An instance of the darkening and discoloration of a concrete surface by a polluted atmosphere is illustrated in Fig. 26.1. Except for the reeded portion in the centre, the surfaces were wire-brushed while the concrete was green. Differences in porosity at each lift have been accentuated by the deposits of grime, and the construction joints show as dark lines.

Of the gases present in the atmosphere, only carbon dioxide and sulphur dioxide cause material damage to concrete surfaces. Carbon dioxide dissolves in rainwater forming carbonic acid, which has a solvent action on any lime compounds present in the concrete. Calcium carbonate is removed from the surface in this manner and is replaced by diffusion of calcium hydroxide from the interior. When the calcium hydroxide is precipitated, as calcium carbonate, water held in the pores becomes unsaturated and more calcium hydroxide is taken into solution. The depth to which the calcium hydroxide is changed to calcium carbonate by carbon dioxide depends on the porosity of the concrete, but even for porous concrete it rarely exceeds 30 mm. The process is known as carbonation (see also p. 16).

The white deposit of efflorescence (or 'lime bloom') so commonly seen on concrete surfaces is the result of leaching and subsequent carbonation and evaporation. It sometimes occurs on initial drying out of the concrete when salts are brought to the surface.

The sulphur dioxide in the atmosphere results in very small amounts of sulphurous acid in rainwater and this reacts with lime compounds, forming calcium sulphate. This is only slightly soluble in water and is deposited on or near the surface, forming a skin. Crystallization behind this skin, and differen-

Fig. 26.1 Darkening and discoloration in a polluted atmosphere. Differences in porosity are accentuated by deposits of grime

tial movements between it and the concrete underneath, loosen it and cause disintegration.

The concentration of sulphur dioxide in the atmosphere is highest in industrial areas and varies somewhat with the time of year, December, January and February being the worst months. Exposed concrete in, say, Manchester or Sheffield, would be expected to weather very differently from that in a country district well away from towns. The former would get very dirty with a sulphate-bearing skin which would eventually disintegrate, while in the country district the concrete might be expected to remain clean and any wear would be due to erosion.

The effect of erosion in a polluted atmosphere is shown in Fig. 26.2. The

Fig. 26.2 Erosion of concrete in a parapet wall. Reinforcement which had been given insufficient cover has also corroded and become exposed

surface was originally smooth as it left the mould, but much of the cement and fine material has weathered away, especially on the coping, leaving the coarse aggregate exposed. Spalling due to corrosion of the reinforcement has also occurred. Other examples of the weathering of concrete are given in Figs. 26.3 and 26.4.

The unsightly effects of atmospheric pollution on the appearance of concrete surfaces on a building may be reduced by paying particular attention to the way in which rainwater is discharged. Detailing should aim at either keeping the surfaces relatively dry or uniformly wet in order to avoid streaking.

Frost action

Concrete surfaces may be damaged by frost action either before or after the concrete has set. If the temperature falls below zero before the final set, the expansion of water while freezing exerts a force sufficient to destroy the cohesion between particles of the green concrete. The surface of concrete frozen before the final set has taken place scales badly and becomes pitted.

Fig. 26.3 Weathered concrete to a crane gantry in which a sloping pour plane is in evidence, and also spalling of concrete over reinforcement, due to insufficient cover

The imprint of frost crystals may often be seen on the broken surface. Reference should be made to Chapter 15 for further information on concreting in cold weather.

Disruption of set concrete by frost is caused by the expansion of water in the pores during freezing. Pressures are developed within the material progressively as ice formation continues.

Slow rates of freezing are the least harmful since they are accompanied by appreciable extrusion from the surface with only small extrusional strains. Extrusion into unfilled pores helps to relieve strains in materials which are not saturated, and it is generally found that if the percentage saturation of the pores is below 80 per cent little damage is done, but the effect varies with different materials and with the rate of freezing.

Concrete in exposed damp positions such as road surfaces, reservoirs and dams, which is liable to suffer from frost, should be made with the lowest practicable water content which will give sufficient workability for placing. Concrete with a water/cement ratio of less than 0·5 may be regarded as safe from frost damage; between 0·5 and 0·6 there is a risk in isolated cases; over 0·6 the risk becomes progressively greater. Examples of frost failure are shown in Figs. 26.5 and 26.6.

Where de-icing salts have been used during the winter months to remove ice from road surfaces, spalling and flaking of the surface have proved troublesome.

The frost resistance of concrete can be improved by the addition of a small quantity of an air-entraining agent, usually a neutralized vinsol resin, during the mixing of the concrete. This will produce a very large number of small air

Fig. 26.4 Weathering of a bridge abutment in Scotland. Efflorescence has occurred at two construction joints due to seepage of water through the joints

bubbles in the concrete, and the aim should be a total air content of about 5 to 7 per cent in an average concrete. Air entrainment is more effective at the lower water/cement ratios which, in any case, give improved frost resistance.

Air entrainment has been particularly beneficial in improving the frost resistance of concrete roads and appears to be of the greatest value when frost attack and the effects of salt used for de-icing are combined. It is specified in

Fig. 26.5 Disintegration of concrete kerbs due to frost action on porous aggregate

many countries for concrete paving, at least in the upper layer. Air entrainment is not a substitute for good specification of materials and workmanship. Further reference is made to air entraining under Admixtures, Chapter 5.

Sea action

The deterioration of concrete in a sea environment is caused by a number of agencies, notably chemical attack by the seawater and abrasive action by sand and gravel carried in the water. In areas with cold climates, frost can have a particularly destructive effect on concrete exposed to cycles of wetting and drying. In warm climates, the growth of plant life and the activities of marine borers can be destructive on concrete which is weak and contains soft aggregates such as certain limestones or sandstones. As the concrete ages, its surface becomes rougher and cracking may eventually develop which allows the salt-laden water to reach the reinforcement. This causes corrosion with consequent expansion and disruption of the concrete.

After 30 years of observations the Sea Action Committee of the Institution of Civil Engineers(1) concluded that the deterioration of concrete piles in seawater was primarily due to the corrosion of the steel reinforcement. Salts in the water contributed to the incidence of corrosion but were not the dominant factor.

The chemical constituent of seawater that is mainly responsible for the attack on concrete is magnesium sulphate. This substance reacts with cement

Fig. 26.6 Failure of over-wet concrete in a bridge. Failure was caused by frost action. With a water/cement ratio below 0·6 such failure would be unlikely

to form calcium sulphate and calcium sulphoaluminate. A molecular increase in volume takes place during the reaction adding a physical action to the process, and on crystallizing the further increase in volume results in a disruptive force not only at the surface but at progressively greater depths. Lea[2] reports that the deterioration of concrete in seawater is often not characterized

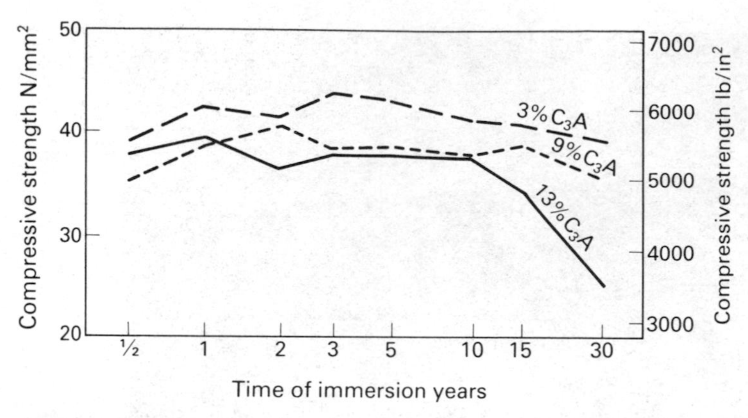

Fig. 26.7 Change of compressive strength of concrete made with cements of varying C_3A content. The water/cement ratio in each case was 0·6

by the expansion found in concretes exposed to sulphate solutions, but takes more the form of an erosion or loss of constituents from the mass. Although laboratory data are conflicting it seems that the presence of chlorides must retard the swelling of concrete in sulphate solutions. The deterioration of concrete exposed to seawater would, therefore, be severe if it were not for the presence of a high chloride content in the seawater.

It has been reported[3] that blocks of concrete submerged partly or totally in seawater off the Los Angeles harbour breakwater were in a reasonably good condition after 67 years. The blocks still showed fairly sharp edges and the original form marks.

Reliance should not be placed solely on the use of cements or cements and admixtures thought to be immune from attack by seawater. Callet[4] quotes the case of large blocks of concrete made with pozzolanic cement, forming part of the original quays at Le Havre, which were burst open by explosions during the war. The broken concrete, which had been in contact with seawater for about four years, showed advanced chemical decomposition.

The long-term durability of concrete immersed in seawater was investigated in Trondheim, Norway over a period of 30 years[5]. Concrete prisms made with different cements generally remained unaffected by the seawater apart from those cements having a high tricalcium aluminate (C_3A) content. The effect on compressive strength at three levels of C_3A content is shown in Fig. 26.7. The cement content in each case was 313 kg/m³, and the water/cement ratio 0·6. It is evident that the concrete made with cement having a C_3A content of 13% eventually (from 10 years onwards) suffered a considerable reduction in compressive strength. This same cement was also investigated in concrete mixes where the amount of cement was varied. Fig. 26.8 shows that when the cement content was increased to 417 kg/m³, there was little change in the compressive strength with time. The effect of variations in the water/cement ratio was also investigated when using the cement with a C_3A content of 13%, and the results are shown in Fig. 26.9. The mixes with water/cement ratios of 0·6 and 0·65 gave a marked decrease in strength with time, and even

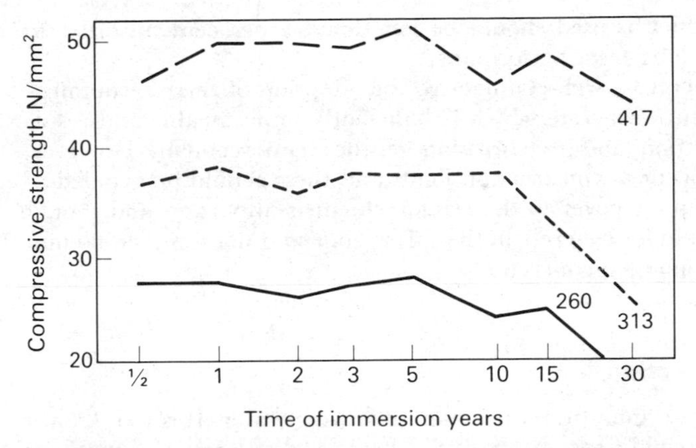

Fig. 26.8 Change of compressive strength of concrete made with varying contents of cement having a C_3A content of 13 per cent (values of cement content given in kg/m³)

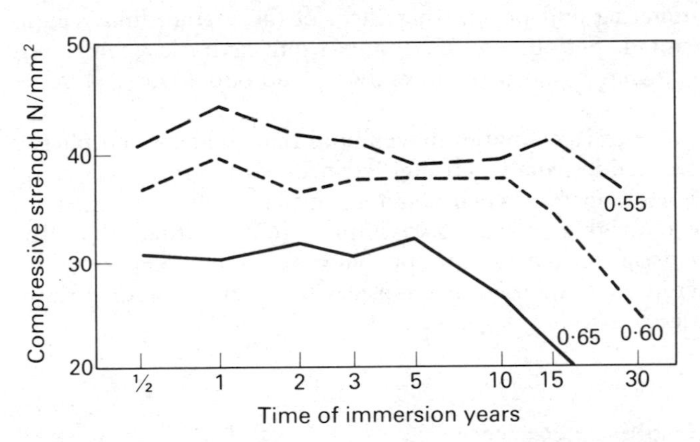

Fig. 26.9 Change of compressive strength of concrete made with different water/cement ratios. The cement had a C_3A content of 13 per cent

the lowest water/cement ratio of 0·55 showed a significant reduction, at an earlier age. The interrelationship between C_3A content, cement content, water/cement ratio and time should be noted.

The FIP Recommendations for the Design and Construction of Concrete Sea Structures [6] state that for high quality concrete of low permeability, the water/cement should be less than 0·45 and preferably 0·4 or less, subject to the attainment of adequate workability. A minimum cement content of 400 kg/m³ is required for concrete in the splash zone, but this requirement is reduced to 360 kg/m³ (corresponding to a maximum aggregate size of 20 mm) for concrete elsewhere. Ordinary Portland cement is considered to be satisfactory for submerged and less exposed atmospheric zones, but it is suggested that in the splash and more exposed atmospheric zones the C_3A content of the ordinary

Portland cement which is used should be less than 12 per cent, in order to achieve satisfactory long-term performance.

Provided the concrete is well-compacted, the adoption of these recommendations should produce concrete which is sufficiently impermeable and resistant to chemical action and to corrosion of the reinforcement. However, trouble is likely to occur at construction joints and these should be avoided as far as possible. Adequate cover to the reinforcement is important and should not be less than 75 mm for concrete in the splash zone and not less than 60 mm for permanently submerged concrete.

Erosion by water action

Serious erosion may result from the action of flowing water. It is well known that under certain conditions of hydraulic flow, a persistent free space or cavitation pocket will occur at some given location, which may under some conditions contain nothing but water vapour and flying particles of water. If this occurs in contact with a solid object such as a concrete surface, damage results from the hammering and penetrating effect of the extraordinary high water pressure generated. Serious erosion results from cavitation, the most notable cases of damage having occurred in spillways and outlet works of large dams.

The American Bureau of Reclamation have found that, under atmospheric conditions, cavitation and destructive erosion begin downstream of relatively small offsets when water velocities reach about 12 m per second.

To minimize the damaging effects of cavitation, it is essential that the concrete surfaces are designed to avoid abrupt changes in slope and curvature. A high quality concrete must be used and finished to a hard smooth surface free from irregularities.

Crack formation in plastic concrete

Cracks occurring before the concrete hardens are usually due to the settlement of the plastic concrete around a rigid obstacle, such as a piece of aggregate or reinforcing bar, or excessive evaporation of water from the concrete surface.

Settlement cracks, which tend to occur more in thick than thin sections, particularly when top reinforcement is present, can best be prevented by using a well-proportioned concrete mix with as low a slump as possible, and by adequate compaction of the concrete.

Plastic shrinkage cracks, caused by water evaporating from the surface of the concrete faster than it bleeds from beneath, are apt to occur when hot drying winds are prevalent (see Chapter 2). Ground and road slabs are susceptible to this type of cracking.

The formation of plastic cracks results in a weakened section of concrete, possibly unable to resist subsequent drying shrinkage stresses, with the result that cracks may develop through the full depth of the section.

Fig. 26.10 Complete disintegration of a concrete breakwater by sea action. Failure occurred primarily at the construction joints, with the result that the concrete broke up into a series of independent masses

Cracks in hardened concrete

The formation of cracks in hardened concrete is not often attributable to any single cause, but besides being unsightly they may allow the leakage of water and lead to corrosion of any reinforcement. Factors which may lead to cracking are:

(1) Drying shrinkage of restrained members, especially thin units.

(2) Changes in length due to changes in temperature. This applies, particularly, to the larger masses of concrete in which there is a fairly considerable temperature rise during setting and hardening, and a subsequent shrinkage as the concrete cools. Cracking in dams and large sections results from these temperature changes. Thin wall sections also suffer from cracking due to the same cause.

(3) Movements in the structure due to settlement of the foundations, resulting in local overstressing.

(4) Expansion of metal, partly embedded in the concrete, or firmly fixed to the concrete. This is especially liable to happen in industrial structures subject to heat from boilers or furnaces.

(5) Overloading, whether due to inaccuracies of construction, or from the imposition of a greater load than that for which the structure was designed.

The risk of cracking can be reduced by the inclusion of joints in the structure, as outlined in Chapter 11. The effect of increasing the amount of reinforcement in a concrete member is to reduce the width of any particular crack along its length, but to increase the number of cracks. With sufficient reinforcement, it is possible to reduce the width so that the cracks are no longer visible, and this method is often applied when the risk of crack formation is foreseen, and would be difficult to avoid by other means (see also p. 284. In concrete pavings, reinforcement is used principally to hold together

Fig. 26.11 Badly stained and cracked reinforced concrete warehouse in London. The surface was originally cast against a hardboard form lining to give a smooth finish. Crazing may be seen about the middle of the column on the right

any cracks which form due to movement in the subgrade, and so preserve the road from further deterioration.

Alkali-aggregate reactivity, which causes 'map' cracking of the concrete and its subsequent disintegration, is referred to on p. 46.

Crazing

Crazing, or the formation of fine hair cracks on concrete surfaces, is a common cause of disfigurement. Crazing is caused mainly by the accentuated drying shrinkage of the surface layers of concrete, to which must be added the effects of carbonation, and differences in temperature and moisture content between the surface and the interior. This leads to the formation of very fine cracks, rarely more than about 10 mm in depth which divide the surface into small areas, about 5–20 mm across.

Crazing is most commonly found with concrete mixes having a high water/cement ratio and is usually associated with the use of formwork which is smooth and of low absorbency. The hair cracks so formed provide a convenient route for disintegrating agents to attack the interior of the concrete.

Sulphate attack on concrete

Sulphates in solution attack concrete and cause expansion, deterioration, and eventual disintegration. In the early stages of attack there is little visual evidence of any change in the concrete, although there may be some reduction in strength as the chemical action proceeds. This is accompanied by slight expansion, which may not be apparent in the concrete itself, but may cause trouble at points of restraint. As the attack proceeds there is usually some change in colour from the normal cement grey, and cracking or spalling occurs, starting from the surface in dense concrete, but penetrating much deeper in more porous concrete. The rate of attack depends on a number of factors, the principal of which are:

(1) *Form in which the sulphate occurs.* Easily soluble sulphates such as those of sodium, potassium, magnesium, and ammonium, react more vigorously than calcium sulphate or gypsum. Insoluble sulphates do not cause any deterioration.

(2) *Concentration.* The higher the concentration of sulphates in solution the more serious the attack. The severity of attack is increased in circumstances in which a flow of sulphate-bearing water brings a continuous supply of the salt into contact with the concrete.

(3) *pH value of the soil or groundwater.* Its temperature may also be a factor at ground level, the activity of sulphate solutions increasing with temperature.

If the pH value is below 6, i.e. in acid soils, the rate of attack is likely to be increased.

(4) *Permeability of the concrete.* Impermeable concrete is attacked only at the surface and the deterioration is therefore slow, whereas a permeable concrete, particularly if one side is in contact with moist soil containing sulphates while the other side is open to the air so that evaporation occurs, or if subjected to a one-sided water pressure, is attacked throughout its thickness resulting in an overall expansion and complete disintegration.

(5) *Formation of cracks.* Attack proceeds along the lines of cracks, particularly when the movement of moisture along any cracks is encouraged by one-sided water pressure or evaporation from a free surface.

Examples of sulphate attack

The severity of sulphate attack depends mainly on the quality of the concrete, the type of cement, and the varying conditions which surround it. Before considering precautionary measures to minimize the effects of sulphate attack, two examples of severe attack may serve to illustrate the inter-relationships of the various factors involved. The first example is of concrete sewers in sulphate-bearing ground which have failed in a number of instances within a year of construction, one-sided water pressure being the cause of attack throughout the thickness of the pipes, although the concrete was of reasonably good quality.

These failures have occurred even in clay soils which might have been thought to be impermeable. However, the disturbance in digging the sewer trench, and probably some porosity in the backfilled clay is known to have resulted in the flow of ground water necessary to carry sulphates in solution into the concrete. It is probable that the failures would not have occurred if the pipes had been of dense impermeable concrete made with sulphate-resisting Portland cement, or if they had been surrounded by a haunching of such concrete. Other precautions might have been to wrap the pipes in a plastic membrane, or to coat them with bitumen. In some cases, the failures might not have occurred if nothing more had been done than to set the pipe on dense concrete, in order to avoid problems in compacting clay underneath the pipes, and then to pun the clay well in, backfilling the sides and over the top of the pipe.

The second example is of failures of oversite concrete forming the ground floor of a number of houses, resulting in lifting by as much as 75 mm, and pushing outwards of the external walls due to expansion. This was attributed to movement of sulphates from the underlying shale fill into the concrete. This movement was aided by cracking and evaporation from the upper surface before completion of the floor surfacing with pitch mastic. Due to the high concentration of sulphates, it is probable that, in the particular instances examined, it would be necessary to use super-sulphated cement in a dense concrete. However, in cases such as this it is usually simpler to provide a plastic membrane at least 500 μm thick immediately below the concrete.

It will be noted that concrete is attacked by sulphate salts only when they are in solution, and that the rate of replenishment is important. It follows that attack is most severe when the soil conditions allow a free movement of ground water, as in the case of sands and gravels, and porous or fissured rocks. Clay is relatively impermeable when undisturbed. Examples where moisture movements result from evaporation at an exposed surface are: oversite concrete laid directly on the soil surface; porous concrete in foundation through which moisture is sucked up into brickwork above; and especially permeable concrete below fireplaces, boiler houses, etc.

Precautionary measures in sulphate-bearing ground

An essential condition in the production of a concrete resistant to chemical attack is that the concrete itself should be of high quality. A lean mix or badly graded aggregate will not give a resistant concrete, whatever type of cement is used. If the conditions are such that in-situ concrete is immediately exposed to the aggressive agent, it is more susceptible to attack than precast units which have been moist cured for a month or more before being placed in a position subject to attack. The compressive strength of concrete is a reliable index of its resistance to attack, since strength, density, and low permeability are all interrelated; the stronger the concrete the more durable it is likely to be.

The importance of dense well-compacted concrete of high quality cannot be over-emphasized, but the resistance to attack can be supplemented by other precautions. However, as the probable severity of attack is largely a matter of judgment based on general guidance, such as that given in Table 26.1, any decision concerning precautions to be taken is likely to be a compromise between what may be accepted as a reasonable risk and economic savings. In cases of doubt, particularly where any failure in the concrete could have disastrous consequences or would be difficult to repair, it is wise to err on the side of excessive precautions. Special consideration should also be given to thin concrete sections, site conditions which may make the production of dense concrete impracticable, and the chances of one-sided water pressure. Changes

Table 26.1 Classification of Sulphate Soil Conditions Affecting Concrete and Recommended Precautionary Measures

Classification of soil conditions			Precautionary measures	
Class	Sulphur trioxide in ground-water parts SO_3 per 100 000	Sulphur trioxide in clay % SO_3	Buried concrete surrounded by undisturbed relatively impermeable clayey soils.	Concrete exposed to one-sided water pressure or evaporation from one face. Buried concrete in permeable ground, including all backfill materials.
1	Less than 30	Less than 0·2	No special measures except that the use of lean concretes (e.g. concrete containing less than 275 kg/m³ cement) inadvisable if SO_3 in water exceeds 20 parts per 100 000. In latter case use good well-compacted concrete made with ordinary Portland cement and a maximum w/c ratio of 0·55.	**Dense fully-compacted concrete made with ordinary Portland cement (see Table 26.2 for cement content), and a water/cement ratio not exceeding 0.50**
2	30 to 100	0·2 to 0·5	Dense fully-compacted concrete made with ordinary Portland cement (see Table 26.2 for cement content), and a water/cement ratio not exceeding 0·50. If sulphates are in highly soluble form use sulphate-resisting Portland cement.	Dense fully-compacted concrete made with sulphate-resisting Portland cement (see Table 26.2 for cement content), and a water/cement ratio not exceeding 0·45. **Super-sulphated cement may be used for the higher concentrations if in doubt.**
3	100 to 250	0·5 to 1·0	Full compaction to give the densest possible concrete essential. Use sulphate-resisting Portland cement (see Table 26.2 for cement content), and a water/cement ratio not exceeding 0·45.	Full compaction to give the densest possible concrete essential. Use super-sulphated cement (not less than 385 kg/m³) and a water/cement ratio not exceeding 0·45.
4	Above 250	Above 1·0	Full compaction to give the densest possible concrete essential. Use super-sulphated (or high-alumina) cement (not less than 385 kg/m³) and a water/cement ratio not exceeding 0·45.	Preference should be given to protection by impervious membrane, see p. 393. Otherwise use of super-sulphated cement as for buried concrete in undisturbed clay, with lowest practicable water/cement ratio for full compaction.

See text for influence of pH value for soil, type of sulphate, and size of section.

in ground water level occur with the seasons, and may also result from changes in conditions as a result of the building of a structure or adjacent structures. Footings and mass concrete buried completely in clay are unlikely to suffer seriously unless the concrete is of poor quality and the sulphate concentration of the ground water high. Little is known of the possibility of attack on concrete piles, mainly due to the difficulty in obtaining evidence. It seems probable, however, that attack does at times occur near the head and when such piles penetrate sulphate-bearing soil, through which there is a continual flow of water.

An assessment of the probable severity of attack is primarily based either on a determination of the percentage of sulphates (as SO_3) present in the soil, or on a determination of the quantity of sulphates (as SO_3) in parts per 100 000, present in the ground water. The analysis should also indicate the metal radicals present in conjunction with the sulphate and the pH value of the soil or ground water. A guide to the precautions necessary may be obtained by reference to Table 26.1.

The table is provided only as a general guide. From the paragraph headed Sulphate attack on concrete, it will be noted that the presence of gypsum (calcium sulphate) is less important than that of the easily soluble sulphates. For mass concrete surrounded by undisturbed and relatively impermeable clay, it is suggested that the presence of gypsum crystals may be ignored provided dense well-compacted concrete is used. Whenever practicable, footings in clay should be cast without shuttering. This avoids backfilling, which may leave a porous zone likely to become saturated by surface water even if the general ground water level is below foundation level. If shuttering cannot be avoided, but mass concrete footings are isolated there may not be much movement of water, so that the precautions given for concrete buried in clay should suffice. For small concrete sections, the guidance given in Table 26.1 should give a reasonable insurance against failure; if spaces between the member and the soil are backfilled the precautions adopted should be those for buried concrete in permeable ground.

If the soil or ground water is acid, i.e. with a pH value of less than 6, it is suggested that the precautions taken should be equivalent to those outlined under class 4, i.e. super-sulphated cement (see Table 26.1), should be used. When the SO_3 content of the soil is high, particularly if the soil is permeable, other precautionary measures, such as those described on p. 393, should be employed. For an SO_3 content above 1%, unless super-sulphated cement is used, such measures are a wise precaution in any soil conditions. The sulphate content of the ground water is a more reliable guide than that of the soil, since the latter is more subject to local variation.

Sulphate resistance and type of cement

The characteristic of Portland cement which has the greatest influence on its resistance to sulphate attack is the tricalcium aluminate (C_3A) content. The effect of the C_3A content is clearly shown in Fig. 26.12 which gives the results of field studies made on concrete specimens exposed to sulphate soils by the Portland Cement Association[7]. The soil in this particular case contained

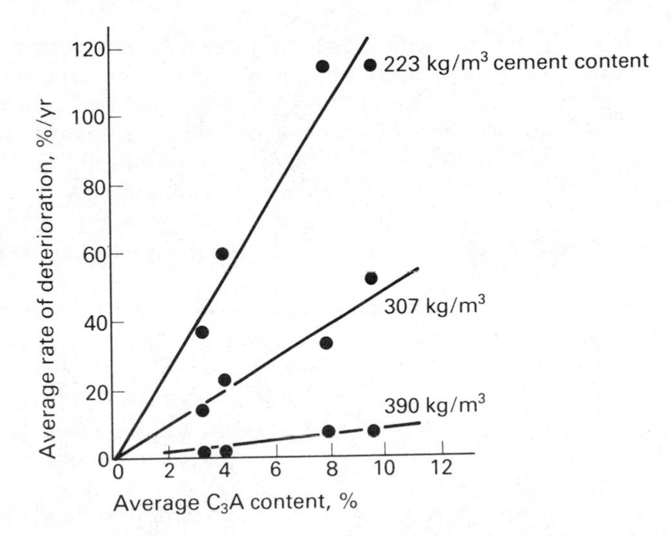

Fig. 26.12 Effect of cement content of concrete and average C_3A contents of different types of cement on rate of deterioration in sulphate soil (the soil contained approximately 10 per cent sodium sulphate

about 10 per cent sodium sulphate. The deterioration of the concrete specimens was evaluated visually, by measurements of strength after various periods of exposure, and by measurement of dynamic modulus of elasticity. It can be seen that for selected concrete mixes with three different levels of cement content the rate of deterioration in each case depended upon the C_3A content of the cement. These results also clearly indicate that the cement content has a very significant influence and it is evident that the quantity of cement is more important than the amount of C_3A it contains.

It is evident that since sulphate-resisting Portland cement has a low C_3A content (less than 3·5%), it is more resistent to sulphate attack than ordinary Portland cement. However, to be effective, sulphate-resisting Portland cement must be used in sufficient quantity in dense concrete. Recommended minimum cement contents are given in Table 26.2 for the more usual irregular and angular aggregates and varying water/cement ratios and workabilities.

Super-sulphated cement is resistant to sulphates, other than ammonium sulphate which causes severe attack. This cement is also resistant to attack by hydrochloric and sulphuric acids in concentrations up to 15 g per litre.

Reference is made to the properties of pozzolanic cements on p. 65.

Physical protection against sulphate attack

As an alternative to the use of one or another of the special cements, an impervious coating to the outer surface of the structure may be considered. In the case of a basement to a building constructed in water-bearing ground, an asphalt tanking is sometimes provided to ensure watertightness of the basement. This tanking provides an adequate protection from attack by sul-

Table 26.2 Recommended Minimum Cement Contents to give Resistance to Sulphate Attack for Different Types of Aggregates, Water/Cement Ratios and Workabilities (Use this Table in conjunction with Table 26.1)

w/c Ratio (effective)	Aggregate	Estimated average 28 day compressive strength N/mm² (lb/in²)*	Cement contents per cubic metre (yard) of compacted concrete for different workabilities kg (lb)			
			CF 0.78 slump 0–12 mm (0–½ in)	CF 0.85 slump 12–25 mm (½–1 in)	CF 0.92 slump 25–50 mm (1–2 in)	CF 0.95 slump 50–100 mm (2–4 in)
0.55	20 mm (¾ in) irregular	34 (4950)	275 (460)	310 (520)	340 (570)	350 (590)
	20 mm (¾ in) angular	40 (5800)	300 (500)	330 (560)	360 (610)	375 (630)
0.50	20 mm (¾ in) irregular	40 (5800)	310 (520)	345 (580)	375 (630)	385 (650)
	20 mm (¾ in) angular	47 (6850)	330 (560)	370 (620)	400 (670)	410 (690)
0.45	20 mm (¾) irregular	46 (6700)	350 (590)	390 (660)	420 (710)	430 (730)
	20 mm (¾ in) angular	54 (7850)	380 (640)	415 (700)	445 (750)	460 (770)

* Based on average strength: w/c ratio relationship.

Table 26.1 gives maximum values for w/c ratios. These should be used to indicate (from Table 26.2) the minimum cement content.

phates on the concrete in the basement structure, and no additional precautions are normally needed. Concrete foundations not subjected to one-sided water pressure can be protected against attack by thick coats of bituminous paint on the outer face, or by a built-up coating using two or three layers of bituminous felt with a bituminous paint tack coat on the concrete surface and between each layer. Bituminous emulsion does not provide a satisfactory coating.

Sulphates in the aggregates

Deterioration of concrete may also occur due to the presence of sulphates in the aggregates used to make the concrete. This problem is met mainly in the Middle East where many of the sources of limestone and natural sand contain widely varying quantities of sulphates, mainly in the form of gypsum. The sulphates slowly react with the cement paste causing expansion which gives rise to stress conditions conducive to bursting off of the outer layers of concrete, or cracking due to structural movements. This may take from a few months to several years to reach a dangerous state. To avoid this danger, the sulphate content of the aggregates should be controlled to within given limits. For work in the Middle East, a common requirement is that the acid soluble sulphate (SO_3) content of both the fine and coarse aggregate shall be no more than 0·4 per cent, by weight of the aggregate. This limit is generally accompanied by the over-riding requirement that the total acid soluble sulphate present in the concrete, including that present in the cement, shall not exceed 4 per cent by weight of cement.

For cases where the presence of sulphates in the aggregates causes concern, it is recommended that sulphate-resisting cement should be used. However, it is worth noting that, in the presence of chlorides, which are often found together with sulphates in the aggregates, sulphate-resisting cement loses some of its resistance to sulphates. Thus, whilst it may delay failure with badly contaminated aggregates, it does not give adequate protection.

Chloride attack on reinforced concrete

Steel reinforcement in concrete is protected from corrosion by the highly alkaline environment of the cement paste. The higher the alkalinity the greater becomes the protective quality of the oxide film that is formed on the surface of the steel. Conversely, when the alkalinity is reduced, the steel becomes potentially more susceptible to corrosion.

A reduction in the alkalinity of the cement paste around the steel bars can be caused by carbonation (see p. 377, but this is a very slow process and is hardly likely to penetrate to a depth of more than about 15 mm in good quality concrete having a low permeability. Carbonation, therefore, only results in corrosion of steel reinforcement in cases where the concrete is very porous or where the cover to the steel is very small.

A far more serious cause of corrosion of the steel reinforcement is the presence of chlorides in the concrete, which changes the protective quality of the environment to the steel. The reason for this is not completely understood but it appears that the chloride ion is a specific destroyer of the protective oxide film on the steel. In addition, the presence of chlorides in solution in the concrete creates an electrolyte of relatively low resistance.

The corrosion of the steel is an electrochemical process, similar to the action in a basic galvanic cell, but much more complicated. Basically, electrodes with differing electrical potentials are established at different points on the steel surfaces, due mainly to variations in chloride and moisture contents of the concrete. Current then flows between the positive electrodes (cathodes) and the negative electrodes (anodes), with the result that at the anodes the metal goes into solution as positively charged ions. These ions then react with water and oxygen to form rust. The corrosion products so formed occupy a volume over twice the original volume of the steel and consequently internal pressures as high as 30 N/mm^2 may be developed. This force is many times greater than the tensile strength of the concrete and is responsible for the cracking and spalling that subsequently occurs. Soluble components of the rust may be leached through the cracks and appear as rust stains on the surface of the concrete.

The severity of the damage depends on the concentration of the chlorides present in the concrete, as well as the variation of the chloride concentration from one point to another. It is essential, therefore, to impose limits on the amount of chloride which may be introduced into the concrete.

One of the ways in which chlorides are introduced into concrete is through the use of an admixture. Cases of overdosage of calcium chloride, used as an accelerating admixture, have resulted in severe cracking and spalling of concrete units within a few years of construction (see p. 80). However, the use of calcium chloride in concrete has now been virtually banned in the UK (see Chapter 5).

Chlorides also become a problem when they occur in the aggregates used to make the concrete. Their presence in marine aggregates is discussed in Chapter 3. In the Middle East, the problem is far more acute since many of the sources of natural sand and limestone are contaminated with common salt (sodium chloride) as well as other chemical salts. (The effect of sulphates has been discussed on p. 395. Some of the natural sands in hot desert countries have very high salt contents. As a result, many concrete buildings constructed with sand from these sources have started to crack within four or five years of construction in spite of the arid conditions. An investigation made in Bahrain in 1972 showed that, in general, corrosion of the reinforcement and resulting cracking of the concrete occurred where the chloride content of the concrete was in excess of 0·12–0·16 per cent, by weight of concrete. Most of this concrete had been made with nominal 1 : 1·5 : 3 mixes.

It is essential, therefore, that the quality of aggregates should be controlled so that the chloride content is kept within reasonable limits. It is recommended that the aggregates should meet the following specified maximum limits for total chloride (as Cl) content:

Fine aggregate (sand):	0.06 per cent, by weight of fine aggregate
Coarse aggregate:	0.03 per cent, by weight of coarse aggregate

Fig. 26.13 Deterioration of a reinforced wall affected by fertilizers

Small adjustments may be made to these limits, if necessary, subject to the over-riding requirement that the acid soluble chloride present in the concrete does not exceed 0·3 per cent by weight of cement.

To comply with these limits, selective digging or thorough washing of the aggregates is required in certain areas. Control of these operations by reliable and experienced engineers is essential if a continuous supply of satisfactory aggregate is to be obtained. It is inadvisable to use sand taken from just above the water table, since it will almost certainly be contaminated with chlorides due to upward capilliary movement of the salty water.

Resistance to other chemical attacks

Gases such as sulphur dioxide and carbon dioxide may, in damp situations, attack concrete. These gases are prevalent in gasworks, railway tunnels, and at various chemical factories. Sulphuric acid from coke, coal or furnace ash is liable to attack concrete in bins, bunkers, or retaining walls. The chemical attack is in these cases likely to be accompanied by abrasion.

Sulphates of sodium, potassium, magnesium and ammonium may cause serious damage to Portland cement concrete if moisture is also present. This begins by expansion within the concrete, which may be enough to cause a general expansion in the member. Cracking and disruption follow. Fertilizers often contain ammonium, potassium and magnesium sulphates. A wall affected by fertilizer stacked against it is shown in Fig. 26.13.

Generally, inorganic acids are destructive to concrete. These may also be released from some salts, such as ammonium chloride and ammonium nitrate[8] by interaction with lime. Leaching then follows.

Organic materials vary in their rate of attack on concrete and their effect may roughly be summarized as follows:

Petroleum oils—none.

Coal tar distillates—none or very slight.

Organic acids such as stearic, oleic, lactic and tannic—all attack concrete.

Lactic substances, which are derived from dairy produce, have a most destructive effect.

Vegetable oils—slight or very slight attack.

Molasses, sugar, syrup, glucose—fair degree of attack.

Acetic acid which occurs in vinegar, and tartaric acid which occurs in some fruit juices all attack concrete.

Oxalic acid—little effect.

Sewage normally has an alkaline reaction and is harmless, but it may become acid by contamination with factory wastes and will then attack concrete.

The concrete along the top of sewers can be severely attacked as the result of hydrogen sulphide gas being evolved from the stale sewage. The hydrogen sulphide is oxidized by anaerobic bacteria to form sulphuric acid which condenses on the walls of the sewer. The attack may be rapid particularly in warm conditions and where ventilation is poor.

Resistance to abrasion

Concrete is sometimes called on to resist abrasion as in the case of road surfaces, silos and bunkers, retaining walls to storage areas, etc. The movement of sea-shore deposits by wave action may also cause serious abrasion of concrete structures. The best insurance is good dense concrete made with a hard tough aggregate in accordance with the principles already described.

REFERENCES

1 *The Durability of Reinforced Concrete in Sea Water*, National Building Studies Research Paper, No. 30. HMSO.

2 Lea, F. M., *The Chemistry of Cement and Concrete*, 3rd ed., 1970. Edward Arnold (Publishers) Ltd., London.

3 Haynes, H. H. and Zuhiate, P. C., *Compressive Strength of 67-year Old Concrete Submerged in Sea Water*. Technical Note N-1308 Naval Civil Engineering Laboratory, Port Hueneme, California, USA. Oct. 1973.

4 CALLET, M. P., Concreting in sea-water. Notes from a paper published in *The Dock and Harbour Authority, Concr. Constr. Eng.*, Nov. 1949, p. 364.
5 GJORV, O. E., *Long-term Durability of Concrete in Sea Water. Proceedings ACI,* **68**, 1971, pp. 60–67. American Concrete Institute.
6 *FIP Recommendations for the Design and Construction of Concrete Sea Structures,* 3rd ed., July 1977. Cement and Concrete Association.
7 VERBECK, G. J., *Field and Laboratory Studies of the Sulphate Resistance of Concrete. Performance of Concrete.* A symposium in honour of Thorbergur Thorvaldson, E. G. Swensen. University of Toronto Press.
8 LEA, F. M., The action of ammonium salts on concrete, *Mag. Concr. Res.*, **17**, No. 52, Sept. 1965. Cement and Concrete Association.

27

Quality Control Inspection and Testing

THE quality control of concrete should aim at producing a uniform material which provides those properties particularly desirable for the work envisaged. At the same time, a careful balance must be maintained to ensure that the quality of concrete is satisfactory for the purpose, without being of unnecessarily high strength, the achievement of which inevitably results in a marked increase in cost. Thus, lean mixes of low strength are usually adequate for mass concrete in foundations and in the cores of dams where, in fact, they are advantageous because the heat generated during hydration is less than that for richer mixes. On the other hand, concrete pipes and haunchings used in sulphate-bearing ground must be of rich dense mixes if rapid and costly failures are to be avoided.

The choice of plant and procedures may have a considerable influence on the results achieved, and if effective quality control is to be obtained, discussions between the design engineer and the contractor should be arranged at the planning stage. From this point the details of control can be developed.

It is fortunate that most of the desirable properties of concrete are enhanced by increasing the compressive strength, and this, therefore, affords a simple measure of the general standard of quality. It should be noted, however, that it is not only necessary to control the quality of the concrete at the mixer, but to provide experienced supervision of all the processes which follow. This cannot be a matter for individual effort only, and any measures envisaged for the control of concrete quality must provide for cooperation and general instruction in good or improved techniques, and in the use of tools and machinery. It is essential that the educational methods should be such that their influence is apparent from the site agent down to each operative involved.

In order to achieve good results, it is necessary to adopt several methods of approach (1), such as:

(1) Direct control by means of a concrete engineer, and assistants if necessary, specially trained in the work and resident on the site, where the quantity of concrete is large, or for some special reason an unusually high standard of concrete quality is desired. This concrete engineer may be backed by a site laboratory. He should be responsible not only for the production of good quality concrete at the mixer, but for supervision of the many operations which ensue.

The principal duties of the concrete engineer or other person responsible for concrete quality may be defined as follows:

(i) He should be familiar with the plans and specifications, especially those sections which will deal with the concrete work under his control, and should see that the work is executed accordingly.

(ii) If any sections of the specification are impracticable, or require modification, he should report to his immediate superior, or to the employing authority, in order that suitable alternative measures can be discussed.

(iii) He should see that the materials are of satisfactory quality, the proportioning of the materials is accurate, the water content is within the limits allowed, and the concrete is of uniform and satisfactory consistency.

(iv) He should acquaint himself with the details of the formwork, reinforcement, handling, placing, consolidation, finishing and curing, and he should point out any errors which occur in the techniques. He should be able to advise on the best method to adopt in any particular circumstance.

(v) He should take samples of the materials, either for site tests or tests at a central laboratory, make slump tests, and prepare concrete cubes for testing.

(vi) He should keep a record of work he has done and should chart results of tests, in order that any deterioration over a period can be observed and measures taken to effect improvements.

The concrete engineer must be certain before making a complaint that it is of sufficient importance to be justified. If it is, he should insist that the defective work or method is remedied within a reasonable period. At the same time he should make every effort to maintain control and cooperation with the agent, site engineer, foreman, ganger, and workmen.

Delegation of the duty of the control of concrete quality on the site must depend on the quantity of work involved. On small contracts it is obviously uneconomic to employ an engineer solely for the purpose of controlling quality of concrete, and it is usual for the agent to delegate the duties to the general foreman or one of the site engineers. In such cases it is advisable to make every effort to impress on the responsible person the importance of the correct technique in making concrete and the correct method of making control tests. Control on large contracts may require one or more concrete engineers, with assistance to help with the preparation of test specimens, working in conjunction with a site laboratory.

(2) Instructions and advice on the site, on both specific and general questions, by a visiting engineer with wide experience of concrete technology. This form of approach is particularly valuable on the smaller sites, provided the concrete engineer is prepared to talk to the site staff, foreman, and gangers in non-technical language which they can understand.

(3) Education of site staffs by means of pamphlets dealing with concreting techniques, copies of technical papers and encouragement to attend any meetings at which they can join in discussions on concrete technology.

(4) Testing of materials and concrete by experienced laboratory engineers. The emphasis on experience is important, since the observations made during testing and the interpretation of the results are just as important as the figures obtained from the testing machine.

The production of a good concrete at the mixer is not sufficient to ensure a good concrete structure since, for instance, a small percentage of voids in the concrete due to air-bubble holes or honeycombing, causes a reduction in crushing strength of 4–6 times that percentage. Further, it is necessary in most concrete structures built in situ to provide construction joints or lift planes at intervals and, unless carefully treated, these provide a plane of weakness at which disintegration may begin, or through which water may seep. It is therefore necessary in considering quality control measures not to place too much reliance on the results of slump tests and cube tests, and to arrange for supervision of all the other aspects of concrete production besides proportioning and mixing. Much depends on the skill and experience of the foreman carpenter, the reinforcement fixer and the placing ganger, or foreman, and unless they fully understand the basic principles of good concreting techniques, they may run into difficulties when faced with some new shape of member or the use of concrete of different mix proportions from that to which they have been accustomed.

Diagnosis of the causes of blemishes and defects of concrete is not always easy, since there may be several possible causes of any one blemish or defect, and the necessary or possible remedies may be even more numerous. This is illustrated in Table 27.1, in which some major causes of blemishes on concrete structures and their possible remedies are tabulated. The table is by no means exhaustive, but an examination of the remedies illustrates the dependence of the finished product upon everyone concerned, including the designer, the agent, engineers, foremen, gangers and others.

Equipment

Specifications usually require that certain testing equipment shall be provided by the contractor. The following items are among those commonly required:

Straightedge. Rule. Tape.

Thermometer (maximum and minimum).

Slump cone and rod.

Cube moulds, base plates and rammer.

Containers of nominal volume 0·03, 0·015, 0·007 and 0·003 m^3 (or of 1, 0·5, 0·25, 0·1 ft^3) for measurement of density of different sized materials.

Set of BS sieves, 37·5 mm down to 150 μm.

Cylinder, 100 cm^3 for field silt test.

Primus, hotplate or gas ring for quick drying of aggregates. An oven may be used on larger contracts.

Frying-pan, metal trays, etc.

Pycnometer, or other apparatus for measuring the moisture content of fine aggregate. Tins for sampling aggregates, cement, etc.

Scales, 5 kg × 1 g, and weights.

Storage tank for storing concrete cubes. This should be thermostatically controlled on larger contracts.

Trowels, shovels, spanners and other tools.

Testing of materials

Methods of testing aggregates and cement for quality have already been described in Chapters 3 and 4. Samples of the aggregate, and in some

Table 27.1 Some Major Causes of Blemishes on Concrete Structures and Possible Remedies[2]

Blemish	Possible causes of trouble	Possible remedy
A Concrete alignment unsatisfactory.	1 Unsatisfactory formwork design.	1 Remedy design.
	2 Faulty erection of formwork.	2 Tighten bolts, see that wedges are in correctly, bracings tightly fixed, etc.
	3 Placing of concrete at too fast a rate for design of formwork.	3 Alter formwork design or adjust rate of production of concrete.
	4 Dropping of concrete from a height into formwork.	4 Improve plant arrangement and placing technique.
	5 Use of vibration in formwork not designed and erected with vibration in mind.	5 Improve design of formwork or erection technique.
B Construction joints in random positions.	1 No attention to formwork design.	1 Show joint positions on formwork detail.
	2 Concreting stopped for a period during lift.	2 (*a*) Improve concreting gang technique to avoid breaks. (*b*) Adjust quantity of concrete between joints to plant output. (*c*) Eliminate plant breakdowns by regular maintenance.
C Construction joints badly made, including honeycombing just above joint.	1 Failure to remove laitance from previously placed concrete.	1 Remove laitance by scabbling, wet grit blasting, etc.
	2 Failure to clean up	2 Clean up thoroughly by air or air-water jets.
	3 Formwork not tight or gives under pressure, allowing loss of cement-sand grout.	3 Remedy design or erection technique of formwork.
	4 Mix too harsh.	4 (*a*) Revise concrete mix design. (*b*) Check proportioning of materials at the mixer.
	5 Concrete allowed to flow along formwork.	5 (*a*) Instruct concreting gang to place evenly round formwork. (*b*) Improve transport method to simplify even placing.
D Honeycombed patches.	1 Loss of cement grout through formwork.	1 Check formwork.
	2 Shortage of sand in some batches.	2 (*a*) Use more care in proportioning. (*b*) Mix more thoroughly.
	3 Change in grading of aggregate.	3 Check grading of coarse and fine aggregates.
	4 Segregation of mixed concrete.	4 (*a*) Change plant used for transport. (*b*) Alter mix design to decrease tendency to segregate.
	5 Segregation during placing.	5 (*a*) Improve placing technique. (*b*) Improve plant used in placing (if any).
	6 Lack of consolidation.	6 Improve compacting technique.

instances of the cement, should be sent to a central laboratory for comprehensive tests before concreting commences. It is often advisable, and sometimes imperative, that one or more trial mixes be prepared at the laboratory; the aggregate samples should therefore be sufficient for this purpose. It is necessary to give the central laboratory as much information as possible concerning the source of supply of the aggregate, the purpose for which the concrete is to be used, and the method by which the concrete is to be compacted, in order that a suitable and practicable mix can be designed. The nominal mix, when specified, should be quoted. When submitting cement for testing, the brand, situation of works, date and time of sampling, method of storage, length of storage, and other relevant details should be quoted.

Whether or not trial mixes have been made in the laboratory, it is essential for trial mixes to be made on site using the materials and the plant which will be used for the construction. Adjustments to the mix proportions may be necessary to obtain both the workability and the average compressive strength for which the mix was designed.

When sampling cement for testing it is advisable, in most cases, to obtain samples from a number of different loads, and to submit them separately in airtight tins.

Determination of surface moisture on the site

Aggregates are usually delivered to the site in a saturated condition and the water drains away gradually after depositing. This results in a saturated layer, about 300 mm or so thick, at the bottom of the stockpile. It is advisable to avoid using material from this layer and to be careful that samples for moisture content are not taken near the bottom of the heap.

Small differences in moisture content coincide with changes in the grading of the aggregate, especially the sand. Samples for testing should, therefore, be collected from different parts of the stockpile and mixed. At least three samples should be taken and tested and the average value found.

In large hoppers drainage occurs from the bottom. It has been found that this results in a low moisture content at the bottom, becoming wetter towards the middle. Allowance must be made for this by adding rather more water than that indicated by the tests for the first few mixes after any prolonged break.

Practical difficulties occur on the site for the various reasons discussed above, and it is rarely economical to make frequent tests for surface moisture when the quantities of concrete produced are small. On those larger works where strict control can be maintained, adjustments to the water added at the mixer on the basis of frequent surface moisture tests, should make it possible to maintain the water/cement ratio within the limits $\pm 0{\cdot}02$. It is suggested that when the quantity of water to be added to the mix has been estimated, a tolerance of ± 1 litre in 20 should be allowed, in order to take care of short-period changes.

An alternative method of strictly controlling the uniformity of the concrete is by frequent checks on the workability, preferably by means of the compacting factor test (see below). While this is only an indirect way of controlling the water/cement ratio, it has been found to be satisfactory, and is particularly valuable in road construction, for which the plant demands very uniform workability. However, checks by surface moisture tests on the aggregates must still be made periodically.

Mention has already been made of the necessity for determining the moisture content of the aggregate in order that the water/cement ratio may be controlled. Several methods have been devised for this purpose, but of these only two will be described in detail.

(1) *Drying Methods*

The surface moisture can be found by drying in two different ways. Either the aggregate can be dried until it is in a saturated surface-dry condition, or it can be dried right out and an allowance made for the loss of absorbed water.

It is difficult to decide just when the aggregate is saturated surface-dry, especially for fine aggregates and careful manipulation is essential if reliable results are to be obtained.

The second method is recommended for use on the site, since it is easy to dry the aggregate completely. The absorption can be determined by preliminary tests in the laboratory, and for all practical purposes it may be assumed at a constant figure.

The method of drying the aggregate depends on the facilities available. The most accurate results are obtained by drying a sample weighing 1 kg to 5 kg, according to the grading, in an oven maintained at 105–110°C. As this takes several hours, the method is too slow for the site control of concrete. Drying can, however, be speeded up by using 1 kg to 2 kg of aggregate, according to grading, in a frying-pan over a primus or other stove. This takes about 10 to 15 minutes, but care must be taken not to overheat the aggregate, and to prevent any small pieces from jumping out of the pan.

It is necessary to make at least two tests, and to find the average, if reliable results are to be obtained for the moisture content.

(2) *Pycnometer Method*

This is useful for dealing with sands and other fine aggregates. The method is described in BS 812 : Part 2 : 1975. It is necessary to determine the bulk relative density of the material from a saturated surface-dry sample. This needs careful manipulation, and it is suggested that at least three determinations should be made and the average found. The surface moisture can then be estimated without repeat determinations of the relative density, unless material is delivered from another source with the possibility of a change in the relative density.

It should be noted that all calculations when using a pycnometer are based on the saturated surface-dry condition, and to obtain results based on a dry condition, a correction for absorbed water must be applied.

Precautions to be taken in the use of the pycnometer are listed below:

(1) Weighing must be very accurate.

(2) The outside of the bottle must be carefully wiped free of any drops of water.

(3) It is not easy to get rid of air bubbles, and any carelessness in this respect upsets the results.

The pycnometer has the advantage that it can be used when no sources of heat for drying are available.

(3) *Other Methods*

Various sets of equipment are available which employ other methods of determining the surface moisture. Some of them are not very accurate or are less convenient than the methods already described. Others use far too small a sample.

The methods include measurement of the change in relative density of acetone or a saline solution when mixed with the wet aggregate, addition of methylated spirit and ignition to dry off the water, use of small cups on finely balanced scales reading off the moisture content directly, a saturation method using a graduated cylinder, the syphon-can method, and measurement of the electrical resistivity.

Sampling of concrete

The sampling of concrete for testing should be done with care in order to ensure that the test results are representative. Concrete may not be homogeneous for a number of reasons, such as, errors in proportioning, non-uniform mixing, segregation during handling and placing, change of consistency during handling, or loss of moisture from the surface or from absorption by contact with an absorbent material.

Samples should be taken either at the mixer or as close as possible to the point of final deposit, in which case they should be taken immediately after the concrete is deposited and before it is compacted. Occasionally tests for consistency should be made on the same concrete both at the mixer and at the point of placing in order to establish the extent of any loss of workability during transport.

The procedure to be used for obtaining samples of fresh concrete is described in BS 1881 : Part 1 : 1970 to which reference should be made. Whenever possible the sampling should be done when the concrete takes the form of a moving stream, as for instance from the discharge chute of a mixer. This does not apply to discharges from lorries or dumpers, when sampling is best done from the stationary vehicle or a heap. BS 1881 requires not less than four increment (samples) when sampling from a moving stream and not less than six increments when sampling from a lorry or heap. These increments are to be taken using 5 kg and 3·5 kg scoops respectively.

The increments are taken at equally spaced intervals and the scoop moved about in the concrete stream to obtain as representative a sample as possible. When sampling from a heap or lorry some effort should be made to obtain increment samples from within the concrete as well as at the surfaces. The increments are mixed to provide one composite sample which is then divided as may be necessary to give the quantity required for any specific test.

Several other points are worthy of note:

(1) Samples should not be taken at the beginning or end of a period of concreting without making special note of the fact.

(2) It is advisable to take samples at irregular intervals and without prolonged preparation which is evident to the mixer operator.

(3) Note should be made of the appearance of the mix, including any tendency toward segregation.

Fig. 27.1 Compaction of 150 mm cube using a standard rammer

(4) Note should be made of the time and date, mix proportions, the batch number, the mixer from which the concrete was delivered, the position in the structure, the method of sampling and the temperature.

Testing for consistency

The slump test is the most convenient test of consistency for site use. This and other tests are fully described in Chapter 6.

Preparation of specimens for compression test

The preparation of test cubes from a representative sample of the concrete, obtained in the manner already described, should be done in accordance with the method laid down in BS 1881: Part 3: 1970, Methods of Making and Curing Test Specimens, and it should be noted that the prime requirement is for the concrete to be fully compacted. Whereas the minimum number of blows per layer on a 150 mm cube is specified as 35, mixes of low workability

Fig. 27.2 Use of electric hammer to compact a stiff mix which could not be compacted by hand

with less than 25 mm slump are likely to require more than the minimum. Very dry mixes with zero slump may require over 100 blows per layer to achieve full compaction, and are best compacted using vibratory methods. Poker vibrators are not suitable.

Particular attention should be given to the ramming of the concrete into the moulds to obtain full compaction, Figs. 27.1 and 27.2, and to protection to prevent premature drying out.

The curing of cubes has an important influence on the strength attained. Particular attention should be given to the avoidance of any possible drying out, and to the curing temperature. BS 1881 : Part 3 : 1970 gives details for the curing of test cubes. For specimens made on site it is required that the cubes shall be stored immediately after making, under damp matting or other suitable damp material covered with polythene at a temperature of 20 ± 5°C for 16–24 hours. As soon as they are demoulded they are submerged in a tank maintained at a temperature of 20 ± 2°C until they are transported to the testing laboratory.

A difficulty arises with the curing of cubes for testing at 7 days when these have to be transported to a distant laboratory for testing; it is that the cubes must be packed two or three days after casting and may be in transit for the next two or three days. During transit the temperature cannot normally be controlled, and the effect on the crushing strength may be considerable in very cold weather.

Curing is much easier in the laboratory, and BS 1881 requirements are somewhat more stringent. The initial curing in the moulds for 16–24 hours is to be in moist air of at least 90% relative humidity. A moist curing room or cabinet is usually used, but similar methods to those specified for site use may be adopted, provided the correct temperature is maintained. After demoulding the specimens are submerged in a tank of water maintained at 20 ± 1°C, until they are taken out immediately before testing.

Specimens from site are also stored in the tank, at 20 ± 1°C, from the time they arrive in the testing laboratory.

There is no provision in BS 1881 for 'field curing' of specimens, which is often a useful method for providing an indication of the strength attained by the concrete in the structure. This information provides a useful guide to the stripping of formwork and supports.

Specimens intended to be cured under field conditions should be placed in contact with the forms and given the same protection as the concrete structure with a little additional insulation to compensate for the fact that they are cast in steel moulds and have greater relative surface area from which heat losses can occur. Some indication of the quality of the concrete can be obtained by making careful note of the temperatures during the curing period, and then making an allowance for any reduction in early strength using the curves given in Fig. 4.7. A fair approximation can be obtained in the following manner:

Assume average daily temperatures are as tabulated below:

Day	1	2	3	4	5	6	7
Average daily temperature, °C	5	5	0	10	10	15	5
Ordinary Portland cement in use							

Then, start off on the 5°C line and continue for 2 days, then move horizontally to the 0°C line (2—2) and move up this line for a one day period, reaching point 3. Move horizontally back to 3 on the 10°C line, and move along this for a two day period (days 4 and 5), to point 5. Continue to point 7 on the 5°C line. A strength of 52 per cent of that of concrete cured at 20°C for 28 days is indicated, compared with 68 per cent at 7 days on the 20°C line. Thus, concrete cubes cured for 7 days at the temperatures indicated, might be expected to reach $\frac{52}{68}$ = 76 per cent of that of similar concrete cured at 20°C for 7 days.

Cores, usually about 150 mm diameter, may be cut from the finished concrete by diamond drills or by the grinding action of steel shot under a rotating steel tube, a supply of water being used to wash away the slurry formed.

Information on the drilling, examination and testing of cores is given in BS 1881: Part 4: 1970, see also p. 411.

DETAILS FOR COMPRESSION TESTS ON CONCRETE CUBES

CONTRACT
Name ....Ledger No. ....
Full address ....
MATERIALS
Coarse agg. Supplier ....Size...
Pit ....
Fine agg. Supplier ....
Pit ....Zone ....
Cement type and trade name..Works ....

MIX DETAILS
Ready mix supplier (if applicable) ....
Ready mix plant location ....
Class of concrete...Mix proportions (wt/vol)* ....
Specified strength.......................................@ 7 days.......................................@ 28 days
Mixer size.............................Slump (mm).............................
Where concrete placed ....
Batch quantities: Cement...........kg Fine agg............kg Coarse agg............ Water...........l
(dry/damp)*
CUBE DETAILS
Date cast......................................Cubes made by..................Status...
Cube numbers ....Despatch date ....
Testing Laboratory ....
Age at which to be tested ....
Site remarks...
....

* Delete whichever is inapplicable.

Fig. 27.3

Despatch of materials and specimens for testing

Samples of material or specimens of concrete should be accompanied by complete information regarding the conditions under which they were obtained or made. The name of the site and other means of identification should be clearly marked on strong labels attached to each article.

Cement should be packed in airtight tins holding not less than 5 kg.

Samples of aggregate should be sent in strong polythene or closely woven jute bags, since the loss of fine particles in transit can materially change the grading, more particularly with sands.

Strong labels should be used with an identification number which can be quoted on an information form sent separately.

Concrete cubes should be packed in wet sand or wet sacks, and enclosed in a strong polythene bag, before placing in a wooden box or other suitable container for transport to the laboratory. The information to be sent with them should include the date of casting, the source of supply and sizes of the aggregate, the brand of cement, the proportions of the mix, the consistency, the location from which the sample was taken, the curing temperatures, the method of curing, and the age at which the test is to be made. A suitable form to use for such a purpose is shown in Fig. 27.3. It is preferable for the completed form to be sent by post and for similarly numbered counterfoils, with the name of the site and the date on which the cubes were cast, to accompany the cubes.

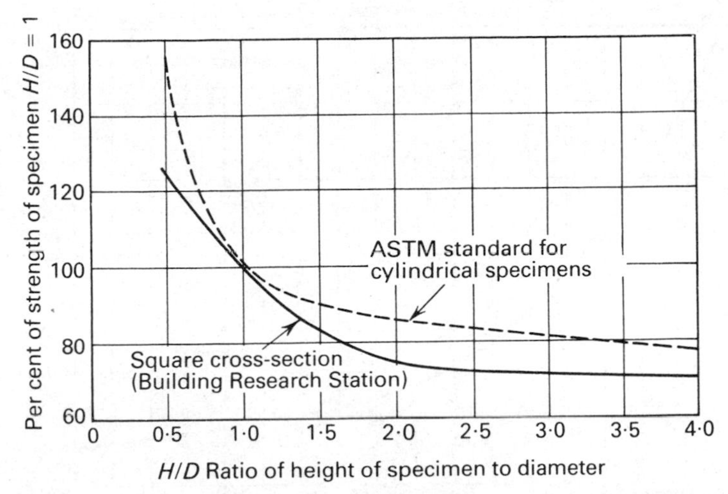

Fig. 27.4 Relation of height and diameter of specimen to compressive strength

Testing for compressive strength

Tests on concrete cubes (or cylinders in some countries) for compressive strength are widely accepted as the most convenient means of quality control of the concrete produced whether on the site or at a ready-mixed concrete plant. The cube test is, however, no more than a measure of the quality of the concrete produced at the mixer, and as the amount of concrete used in cube making is only a very small proportion of the quantity of concrete placed in the structure, and is only taken from a few of the batches mixed, individual tests should only be regarded as giving a general indication of the quality of the concrete. The main causes of variation in cube strength are summarized on p. 416. While cube strengths are accepted or rejected on the basis of passing or failing certain limits, whether these are related to characteristic strengths or minimum strengths, they should also be examined in relation to the mean strength expected from the mix design calculations (see Chapters 6 and 7).

When cube strengths are in the region of the mean strength it may be accepted that the concrete being produced at the mixer is of reasonable standard, but when cube strengths approach the upper or lower limits, then an investigation should be made of conditions at the mixer. Further information is given on p. 415 under Variation in cube strengths.

It should be remembered that procedures for making and testing cubes have a significant bearing on the accuracy of the results.

Testing of concrete cubes, cylinders and cores

Compression tests are usually made on 150 mm cubes in Great Britain, whereas in the United States of America cylindrical specimens, 300 mm high and 150 mm in diameter, have been adopted as standard. Unfortunately, the height of the test specimen in relation to its width greatly influences the result obtained, as is illustrated in Fig. 27.4, and the crushing strengths of the cylindrical test specimens used in the United States can only be compared with the

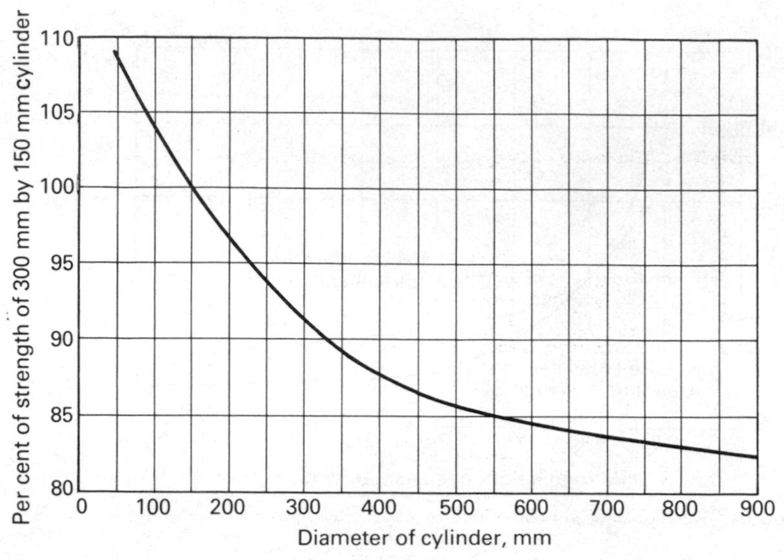

Fig. 27.5 Effect of cylinder diameter on crushing strength

results obtained from 150 mm cubes in this country by applying a suitable correction factor. BS 1881 : Part 4 : 1970 gives correction factors for the calculation of the cube strength (150 mm) equivalent to cylinders or cores with various length/diameter (L/D) ratios. The necessary correction may be simply expressed in the following formula:

$$\text{Correction factor, } C = \frac{L}{10D} + 1.05$$

where C is the factor by which the cylinder strength is multiplied to obtain the equivalent cube strength.

The size of specimen also affects the crushing strength; for instance, compared with a 150 mm cube, a 100 mm cube of the same concrete is approximately 4 per cent stronger. Again, a 200 mm cube is approximately 8 per cent weaker, and a 250 mm cube 10 to 12 per cent weaker than a 150 mm cube. The relationship between diameter and strength of concrete cylinders is given in Fig. 27.5 by a curve based on the results of a large number of tests made in the United States.

The advantage of the cube form of specimen used in British practice is that the sides can be made accurately plane and parallel by using carefully machined steel or cast-iron moulds. The cubes can then be tested on their sides without any special preparation of the test specimen or any necessity for bedding material, which results in lower strengths, and unless carefully done, leads to unreliable results. For this reason the use of wooden moulds is unsatisfactory, since it is impossible to maintain their accuracy, thus necessitating some form of bedding material. As an illustration of the influence of end packings on the crushing strength, a reduction of 30 to 50 per cent may be obtained by the insertion of lead or soft millboard.

Fig. 27.6 Hydraulic testing machine with a capacity of 300 tonnes

In the United States the difficulty of trowelling accurately the top surface of the cylindrical specimen is overcome by capping with neat cement an hour or so after filling with concrete. Capping of cylinders is also often done with a heated mixture of sulphur and finely divided filler. The advantage of this material is that it develops a high early strength. However, it does appear that capping introduces an additional source of error.

The rate of applying the load to the specimen affects its crushing strength, especially for high rates of loading. Below a rate of loading of 17 N/mm^2 per minute, however, there is little change. It will be noted that a rate of 15 N/mm^2 per minute is specified in BS 1881 : 1970.

A point which may be missed is that cubes should be tested in a saturated condition in order to obtain uniformity. Dry cubes give higher results, but since there are different degrees of apparent dryness, it is not possible to ensure agreement.

Crushing tests are made in various machines either of the lever-arm type or adaptations of hydraulic presses. A machine of the hydraulic-press type with a maximum capacity of 300 tonnes, is illustrated in Fig. 27.6. The machine consists of two parts, namely, the press itself and the dynamometer which measures the load. The press consists of a ram, a crosshead, and upper and lower compression plates between which the specimen is crushed. The lower compression plate is scribed to facilitate centering of the specimen, and rests on the ram of the press. A spherical seating provides sufficient rotation for any lack of parallelism in the sides of the specimen to be accommodated.

The ram is raised by pumping oil into the press cylinder from a reservoir at the base of the dynamometer. The oil pressure acts on a small piston in the dynamometer, which in turn rotates the pointer on the dial. A second pointer having a light friction fitting to the glass front, is engaged by the indicating pointer as it ascends, and remains in the highest position, thus recording the maximum load obtained.

Some hydraulic machines are hand operated, but this results in a jerky rate of loading, and is tiring if a large number of specimens have to be tested. Ordinary pressure gauges are also commonly used to indicate the load instead of a dynamometer. They are less expensive, but are not particularly reliable or accurate.

Inaccuracies may occur in making crushing tests for various reasons, such as errors in centering the cubes, wear of the platens, friction in the spherical seating of the upper compression plate, and inaccurate calibration of the machine itself. It has been found that considerable discrepancies occur between the results of crushing tests made with different designs of machine, although when calibrated with standard gauges these machines agree with one another to within 0·5 per cent. This is due to the imposition of different strain distributions on the cubes by different machines. The cause of these discrepancies, which may result in cube strength variations of 10% or even more, has been the subject of much research in recent years[(3),(4)]. An amendment slip to BS 1881 : 1970 : Part 4[(5)] should be consulted for detailed requirements concerning testing machines and the cube testing procedure.

When a cube has been tested it should ideally break to the shape of an inverted pyramid supported by an equal pyramid on its normal base. A cube which has sheared diagonally is indicative of some eccentricity of loading. If this happens repeatedly the cause should be sought.

It appears advisable that all cubes from a particular site should be tested in the same machine in order to get comparable results.

Number of tests

Concrete is used in such widely varying circumstances that the amount and the type of tests made are matters which need to be determined by the engineer. Inspection of the concrete production on site and of the cast concrete is just as important, if not more so, than test results. Serious problems can arise from badly placed reinforcement; poor compaction, especially near construction joints; honeycombing and segregation. These problems are not revealed in the results of the usual concrete tests. The cost of testing cubes and of analyzing fresh concrete for cement content can become a significant proportion of the cost of the concrete itself. The engineer, therefore, needs to make a careful appraisal of all the factors when deciding the level of inspection and testing to

Table 27.2 Sampling Rates Suggested for Concrete for Test Cubes

CP 110 : 1970		BS 5328 : 1976	
Type of Structure	Volume in m³ which sample represents	Type of Structure	Volume in m³ which sample represents
Highly stressed structural elements	10	Critical structures, e.g. masts, cantilevers, columns	10
Ordinary structural work	20	Intermediate structures, e.g. beams, slabs, bridge decks	50
Mass concrete not subject to high stress	50	Heavy concrete construction, e.g. breakwaters, solid rafts	100

Note: CP 110: 1970 also suggests that the sample should represent 10, 20 or 50 batches if this is less than the volume in m³. Higher rates are suggested at the start, to be reduced when high quality has been established.

adopt for the particular work to be undertaken. The following information is given for guidance.

Slump tests are easily made on site and provide a useful check on the consistency of the mix. They may be made whenever a change becomes discernible. Other tests for workability are generally more appropriate to laboratory testing (including site laboratories) and, therefore, the rate of testing will be directed more closely by the engineer. Usually they will be made only on the larger structural works.

Sampling rates for concrete for test cubes suggested in CP110 : 1970 and in BS 5328 : 1976 are different (see Table 27.2) illustrating the problem of giving guidance on sampling rates.

From any one sample of concrete BS 5328 : 1976 requires that two test specimens shall be made, both specimens being cured and tested at 28 days, the average of the two results to be taken as the test result. Provision is also made for the use of other curing regimes if it is agreed that they are capable of predicting the strength at 28 days. Again the result is based on the average of two results. CP 110 : 1970 suggests that only one cube should be made from one sample of concrete. The BS 5328 : 1976 arrangement is the better one. The authors have, however, always preferred to make three cubes, one for test at 7 days as a guide, and the other two for test at 28 days.

Variation in cube strengths

It is well known that variations occur in the strengths of dense concrete test cubes due to a variety of factors. Except for testing errors, which should be small, and variations in the rate of hardening of the cement, the variations are dependent on the care and attention given to the manufacture of the concrete and treatment of the specimens on the site. The main causes of variation in concrete cubes may be summarized as follows:

(1) Inaccuracy in the proportioning of the gravel, sand and cement. This is probably the largest single cause of site variations.

(2) Variations in the water/cement ratio. Such variations are aggravated by the need to obtain a reasonable workability for placing when using a mix in which the proportions of gravel, sand and cement are varying widely.

(3) Variations in the grading of aggregates which necessitate changes in the water/cement ratio if a uniform workability is to be maintained.

(4) Insufficient compaction. Very small percentages of air voids cause a large reduction in strength.

(5) Unsatisfactory curing. If the cubes are allowed to dry out during the first 24 hours, the loss in strength may be up to 50 per cent, which is not completely recovered by subsequent wetting.

(6) Variations in temperature. These have a most marked effect at early ages.

(7) Variations in cement quality.

Fortunately these variations are not additive, since some factors tend to improve the concrete, while others tend to cause a lowering of quality. The result is that concrete cube strengths vary about the average in such a way that while most of the values show little deviation above or below, a gradually decreasing number vary by greater amounts. If the number of cubes differing from the average by certain amounts, say, 0–5, 5–10, 10–20, 20–30, etc., N/mm^2 are plotted a histogram is obtained such as that given in Fig. A.1, p. 421. An envelope of such a histogram has the form shown for a Gaussian or normal distribution.

It is sometimes desirable to summarize the test data obtained over a period. This may be done by taking an average and also indicating the maximum and minimum values, but such figures give little indication of the way in which most of the results departed from the average. An index which provides a measure of the extent of the variations from the average is, therefore, useful. The usual index is the 'standard deviation' or the coefficient of variation, expressed as a percentage of the average value. These terms are defined in Appendix A.

On most sites, changes in the grading of the sand, differences in workability required for differences in placing conditions, and other factors make some changes in the water/cement ratio necessary from time to time, and these are not normally recorded with sufficient accuracy for them to be taken into account in any statistical analysis of the cube strengths. The number of cubes made is also too small on most sites to be of much use statistically; the standard deviation calculated on a few results is likely to depart considerably from the true value, which can only be obtained with reasonable accuracy from a large number of cube results, say 40 or more. With such variants unaccounted for, the standard deviation would be expected to be least on works requiring large quantities of concrete of uniform workability, such as road or airfield pavings, and this is borne out in practice. Even then, an upward or downward trend in the concrete strengths over a period, whilst increasing the standard deviation or coefficient of variation, may be caused by a similar trend in cement strengths, and not by a weakening of quality control.

Although these limitations on the value of statistical analyses must not be overlooked, such analyses, if carefully interpreted, are useful as an indication of the variations in cube strengths which might be deemed reasonable under a variety of site conditions. This knowledge is of importance both in mix design and in day-to-day control on the site.

The information given in Table 7.1 is based mainly on an analysis of cube results from nearly two hundred sites, ranging from small works on which only a few cubes were made, to London Airport where the number tested by the contractor was nearly 30 000. In the Table it will be noted that the variation in strength is taken to be proportional to the average strength for strengths below about 21 N/mm². Above this value the variation is indicated to be independent of the average strength, and is related more to the standard deviation than to the coefficient of variation. There is considerable evidence [1] to support this view as being approximately correct, both theoretically and from analysis of site cube test results.

These conclusions are based on the use of the 'minimum strength' as the measure of control, which remains the only practical approach for the many sites where the total number of cubes tested is small. It is assumed that the engineer wishes to use cube strengths as a control measure and is not prepared to rely on prescribed mixes (BS 5328 : 1976) without site testing. He can, of course, fix his own minimum strength and specify it for a prescribed mix.

On those sites where large numbers of cubes are made from one quality of concrete etc. (see Chapter 7, p. 108, for further details on CP 110 requirements for designed mixes) a statistical approach becomes practical. This also applies to ready-mixed concrete plants, providing they turn out sufficient of any one quality of concrete, etc.

Whichever approach is adopted a low, or high, result should never be ignored. It may have been caused by a blockage or other fault at the weigh batcher which has gone undetected, with the result that the concrete has been seriously short of cement, or far too rich in cement, or otherwise unsatisfactory. Similarly, a low, or high, result may result from a delivery error in ready-mixed concrete or failure to make the necessary changes in proportions when changing from one grade to another.

Another possibility is that the water/content ratio has been increased beyond that intended, in order to increase the workability of the concrete, either to facilitate placing in a difficult position, or possibly because the plant being used will not deal satisfactorily with the original consistency. An instance of this is the attempt to use a transporting hopper with too narrow an opening for the discharge of stiff concrete, with the result that the concrete gang quite rightly calls for wetter concrete in order to get on with the job. A study of such plant difficulties can often lead to a substantial improvement in concrete quality, and sometimes to a saving in cost at the same time.

Assessment of cube test results

For compliance with the 'characteristic strength', which provides for a failure rate of 5 per cent, BS 5328 : 1976 requires two provisions to be met:

(1) The average strength determined from any group of four consecutive results exceeds the characteristic strengths by

3 N/mm^2 for concretes of grade C20 and above
2 N/mm^2 for concretes of grade C15 and below.

(2) The strength determined by any test result is not less than the specified characteristic strength minus

3 N/mm^2 for concretes of grade C20 and above
2 N/mm^2 for concretes of grade C15 and below.

The specification also states that the quantity of concrete represented by any group of four consecutive test results shall include the batches from which the first and last sample were taken together with all intervening batches. When a test result fails to comply with (2), above, only the particular batch from which the sample was taken is at risk. It may be noted that provision (2) is virtually a 'minimum strength'. This is a considerable change from the requirements of CP 110:1970, which are unnecessarily onerous.

Provision is made in BS 5328 : 1976, Methods of Specifying Concrete, for conditions where there are not four, or more, consecutive test results. The purchaser is then advised to specify his requirements when:

(1) There may be a lapse of two weeks or more between successive pours of the same grade.

(2) Different sampling rates are specified for the same grade.

(3) The purchaser requires concrete to be supplied from more than one source.

Where individual results are specified for compliance the purchaser is advised to state his own requirements.

The above provision is important since it will apply to many sites, especially the smaller civil engineering works and most building works. The 'minimum strength' method of assessment which for many years has been based on a notional 2 per cent failure rate will usually be found most appropriate for quality control. An alternative is to adopt provision (2), above, as the criterion for the minimum strength for single results. Each result (an average of two cube results may be used, see Number of tests) then provides the indication of the cube strength for the concrete the sample is taken to represent, and action can be taken quickly if the strength is higher or lower than expected.

Records

Tabular forms greatly simplify the work of recording and reporting, and also form a check of the items covered. Forms may be prepared for calculations of the batch quantities of aggregates, the recording of information when submitting aggregates or cubes for testing, records of slump tests, and records of cube test results. Charts give a useful summary of test data in a form which can be readily appreciated, and from which the trend of results can be quickly assessed. Reference may usefully be made to ACI recommendations[6].

Flexural tests

Tests on beams to determine the flexural strength of the concrete are described in Chapter 2. Two-point loading is called for in BS 1881, which also provides for compression testing of the broken halves of the beam to obtain the equivalent cube strength.

Indirect tensile test

An indirect tensile test is sometimes specified. This is made by loading a 150 mm diameter cylinder placed horizontally in the testing machine, so that it is split along its length. The indirect tensile strength, f_t, is then calculated from the formula

$$f_t = \frac{2P}{Dl}\ \text{N/mm}^2$$

where D = diameter of cylinder in mm
l = length of cylinder in mm
P = maximum load applied to the cylinder in Newtons.
Details of the test are given in BS1881 : Part 4 : 1970

Analysis of fresh concrete

A method of analyzing fresh concrete to determine the proportions of the aggregate, cement and water is given in BS 1881 : Part 2 : 1970. It is too elaborate for ordinary site use. Machines have been devised to give an answer by a fairly simple procedure but, due to their cost and the importance of operation by a trained technician, are unlikely to come into common use. The analysis needs to be commenced within a few minutes of discharge of the concrete from the mixer. The methods so far devised are not particularly accurate; they are dependent to a considerable degree on knowledge of the properties of the constituents of the fresh concrete, particularly the silt content of the fine aggregate.

REFERENCES

1 MURDOCK, L. J., The control of concrete quality, *Proc. Inst. Civ. Engrs*, Part 1, July 1953.
2 MURDOCK, L. J., Methods of achieving control of quality, *Symposium on Mix Design and Quality Control of Concrete*, London, 11–13 May 1954. Cement and Concrete Association.
3 NEWMAN, K., Concrete control tests as measures of the properties of concrete, *Proc. Symposium on Concrete Quality*, Nov. 1964. Cement and Concrete Association.
4 *The Performance of Existing Test Machines*, A Concrete Society Working Party Report, July 1971. The Concrete Society.
5 BS 1881 : Part 4 : 1970, *Methods of Testing Concrete for Strength*. British Standards Institution.
6 ACI 214–65, *Recommended Practice for Evaluation of Compression Test Results of Field Concrete*. Adopted as Standard by American Concrete Institute, 1965.

Appendix A

Variations in Cube Strengths

VARIATIONS in cube strengths result from many different factors such as the strength characteristics of the cement, the water/cement ratio, changes in grading of the materials, air-voids content of the compacted concrete, curing, temperature, and testing errors. For any particular sample of concrete, some of these factors will tend to reduce the compressive strength, while others will tend to increase it. The compressive strength actually obtained will then depend on the balance between the positive and negative influences, and the extent of deviations from an average value will depend on the amount by which the total of the positive influences exceed the total of the negative influences, or vice versa. In general, it is found that the bulk of the results lie near the average value with a decreasing number showing greater deviations, following the distribution shaded in Fig. A.1.

Analyses of cube strengths obtained on large numbers of sites have shown that the distribution of values is in accordance with the theory of probability, i.e. they have a normal or Gaussian distribution which is represented by the curve in Fig. A.1. A Gaussian distribution has the property of being wholly defined by the average value of the set of results and their standard deviation.

The standard deviation of an infinite set of numbers is defined by:

$$s = \sqrt{\frac{\Sigma(x - m)^2}{n}}$$

where s is the standard deviation of the set of numbers,
n is the number of values in the set,
x is any value in the set of numbers,
m is the average of the set of numbers.

However, when examining concrete cube results the main interest is in an estimation of the variation of the compressive strength of the whole consignment of concrete, rather than the variation in the small number of samples which have been taken and tested from the entire consignment. A better estimate of the standard deviation for the whole consignment is given by the following formula:

$$s = \sqrt{\frac{\Sigma(x - m)^2}{n - 1}}$$

The coefficient of variation is obtained by dividing the standard deviation (s) by the average m and is usually expressed as a non-dimensional percentage:

$$v = \frac{s}{m} \times 100$$

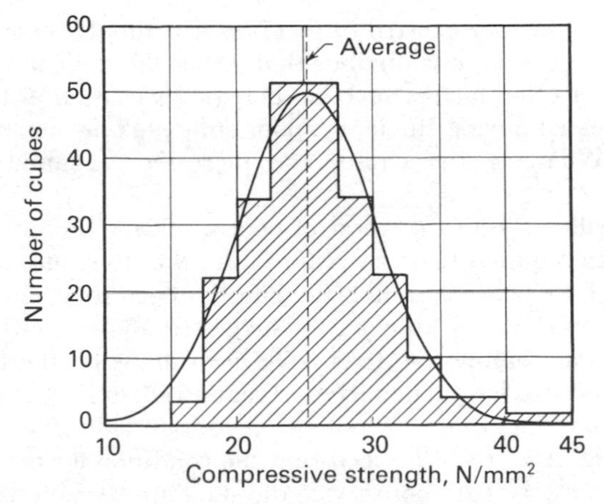

Fig. A.1 Example of histogram and a normal probability curve

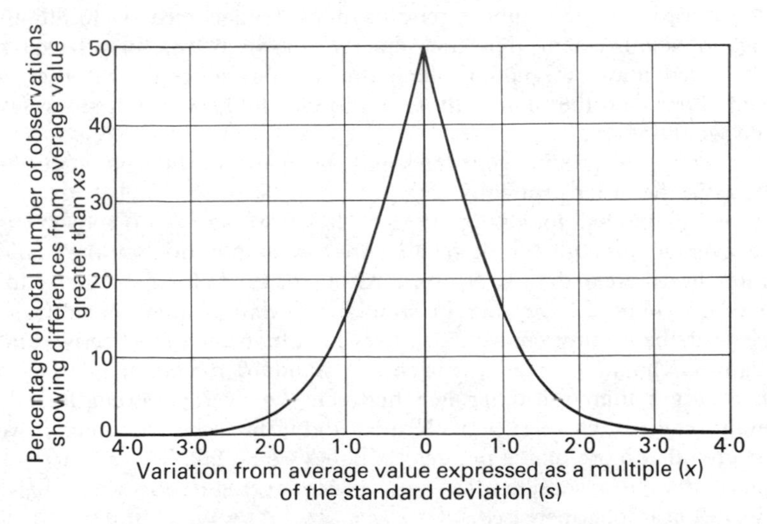

Fig. A.2 Distribution of observations deviating from the average value due to operation of numerous chance causes

The normal distribution curve may be used to calculate the proportion of cubes differing from the average by any multiple of the standard deviation. Figure A.2 has been plotted to show the relationship in convenient form.

The standard deviation provides a useful basis for quality control when large numbers of cubes are made from one quality of concrete and are tested in a reasonably short period, particularly during the early stages of construction. As an example, assume that thirty or more cubes are made and tested during the first few days of construction (or if economically justifiable from trial mixes before commencing concreting operations), and that the standard deviation is calculated from the results obtained. It is then possible to form an estimate of the future distribution of results. Should these show greater variability as time goes on, it is obvious that either the site control is showing some relaxation, or other factors such as increased variation in the quality of the cement, decreased mixing efficiency, poor cube-making, etc. are having an adverse effect. Whatever the cause, the increased variability should be investigated.

An example will make this method of control clear. Assume the specified minimum strength required to be 21 N/mm² and that to ensure that not more than 2 per cent of the cubes shall fail, the aim has been to achieve an average strength of 30 N/mm². (See Table 7.1 for suggested allowances.)

Suppose now that compressive tests have been made on the first 40 cubes cast on the job, and two have failed; the average strength is 29 N/mm² and the standard deviation (s) 4 N/mm².

Referring to Fig. A.2, it will be seen that the condition for not more than 2 per cent cubes failing, is that about twice the standard deviation ($1 \cdot 96\ s$) shall not exceed the allowable variation from the average, in this case $29 - 21 = 8$ N/mm². As twice the standard deviation is $2 \times 4 = 8$ N/mm² this condition is just fulfilled and the cube strengths may be accepted as fulfilling the specification requirement. The fact that two cubes of the forty tested have already failed may be ignored, since the analysis suggests that they were chance failures, and that if an infinite number of cubes were tested less than 2 per 100 would fail.

Suppose now that testing continues and the results of the next 20 cubes are as follows in the order printed:

24, 31, 34, 36, 22, 23, 39, 28, 33, 32, 41, 28, 32, 21, 37, 32, 31, 29, 41, 27.

The average strength for these 20 cubes is 31 N/mm² but the standard deviation has increased to 6 N/mm². Although no failures occurred in the batch of 20 cubes the increase in standard deviation alone is sufficient to suggest that the quality control measures have not been so effective and an investigation should be made. Twice the standard deviation is 12 N/mm² which is larger than the difference between the average strength and the minimum required, $31 - 21 = 10$ N/mm² and if the trend continued it would be expected that more than 2 per cent of cubes would fail. This fact would not be apparent from a consideration only of the average cube strength which had, in this example, increased.

This example of the use of the statistical approach to quality control indicates the usefulness of the method in assessing whether cube strengths are being maintained at a sufficient level to comply with specification requirements. It should be noted, however, that an analysis based on only 20 to 30 results is not usually sufficiently accurate for investigational purposes, and it is desirable to base any conclusions on a very much larger number, say, a hundred or more cube tests.

The curves in Fig. A.2, may also be used to estimate the proportion of cubes differing from the average by multiples of the standard deviation. For instance, 16 per cent of the cubes are likely to deviate from the average by a value equal to the standard deviation, x being equal to 1. It will be noted that as many cubes are above the average as are below it, and that the deviation above the average is likely to be as great as that below it.

Table A.1 Example of Calculations for Standard Deviation and Coefficient of Variation

No.	Crushing strength of cubes, N/mm^2	Diff. from average $(x - m)$	$(x - m)^2$
1	28	0	0
2	27	−1	1
3	22	−6	36
4	35	7	49
5	31	3	9
6	28	0	0
7	25	−3	9
8	34	6	36
9	29	1	1
10	22	−6	36
Average (m)	28	$\Sigma(x - m)^2 = 177$	

$$\frac{\Sigma(x - m)^2}{n - 1} = \frac{177}{9} = 20$$

$$\text{Standard deviation} = \sqrt{20} = 4{\cdot}5\ \text{N/mm}^2$$

$$\text{Coefficient of variation} = \frac{4{\cdot}5}{28} \times 100 = 16 \text{ per cent.}$$

An example of a histogram plotted to show the numbers of cubes falling into 5 N/mm^2 intervals in cube strength is given in Fig. A.1. The outline of such a histogram should generally follow the form of the normal distribution curve. Any substantial departure should be a matter of investigation. In the example given in Fig. A.1 it will be seen that there is a somewhat sudden cut-off at the lower end, whereas there are a few more high results than would be expected from a normal distribution. It would be expected that the tails of the histogram would stretch to roughly the same range above and below the average. The sudden cut-off should be investigated. Low cube strengths may not have been reported, cubes of poor appearance may have been 'lost', or there may have been some error. Other anomalies should also be investigated; for instance, too many high results may be due to selection of concrete from batches of lower workability than the average for the purpose of cube making.

The above is based on 'minimum' strength, which for most sites provides the most convenient means of quality control (see Chapter 27). The term 'minimum' is not strictly correct since it is normally accepted that 2 per cent of the cubes are likely to fail. When testing relatively small numbers of cubes, the engineer, while hoping that all cubes will be above the minimum, is called upon to give a final judgment if one or two fail by a narrow margin. Attempts to eliminate this situation have led to the concept of 'characteristic' strength, in which the acknowledged failure rate is increased to 5 per cent. This has entailed a more complicated acceptance procedure (as outlined in Chapter 27) and is only applicable to those sites where large numbers of cubes from each different mix and from each mixer set-up are justified. The general principles of the statistical approach described in this appendix apply to both the 'minimum' and the 'characteristic' strength approaches to quality control.

It is useful to note that the difference between the mean value for cube strength and the characteristic strength (5 per cent failure rate) is 1·64 × standard deviation, compared with 1·96 × standard deviation for a 2 per cent failure rate.

For further information on the subject, reference should be made to standard textbooks[1]. Much useful information on the utilization of statistical methods for the control of quality is given in British Standard 2564 : 1955[2].

REFERENCES

1 McIntosh, J. D., *Concrete and Statistics*, 1963. C. R. Books Ltd., London.
2 B.S. 2564 : 1955, *Control Chart Technique when Manufacturing to a Specification*, P. B. Dudding and W. J. Jennett. British Standards Institution.

Appendix B

Metric, Imperial and American Equivalents

The following brief selection of equivalents are those most likely to be of use to the reader.

Metric	*Imperial*	*Reciprocal*
Length		
1 mm	= 0·0394 in	25·4
1 m	= 3·28 ft	0·305
Area		
1 mm^2	= 1·55 × 10^{-3} in^2	645
1 m^2	= 10·76 ft^2	0·0929
Volume		
1 mm^3	= 0·0610 × 10^{-3} in^3	16387
1 m^3	= 35·3 ft^3	0·0283
1 m^3	= 1·31 yd^3	0·765
1 litre (l)	= 0·22 gallon	4·5436
Mass		
1 kg	= 2·205 lbs	0·4535
1 tonne (1000 kg)	= 2205 lbs	0·454 × 10^{-3}
Stress, pressure		
1 N/mm^2	= 145·04 lbf/in^2	0·0069
1 N/mm^2	= 0·029 lbf/ft^2	47·9
1 kgf/cm^2	= 14·2233 lbf/in^2	0·0703
1 kgf/m^2	= 0·205 lbf/ft^2	4·882
Density		
1 kg/m^3	= 1·685 lb/yd^3	0·593
1 kg/m^3	= 0·0624 lb/ft^3	16·02
Force		
1 N	= 0·225 lbf	4·45

Covering capacity

1 m^2/l	= 5·44 yd^2/gallon	0·18

Mass per unit area

1 kg/m^2	= 0·205 lb/ft^2	4·882

Other useful data:

1 Imperial ton	= 2240 lb	1 gallon (Br)	= 10 lb water
1 American ton	= 2000 lb	1 gallon (Am)	= 8·33 lb water
1 litre	= 0·264 American gallon		

1 bag of cement (Br)	= 50 kg (110 lb)
1 bag of cement (Am)	= 94 lb

Subject Index